Piezoelectric Materials

Piezoelectric Materials

From Fundamentals to Emerging Applications

Volume 1

Edited by Jiagang Wu

WILEY-VCH

The Editor

Prof. Jiagang Wu
Sichuan University
No. 24 South Section 1
Yihuan Road
Chengdu
CH, 610065

Cover Image: © Yuichiro Chino/Getty Images

Library of Congress Card No.: applied for

British Library Cataloguing-in-Publication Data
A catalog record for this book is available from the British Library.

Bibliographic information published by the Deutsche Nationalbibliothek
The Deutsche Nationalbibliothek lists this publication in the Deutsche Nationalbibliografie; detailed bibliographic data are available on the Internet at <http://dnb.d-nb.de>.

Print ISBN: 978-3-527-35470-2
ePDF ISBN: 978-3-527-84121-9
ePub ISBN: 978-3-527-84122-6
oBook ISBN: 978-3-527-84123-3

Typesetting Straive, Chennai, India
Printing and Binding CPI Group (UK) Ltd, Croydon, CR0 4YY

C9783527354702_020524

Contents

Volume 1

Preface

Piezoelectricity, a phenomenon wherein a material generates electrical charges on its surfaces in response to applied mechanical stress or, conversely, yields mechanical strain when subjected to an electric field, serves as a critical bridge between mechanical and electrical energies. This unique functionality has spurred the widespread adoption of piezoelectric materials across a broad spectrum of modern technologies. Examples of these technologies abound and pervade our daily lives. In the realm of consumer electronics, piezoelectric sensors and actuators are integral components of smartphones, enabling precision touch-screen interactions. In the medical field, piezoelectric probes are indispensable in ultrasonic diagnostic equipment, facilitating noninvasive imaging and diagnostic procedures. Additionally, high-precision positioning units in scientific research equipment, ignitors for lighters, and pressure sensors, all leverage the properties of piezoelectric materials to function effectively.

Piezoelectric materials can be categorized based on the form of matter, with classifications including piezoelectric ceramics, single crystals, thin films, and organic materials. Each category boasts specific characteristics, making them ideal for particular applications. Moreover, in alignment with escalating global concerns for environmental protection and sustainable development, piezoelectric materials are often classified as lead-containing or lead-free. This classification underscores the ongoing pursuit of environmentally friendly materials that can deliver the advantages of piezoelectricity without compromising sustainability. The evolving research and development in this field promise exciting advancements and wider adoption of piezoelectric materials in future technologies.

Over the past two decades, the field of piezoelectric materials and associated devices has undergone swift and significant advancements. Both phase-field simulations and neutron scattering experiments have substantiated the exceptional influence of polar nanoregions, which have been instrumental in enhancing the piezoelectric properties of lead-based ceramics and single crystals. Innovative approaches, including new phase boundary engineering and compositional design, have equipped lead-free piezoelectric materials with superior piezoelectricity, which is comparable to some of their lead-free counterparts. The adoption of a multi-scale structural design strategy has consistently shattered previous records in dielectric energy storage, signaling the potential for further advancements in this area. Concurrently, the evolution of microfabrication technology has facilitated the emergence of piezoelectric materials' enticing transduction capabilities in a variety of cutting-edge electronic devices. These include biomedical devices, piezoelectric nanogenerators, and piezoelectric flexible devices, which embody the intersection of technological progress and practical application. Additionally, the utilization of advanced in situ characterization techniques, including transmission electron microscopy, neutron scattering, and high-energy synchrotron radiation, has empowered the researchers to unravel the complex physical mechanisms underlying the observed high performance. These advancements collectively signal a bright future for the continued development and application of piezoelectric materials.

Motivated by the aforementioned advancements, I contend that the present moment provides an opportune juncture for reviewing the research landscape of piezoelectric materials and devices. Such a book proves imperative, not only for early career researchers seeking an informed entry point into the field but also for established scholars aiming to conceptualize and implement novel strategies for enhancing the performance of piezoelectric materials and devices. Although numerous high-quality reviews and books have been previously published, the majority concentrate on specific topics within individual material systems. To obtain a holistic understanding of the field of piezoelectric materials and devices, it is essential to encompass both the fundamental knowledge of piezoelectricity and the state-of-the-art research progress in piezoelectric materials and devices. Bearing this objective in mind, I have undertaken the compilation of this book. My aspiration is that this work will serve as a comprehensive resource, illuminating the path for future research and development in the realm of piezoelectric materials and devices.

This book comprises 20 chapters that delve into various facets of piezoelectricity, a dynamic and evolving field. Chapter 1 provides a solid foundation in the basic principles of piezoelectricity, delineating piezoelectric coefficients, describing piezoelectricity via third-rank tensor, and discussing symmetry considerations. This opening chapter serves as an invaluable resource for readers unfamiliar with piezoelectricity or those seeking to strengthen their foundational understanding. Chapter 2 explores the probing of piezoelectric effects using X-ray diffraction and the observation of the piezoelectric effect in multidomain ferroelectrics. Chapter 3 offers a succinct summary of the evolution and research advancements in lead-based piezoelectric materials, with emphasis on historical progress and modern-day applications. Chapters 4 and 5 deliver a comprehensive overview of lead-free piezoelectric materials, concentrating on four significant material systems: (K, Na)NbO_3, $BaTiO_3$, (Bi, Na)TiO_3, and $BiFeO_3$. Chapter 6 presents an overview of piezoelectricity in molecular ferroelectrics, focusing on piezoelectricity in different molecular ferroelectrics and piezoresponse force microscopy (PFM)-aided characterization and design. Chapters 7 and 8 review the development and research of $BiScO_3$–$PbTiO_3$ and $Bi_4Ti_3O_{12}$-based high-temperature piezoelectric ceramics, covering aspects like characteristics, fabrication, domain engineering, electrical properties, endurance, and high-temperature applications. Chapter 9 delves into the newly developed texturing method, concentrating on textured piezoceramics, texture–property relationships, and device applications of textured piezoceramics. Chapter 10 reviews the strides made in piezoelectric thin films, focusing particularly on the growth and piezoelectric coefficient measurements of these thin films. Chapter 11 reviews the development of piezoelectric and electro-optic properties in lead-free single crystals, with a focus on (K, Na)NbO_3 and K(Ta, Nb)O_3-based single crystals. Chapter 12 summarizes the electrostrictive effect, touching on its fundamentals and history, reviewing the electrostrictive effect and strain in various ferroelectrics, and providing examples of their applications. Chapters 13 and 14 review ferroelectric materials for dielectric energy storage, including the fundamentals, design, and performance of inorganic, organic, and composite dielectric materials, as well as their applications. Chapter 15 reviews the advances of the electrocaloric effect, concentrating on its occurrence in ceramic films, bulk ceramics, single crystals, and multi-layer ceramics (MLCs), and the designs and demonstrations of ceramics-based cooling devices. Chapter 16 reviews piezo-bioelectronic materials and emerging applications, including materials, fabrication, and emerging biomedical applications of piezo-bioelectronic devices. Chapter 17 reviews the advances of porous piezoelectric materials for energy technologies, concentrating on the types, fabrication techniques, modeling and characterization, and applications of porous piezoelectric materials. Chapter 18 reviews piezoelectric flexible devices, including their working mechanism, materials, structure design, and applications. Chapter 19 reviews the advances of ferroelectrics under compression, including lead-based and lead-free ferroelectrics and their applications. In Chapter 20, the advances of piezoelectricity in biomedical applications are reviewed, concentrating on several representative applications including implantable energy harvesters, ultrasound imaging transducers, bone/tissue regeneration, neural stimulation, and cancer therapy. Each chapter concludes with a section on the future perspective and concluding remarks.

Finally, it is my sincere hope that this book will serve as a valuable resource for those interested in the fascinating world of piezoelectric materials and devices. The broad array of topics covered, from fundamental concepts to cutting-edge research, aims to provide readers with a comprehensive understanding of the field. This book is a testament to the collective efforts of numerous researchers, scientists, and scholars whose contributions have significantly advanced our understanding of piezoelectric materials and their multifaceted applications.

In closing, I would like to emphasize my deep appreciation to all those who have contributed to this endeavor. The work presented herein represents the synthesis of many minds and the culmination of considerable effort. I trust that this book will provide readers with a compelling and enlightening exploration of the world of piezoelectrics.

February 26, 2024

Jiagang Wu
Sichuan University

1

Piezoelectric Coefficients and Crystallographic Symmetry

Semën Gorfman[1] and Nan Zhang[2]

[1] *Tel Aviv University, The Department of Materials Science and Engineering, Ramat Aviv, Tel Aviv 6997801, Israel*
[2] *Xi'an Jiaotong University, School of Electronic Science and Engineering, Electronic Materials Research Laboratory, Key Laboratory of the Ministry of Education & International Center for Dielectric Research, 28 West Xianning Road, Xi'an 710049, China*

1.1 Introduction

Piezoelectricity is the ability of some crystals to convert mechanical and electrical energy into each other. The direct piezoelectric effect describes the linear dependence of electric polarization in a crystal on applied external mechanical stress. The converse piezoelectric effect then describes the linear dependence of the mechanical deformation in a crystal on an applied external electric field. Piezoelectricity possesses many useful applications, such as pressure sensors/actuators [1], transducers [2], piezoelectric motors [3], sonars, and microelectromechanical systems (MEMs) [4]. The market for piezoelectric devices is projected to reach US$41.0 billion by 2027. Over the last few decades, piezoelectricity has been the subject of considerable academic research, review papers, and monographs [5, 6]. In this chapter, we introduce the anisotropy of piezoelectric coefficients and their relationship with the crystallographic symmetry of materials.

1.2 On the Anisotropy of Piezoelectricity and Its Role in the Understanding of Structures of Crystalline Materials

Incidentally, the discovery of piezoelectricity played a key role in the evolution of ideas concerning the periodic atomic structures of crystals. Notably, such periodicity was only a hypothesis at that time and was experimentally proven later in 1912 by Max von Laue [7]. Piezoelectricity presented one of those first physical properties that highlighted the anisotropy (directional dependence) of single crystals. Indeed, an isotropic material cannot be piezoelectric: the development of polarization under applied stress means that at least the opposite ends of the polarization

Piezoelectric Materials: From Fundamentals to Emerging Applications, First Edition. Edited by Jiagang Wu.

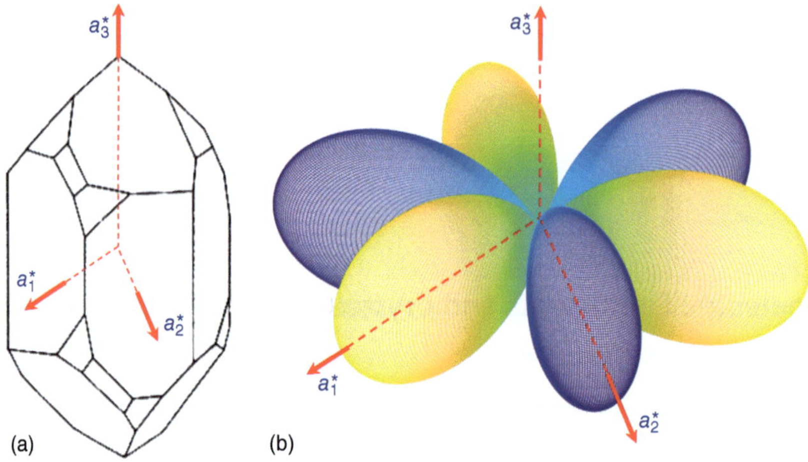

Figure 1.1 Schematic illustration of the anisotropy of piezoelectric effect in α-quartz: (a) typical natural shape of quartz crystal with three non-coplanar directions along the basis vectors of the reciprocal lattice $\boldsymbol{a}_1^*, \boldsymbol{a}_2^*, \boldsymbol{a}_3^*$, (b) the directional dependence of the magnitude of longitudinal piezoelectric coefficient (see the definition in the next paragraph). Remarkably, α-quartz develops no polarization under uniaxial strain along $\boldsymbol{a}_3/\boldsymbol{a}_3^*$ axis (parallel to the threefold symmetry axis) and does not deform under electric field parallel to $\boldsymbol{a}_3/\boldsymbol{a}_3^*$ axis. Source: Semën Gorfman.

vector are different; hence, a crystal must be anisotropic in order to be piezoelectric. The performance of a piezoelectric device strongly depends on its orientation relative to the lattice basis vectors. Figure 1.1 presents one particular example of the anisotropic longitudinal piezoelectricity in α-quartz.

In homogeneous single crystals, the anisotropy of properties must originate from the intrinsic asymmetry of atomic structures maintained over macroscopically long distances. This can only be possible if the structure is long-range-ordered. This is why anisotropy of crystals in general and piezoelectricity of crystals in particular are the key ingredients for the idea of periodicity/long-range order of crystals.

1.3 Simplified Definitions of Piezoelectric Coefficients

As for any anisotropic physical property, the description of piezoelectricity requires handling several property coefficients. There are very few classes of piezoelectric crystals where only one independent coefficient is sufficient (e.g. $Bi_{12}SiO_{20}$ [8]). However, prior to introducing the proper tensor description of the piezoelectricity, we will inspect the simplified didactical definitions and understand the *physical meaning* of piezoelectric coefficients. Specifically, we will assume that a piezoelectric device has been prepared as a parallel plate capacitor, whereas mechanical stress, electric field, polarization, and deformation appear along the plate normally only (Figure 1.2).

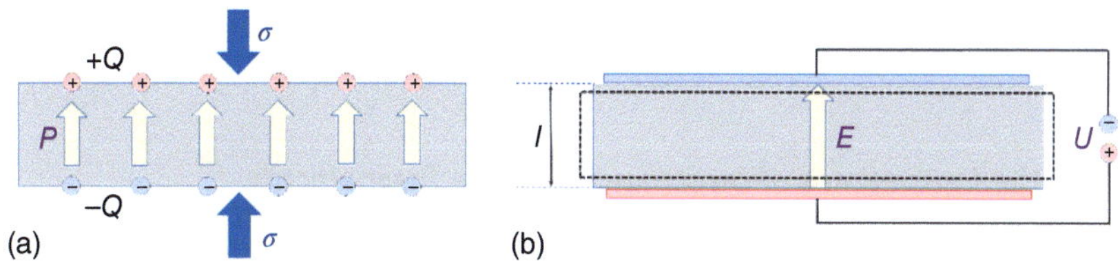

Figure 1.2 Simplified definitions of direct and converse piezoelectric effect: (a) the uniaxial stress σ is applied normally to the piezoelectric plate and causes polarization $\Delta P = d \cdot \sigma$ with the positive and negative charges emerging outside the plate face, (b) electric field is applied normally to the piezoelectric plate with the change of the plate thickness so that $x = d \cdot E$. Here, $x = \Delta l/l$. The shape of the deformed plate is shown by the dashed rectangle. Source: Semën Gorfman.

1.3.1 Direct Piezoelectric Effect

The simplified definition of direct piezoelectric effect is:

$$\Delta P = d \cdot \sigma \tag{1.1}$$

here, ΔP is the projection of the change in the polarization vector to the plate normal and σ is the pressure of the uniaxial forces normally applied to the plate. σ is positive/negative for extending/compressing forces. d is the direct piezoelectric coefficient. The units of polarization and stress are $[P] = \frac{C}{m^2}$ and $[\sigma] = \frac{N}{m^2}$; therefore, the units of the piezoelectric coefficient are $[d] = \frac{C}{N}$ or (practically used) $\frac{pC}{N} = 10^{-12}\frac{C}{N}$. This magnitude becomes more intuitive if we rewrite (1.1) as:

$$Q = d \cdot F \tag{1.2}$$

Here, $F = \sigma \cdot S$ is the magnitude of the force applied to the plate, $Q = \Delta P \cdot S$ is the electric charge outside the plate due to the increment of the polarization. $\boldsymbol{S}$ is the area of each face. According to (1.2), direct piezoelectric coefficient is the amount of electric charges accumulated outside the crystal plate per force applied to the plate. Thus, 1 N applied to a 1 pC/N piezoelectric plate will release 1 pC on the plate faces (positive on one side and negative on the opposite side).

1.3.2 Converse Piezoelectric Effect

Similarly, the simplified definition of the converse piezoelectric effect is:

$$x = d \cdot E \tag{1.3}$$

Here, E is the projection of the electric field vector to the plate normal, $x = \Delta l/l$ is the relative expansion of the plate. x is positive/negative for expansion/contraction. l is the thickness of the crystal plate. d is the converse piezoelectric coefficient. The units of converse piezoelectric coefficient are $[d] = \frac{m}{V}$ or, in practical terms, $\frac{pm}{V} = 10^{-12}\frac{m}{V}$. To get an impression about this unit, we can reformulate (1.3) as:

$$\Delta l = d \cdot U \tag{1.4}$$

Here, $U = E \cdot l$ is the voltage falling between the faces of the plate. Thus, converse piezoelectric coefficient is the elongation of the plate per applied voltage. For example, the application of 1000 V to a 1 pm/V piezoelectric crystal plate expands or shrinks by 1 nm.

1.3.3 On the Equivalence of Direct and Converse Piezoelectric Coefficients

It can be easily shown that the units of $\frac{C}{N}$ are equivalent to $\frac{m}{V}$. This equivalence raises the question of whether there exists any relationship between the direct and converse piezoelectric coefficients. Indeed, these coefficients are not just related; if defined exactly as in Eqs. (1.1) and (1.3), they are *numerically equal to each other*. This is the reason why the same symbol d for the direct and converse piezoelectric coefficients can be retained. However, only for this paragraph, we will temporarily introduce the notations $d^{(d)}$ and $d^{(c)}$ for the direct and converse piezoelectric coefficients, correspondingly.

Here, we will look back at Lippmann's original theory [9] for the longitudinal direct Eq. (1.1) and converse Eq. (1.3) effects only. The interested reader may refer to Nye's book [10] for a comprehensive description of the matter. We assume that a small variation of external mechanical stress, $d\sigma$, and electric field, dE, results in the increments of mechanical strain, dx, and polarization, dP. The first law of thermodynamics expresses the conservation of energy during the electric polarization and mechanical deformation of the material as:

$$dU = \delta A + \delta Q \tag{1.5}$$

Here, dU is the change in the internal energy of the crystal, δQ is the amount of energy received by the crystal in the form of *heat*, δA is the amount of energy received by the crystal in the form of *work* by the external electric ($\delta A_E = E \cdot dP$) and mechanical ($\delta A_M = \sigma \cdot dx$) forces [10]. All the quantities are expressed per unit volume of the material. Assuming no heat exchange with the environment ($\delta Q = 0$ otherwise, further caloric physical properties would be involved), we can rewrite (1.5) as:

$$dU = E \cdot dP + \sigma \cdot dx \tag{1.6}$$

Introducing $\Phi = U - E \cdot P - \sigma \cdot x$, we can rewrite (1.6)

$$d\Phi = -P \cdot dE - x \cdot d\sigma \tag{1.7}$$

Because Φ is a full differential, the polarization and strain can be expressed as partial derivatives:

$$P = -\frac{\partial \Phi}{\partial E}, x = -\frac{\partial \Phi}{\partial \sigma} \tag{1.8}$$

On the other hand, equations (1.1) and (1.3) can also be rewritten in the form of partial derivatives, where $d^{(d)} = \frac{\partial P}{\partial \sigma}$ represents the direct piezoelectric effect and $d^{(c)} = \frac{\partial x}{\partial E}$ represents the converse piezoelectric effect. The equivalence between $d^{(d)}$ and $d^{(c)}$ follows immediately as:

$$d^{(d)} = -\frac{\partial^2 \Phi}{\partial \sigma \partial E} \text{ and } d^{(c)} = -\frac{\partial^2 \Phi}{\partial E \partial \sigma} \tag{1.9}$$

Table 1.1 Typical orders of magnitude of piezoelectric coefficients.

Material	*d* (pC/N)	Refs.	Material	*d* (pC/N)	References
α-quartz (SiO_2)	2	[11]	$BaTiO_3$	150	[12]
PVDF	20	[13]	$PbZr_{0.52}Ti_{0.48}O_3$ (PZT ceramics)	200–400	[14]
$Al_{0.67}Sc_{0.33}N$	26	[15]	PMN–PT single crystals	2000	[16]
$Bi_{12}SiO_{20}$	40	[8]	Sm-doped PMN–PT single crystals	3400–4100	[17]

This is because the second derivative does not depend on the order of variables.

This remarkable feature, appearing in the form of "conjugated" pairs, also holds for many other properties. For example, pyroelectricity is conjugated with the electrocaloric effect and thermal expansion is conjugated with the piezocaloric effect.

1.3.4 Typical Values of Piezoelectric Coefficients

Following our understanding of the piezoelectric coefficient units, it is worth observing their typical values. Table 1.1 lists the piezoelectric coefficients of a few selected materials. Notably, the table here is merely for the purpose of obtaining an impression regarding the typical magnitude of piezoelectric coefficients. As will be discussed in the following chapters, piezoelectric coefficients are described by tensors, and each material may have few coefficients that may differ from one another by an order of magnitude. Still, it is clear from the table that piezoelectric coefficients of most known materials are below 2000 pC/N. Only in so-called relaxor ferroelectric single crystals of $PbMg_{1/3}Nb_{2/3}O_3$–$PbTiO_3$ (PMN–PT) [16, 18] and their derivatives [17], piezoelectric coefficients may exceed such high values.

1.4 Description of Piezoelectricity by Third-rank Tensor

1.4.1 Tensor of Piezoelectric Coefficients

Piezoelectricity of single crystals must be described by tensors. For the direct piezoelectric effect:

$$\Delta P_k = d_{kij}\sigma_{ij} \tag{1.10}$$

Here, ΔP_k represents the components of the change in the polarization vector, σ_{ij} represents the components of the second-rank stress tensor, and d_{kij} represents the components of the third-rank tensor of direct piezoelectric coefficients. More information about the meaning of the stress and strain tensors can be found in [10, 19, 20]. The indices i, j, k run between 1 and 3, and the Einstein summation rule applies to every repeated index (i and j in this case). The vector and tensor components are expressed as relative to a Cartesian coordinate system $\boldsymbol{e}_1$, $\boldsymbol{e}_2$, $\boldsymbol{e}_3$. Later, we will elaborate on the crucial role of this Cartesian system.

According to Eq. (1.10), d_{kij} describes the polarization increment along the Cartesian $\boldsymbol{e}_k$ axis due to the force applied along the e_i to the surface, which is normal to $\boldsymbol{e}_j$. For example, if the piezoelectric coefficient d_{233} is nonzero, then the polarization along the $\boldsymbol{e}_2$ axis will develop in response to the *tensile* mechanical stress on the face, normal to $\boldsymbol{e}_3$. If, for example, the piezoelectric coefficient d_{123} is nonzero, then the polarization along the $\boldsymbol{e}_1$ will develop due to the *shearing* stress on the face normal to the $\boldsymbol{e}_3$ but in the direction of $\boldsymbol{e}_2$.

For the converse piezoelectric effect:

$$x_{ij} = d_{kij}E_k \tag{1.11}$$

Here, x_{ij} represents the components of the second-rank strain tensor and E_k represents the components of the electric field vector. In this case, the Einstein summation applies over the repeated index k. According to (1.11), the converse piezoelectric coefficient, d_{kij}, describes the strain component, x_{ij}, in response to the electric field along the Cartesian $\boldsymbol{e}_k$ axis. For example, the converse piezoelectric coefficient d_{233} stands for the *elongation* along the $\boldsymbol{e}_3$ in response to an electric field applied along $\boldsymbol{e}_2$. The converse piezoelectric coefficient d_{123} describes the *shear* strain in the $\boldsymbol{e}_2\boldsymbol{e}_3$ plane due to electric field along $\boldsymbol{e}_1$.

The thermodynamics of the piezoelectric effect remains similar to the simplified case considered earlier in this chapter. Specifically, if expressed in the same Cartesian coordinate system, *all* the components of the direct piezoelectric tensor are numerically equal to those for the converse piezoelectric tensor. However, it is crucial that the tensor components are defined exactly as in (1.10) and (1.11), i.e. the first index is "electric," while the second and the third indices are "mechanical."

1.4.2 Voigt/Matrix Notation for Piezoelectric Coefficient

Any third-rank tensor has a maximum of $27 = 3^3$ independent components. However, their real number is smaller because of the symmetry. There are two types of symmetries: the first is *physical* and appears as a consequence of energy, charge, and momentum conservation laws. The second is *crystallographic*, which is inherited from the geometrical symmetry of crystal structures and expressed in the form of the Neumann principle. For piezoelectricity, the *physical* symmetry results in the invariance with respect to the interchange of "mechanical" indices:

$$d_{kij} = d_{kji} \tag{1.12}$$

Equation (1.12) is valid for any piezoelectric material, regardless of which atomic structure it contains. This property is intuitively related to similar symmetry properties of the strain and stress tensors $x_{ij} = x_{ji}$, $\sigma_{ij} = \sigma_{ji}$. This symmetry means that the pair ij may cover six different variants: $11, 22, 33, 23 \equiv 32, 13 \equiv 31, 12 \equiv 21$. For writing compactness reasons, a special indexing scheme is commonly used (first introduced

Table 1.2 The substitution of "mechanical" indices using Voigt notation.

ij (tensor)	11	22	33	23 ≡ 32	13 ≡ 31	12 ≡ 21
m (Voigt)	1	2	3	4	5	6
Strain, x_m=	x_{11}	x_{22}	x_{33}	$2x_{23}$	$2x_{13}$	$2x_{12}$
Stress, σ_m=	σ_{11}	σ_{22}	σ_{33}	σ_{23}	σ_{13}	σ_{12}
Piezo, d_{km}=	d_{k11}	d_{k22}	d_{k33}	$2d_{k23}$	$2d_{k13}$	$2d_{k12}$

The first row of the table shows six different pairs of indices ij. The second row shows the Voigt index m, corresponding to each pair. The third, fourth, and fifth rows show the relationship between the components of the strain, stress, and piezoelectric tensors in Voigt and index notations.

by Voigt [21]). The pair ij can be replaced by a single index $m = 1\ldots6$, according to Table 1.2.

Such substitution is used for strain and stress tensors too $x_m \leftarrow x_{ij}$ and $\sigma_m \leftarrow \sigma_{ij}$, and is shown in the next two rows of Table 1.2. The Voigt substitution of indices for the piezoelectric coefficients follows as $d_{km} \leftarrow d_{kij}$, where $ij = 1\ldots3$ and $m = 1\ldots6$, and is shown in the last row of the same table. Then, equations (1.10) and (1.11) can be rewritten in a shorter form:

$$P_k = d_{km}\sigma_m \tag{1.13}$$

$$x_m = d_{km}E_k \tag{1.14}$$

The piezoelectric coefficients are written in the form of 3 × 6 matrix:

$$\begin{pmatrix} d_{11} & d_{12} & d_{13} & d_{14} & d_{15} & d_{16} \\ d_{21} & d_{22} & d_{23} & d_{24} & d_{25} & d_{26} \\ d_{31} & d_{32} & d_{33} & d_{34} & d_{35} & d_{36} \end{pmatrix} = \begin{pmatrix} d_{111} & d_{122} & d_{133} & 2d_{123} & 2d_{113} & 2d_{112} \\ d_{211} & d_{222} & d_{233} & 2d_{223} & 2d_{213} & 2d_{212} \\ d_{311} & d_{322} & d_{333} & 2d_{323} & 2d_{313} & 2d_{312} \end{pmatrix} \tag{1.15}$$

The maximum number of independent piezoelectric coefficients is 18. This number can be reduced if we consider the geometrical symmetry of the crystal or medium.

1.5 Symmetry Considerations

1.5.1 Neumann Principle for the Piezoelectric Effect

As for any physical property, the Neumann principle [22] holds for the piezoelectricity as well. This principle states that: "*if a crystal is invariant with respect to certain symmetry operations any of its physical properties must also be invariant with respect to the same symmetry operations.*" In other words, the components of the piezoelectric tensor should be invariant with respect to the transformation between two symmetry-equivalent Cartesian coordinate systems:

$$d'_{ijk} = d_{ijk} \tag{1.16}$$

Here, d'_{ijk} represents the components of the piezoelectric coefficient tensor relative to the new Cartesian coordinate system $\boldsymbol{e}'_1, \boldsymbol{e}'_2, \boldsymbol{e}'_3$, which is symmetry-equivalent to the original Cartesian coordinate system $\boldsymbol{e}_1, \boldsymbol{e}_2, \boldsymbol{e}_3$. Allow us to introduce the matrix $[T]$, which enables the transformation between these coordinate systems using the following formal matrix equation:

$$(\boldsymbol{e}_1\ \boldsymbol{e}_2\ \boldsymbol{e}_3) = \left(\boldsymbol{e}'_1\ \boldsymbol{e}'_2\ \boldsymbol{e}'_3\right) \begin{bmatrix} T_{11} & T_{12} & T_{13} \\ T_{21} & T_{22} & T_{23} \\ T_{31} & T_{32} & T_{33} \end{bmatrix}, \text{or } \boldsymbol{e}_i = \boldsymbol{e}'_k T_{ki} \tag{1.17}$$

Here, the columns of the transformation matrix represent the vectors' coordinates of the old coordinate system $\boldsymbol{e}_i$ relative to the new coordinate system $\boldsymbol{e}'_i$. It can be shown [10, 20] that the transformation of the tensor components from the coordinate system $\boldsymbol{e}_i$ to the coordinate system $\boldsymbol{e}'_i$ is:

$$d'_{ijk} = T_{ii_1} T_{jj_1} T_{kk_1} d_{i_1 j_1 k_1} \tag{1.18}$$

If $[T]$ describes the symmetry operation, then

$$d_{ijk} = T_{ii_1} T_{jj_1} T_{kk_1} d_{i_1 j_1 k_1} \tag{1.19}$$

It is important to note that equations (1.18) and (1.19) can be applied only to the full tensor form. Obviously, the Voigt 3×6 matrix cannot be directly used here. In the following, we will consider how the presence of one or another symmetry operation in the structure of a crystal reduces the number of piezoelectric coefficients.

Example 1.1 ***Center of Inversion***

Let us assume that a crystal structure is centrosymmetric. The following transformation matrix relates two symmetry-equivalent Cartesian coordinate systems:

$$[T] = \begin{pmatrix} \bar{1} & 0 & 0 \\ 0 & \bar{1} & 0 \\ 0 & 0 & \bar{1} \end{pmatrix} \text{ or } T_{ij} = -\delta_{ij} \tag{1.20}$$

Substituting (1.20) into (1.19) and using the property of the Kronecker symbol, $\delta_{ij} x_j = x_i$, we obtain

$$d_{ijk} = (-1)^3 d_{ijk} \text{or equivalently } d_{ijk} = 0 \tag{1.21}$$

In other words, all piezoelectric coefficients are zero, so that centrosymmetric material cannot be piezoelectric. This property instantly removes the crystals of 11 centrosymmetric (Laue) classes from the list of potential piezoelectric materials. Conversely, observing piezoelectricity in a given material indicates that it is non-centrosymmetric. This ability to identify the absence of the inversion center is useful in X-ray structure analysis [23, 24].

Example 1.2 ***Mirror Plane***

Let us assume that the structure of a crystal has a mirror plane perpendicular to $\boldsymbol{e}_3$. Thus, the following transformation matrix relates two symmetry-equivalent

Cartesian coordinate systems:

$$[T] = \begin{pmatrix} 1 & 0 & 0 \\ 0 & 1 & 0 \\ 0 & 0 & \bar{1} \end{pmatrix} \tag{1.22}$$

Substituting (1.22) in (1.19), we can rewrite (1.19) as:

$$d_{ijk} = (-1)^n d_{ijk} \tag{1.23}$$

Here, $n \leq 3$ is the number of cases where any of the three indices is equal 3 (e.g. $n = 0$ for d_{122}, $n = 1$ for $d_{123,}$ or $n = 2$ for d_{323}). Accordingly, the tensor of piezoelectric coefficients in the matrix (Voigt notation) will take the form:

$$[d] = \begin{pmatrix} d_{11} & d_{12} & d_{13} & 0 & 0 & d_{16} \\ d_{21} & d_{22} & d_{23} & 0 & 0 & d_{26} \\ 0 & 0 & 0 & d_{34} & d_{35} & 0 \end{pmatrix} \tag{1.24}$$

with 10 independent piezoelectric coefficients.

1.5.2 The Choice of the Cartesian Coordinate System

According to (1.18), the values of piezoelectric coefficients depend on the choice of the coordinate system. Therefore, every time when piezoelectric coefficients of a material are reported or retrieved from a literature, it is crucial to know how the coordinate system is chosen. All physical properties are described in Cartesian coordinate systems and are commonly referred to as the crystal physical coordinate system. It is imperative to understand that this system is not related to the experimental setup but rather to the basis vectors of the crystal itself.

For single crystals, it is convenient to relate the axes of the crystal's physical coordinate system $\mathbf{e}_1$, $\mathbf{e}_2$, $\mathbf{e}_3$ to the axes of the crystallographic, $\boldsymbol{a}_1$, $\boldsymbol{a}_2$, $\boldsymbol{a}_3$ and the reciprocal crystallographic coordinate systems $\boldsymbol{a}_1^*, \boldsymbol{a}_2^*, \boldsymbol{a}_3^*$. The basis vectors of the reciprocal coordinate system are related to that of the crystallographic coordinate system by the scalar products $\boldsymbol{a}_i\boldsymbol{a}_j^* = \delta_{ij}$. Table 1.3 summarizes these three types of coordinate systems, including their metric (the lengths and the angles between the basis vectors), their usage, and their conventional choice. The standard relationships between the crystal physical and crystallographic coordinate systems are given, e.g. in [35]. Nowadays, it is common to define the crystal physical coordinate system as in Figure 1.3a, so that $\mathbf{e}_2//\boldsymbol{a}_2$, $\mathbf{e}_1//\boldsymbol{a}_1^*$, and $\mathbf{e}_3 = [\mathbf{e}_1 \times \mathbf{e}_2]$. Nonetheless, for the hexagonal cell setting case, it is more common to define the crystal physical coordinate system as $\mathbf{e}_3//\boldsymbol{a}_3$, $\mathbf{e}_1//\boldsymbol{a}_1$, and $\mathbf{e}_1 = [\mathbf{e}_2 \times \mathbf{e}_3]$. In any case, we discourage the reader from making a blind assumption of one or another convention used and recommend a critical inspection of every individual case. The cases of trigonal crystals are particularly difficult since the crystallographic coordinate system $\boldsymbol{a}_1$, $\boldsymbol{a}_2$, $\boldsymbol{a}_3$ itself has two "standard" cell settings: primitive ($\boldsymbol{a}_1 + \boldsymbol{a}_2 + \boldsymbol{a}_3$ is parallel to the threefold axis) or *R*-centered ($\boldsymbol{a}_3$ is parallel to the threefold axis).

Table 1.3 Coordinate systems used for the description of crystal structures and properties.

Coordinate system	Crystallographic	Reciprocal crystallographic	Crystal physical
Designation for the basis vectors	$\boldsymbol{a},\boldsymbol{b},\boldsymbol{c}$ or $\boldsymbol{a}_1,\boldsymbol{a}_2,\boldsymbol{a}_3$	$\boldsymbol{a}^*,\boldsymbol{b}^*,\boldsymbol{c}^*$ or $\boldsymbol{a}_1^*,\boldsymbol{a}_2^*,\boldsymbol{a}_3^*$	$\boldsymbol{x},\boldsymbol{y},\boldsymbol{z}$ or $\boldsymbol{e}_1,\boldsymbol{e}_2,\boldsymbol{e}_3$
Metric	Depends on the crystal system	Depends on the crystal system	Cartesian $\boldsymbol{e}_i\boldsymbol{e}_j=\delta_{ij}$
Used for	Description of periodic structures of crystals	Description of diffraction pattern and natural shapes of crystalline polyhedrons	Physical properties
The conventional choice, orientation of basis vectors	Aligned to certain symmetry elements of the crystal lattice	Related to the crystallographic coordinate system	Attached to the crystallographic coordinate system

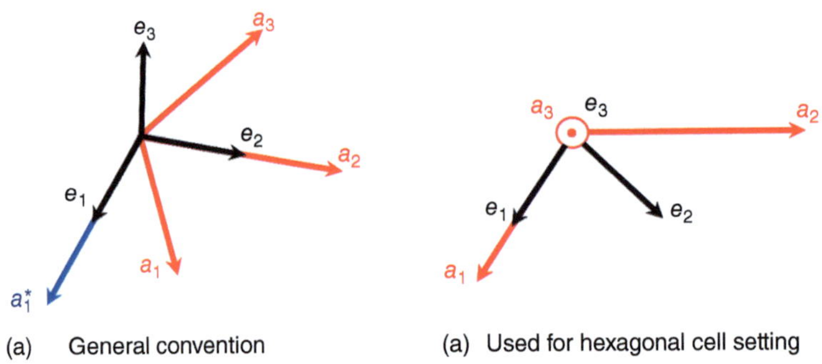

Figure 1.3 Two conventional choices of crystal physical Cartesian coordinate system $\boldsymbol{e}_1$, $\boldsymbol{e}_2$, $\boldsymbol{e}_3$ relative to the crystallographic $\boldsymbol{a}_1$, $\boldsymbol{a}_2$, $\boldsymbol{a}_3$: (a) general three-dimensional case where $\boldsymbol{e}_2 \parallel \boldsymbol{a}_2$ and $\boldsymbol{e}_1 \parallel \boldsymbol{a}_1^*$, (b) special convention, which is frequently used for hexagonal lattice setting (in hexagonal and trigonal crystal classes), where $\boldsymbol{e}_1 \parallel \boldsymbol{a}_1$ and $\boldsymbol{e}_3 \parallel \boldsymbol{a}_3$. Note that special attention should be paid to the trigonal crystal classes 32 and 3*m* (where the two standard orientations of the crystallographic coordinate system itself are possible). Source: Semën Gorfman.

1.5.3 How to Use the Space Symmetry Group Information to Find the List of Independent Piezoelectric Coefficients

The transformation of the coordinate systems, keeping the crystal structure invariant, is contained in the information regarding the type of a space symmetry group. There are 230 types of space symmetry groups describing *three-dimensional periodic crystal structures*, all of which are listed in the International Tables for Crystallography, Volume A [25]. Each symmetry operation includes the "rotation" and "translation" parts. The translation part is irrelevant for macroscopic physical properties: for example, a "screw" axis or "glide" plane acts as a simple rotation axis and a mirror

Table 1.4 The shapes of piezoelectric tensors for 20 piezoelectric crystal classes.

Crystal system	Crystal class	Piezoelectric tensor (Number of independent coefficients)	Conventional choice of Cartesian coordinate system
Triclinic	1	$\begin{pmatrix} d_{11} & d_{12} & d_{13} & d_{14} & d_{15} & d_{16} \\ d_{21} & d_{22} & d_{23} & d_{24} & d_{25} & d_{26} \\ d_{31} & d_{32} & d_{33} & d_{34} & d_{35} & d_{36} \end{pmatrix}$ (18)	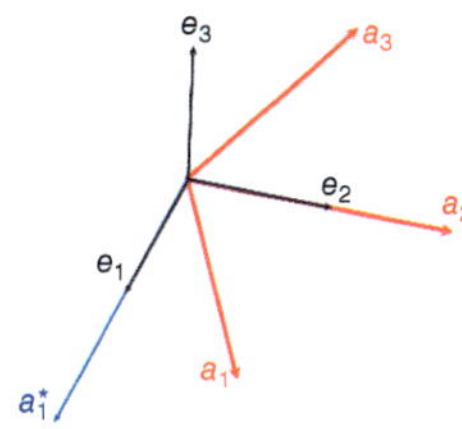
Monoclinic	2	$\begin{pmatrix} 0 & 0 & 0 & d_{14} & 0 & d_{16} \\ d_{21} & d_{22} & d_{23} & 0 & d_{25} & 0 \\ 0 & 0 & 0 & d_{34} & 0 & d_{36} \end{pmatrix}$ (8)	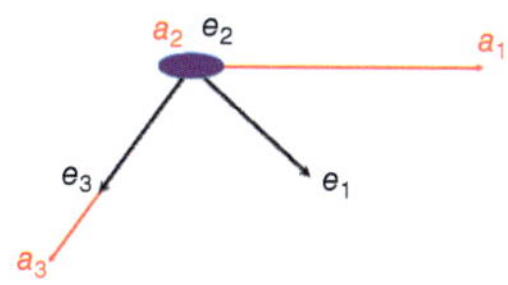
	m	$\begin{pmatrix} d_{11} & d_{12} & d_{13} & 0 & d_{15} & 0 \\ 0 & 0 & 0 & d_{24} & 0 & d_{26} \\ d_{31} & d_{32} & d_{33} & 0 & d_{35} & 0 \end{pmatrix}$ (10)	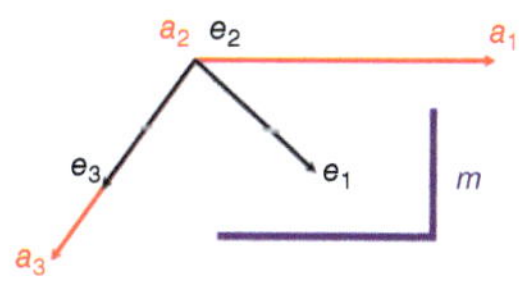
Orthorhombic	222	$\begin{pmatrix} 0 & 0 & 0 & d_{14} & 0 & 0 \\ 0 & 0 & 0 & 0 & d_{25} & 0 \\ 0 & 0 & 0 & 0 & 0 & d_{36} \end{pmatrix}$ (3)	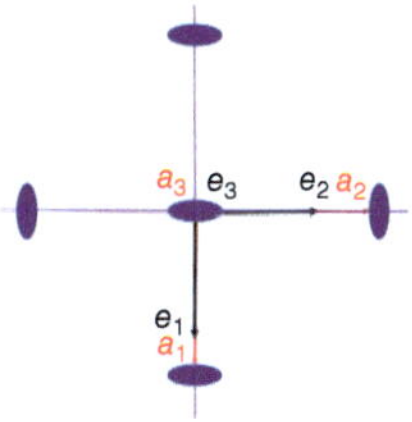
	$mm2$	$\begin{pmatrix} 0 & 0 & 0 & 0 & d_{15} & 0 \\ 0 & 0 & 0 & d_{24} & 0 & 0 \\ d_{31} & d_{32} & d_{33} & 0 & 0 & 0 \end{pmatrix}$ (5)	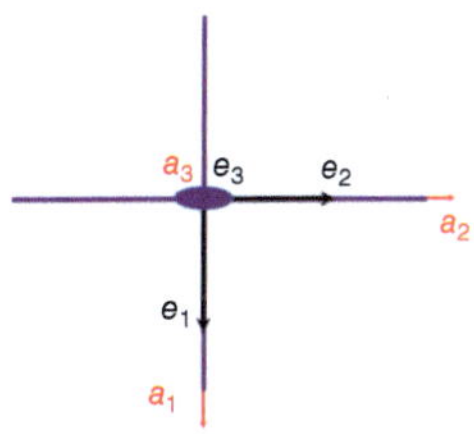
Tetragonal	$4mm$	$\begin{pmatrix} 0 & 0 & 0 & 0 & d_{15} & 0 \\ 0 & 0 & 0 & d_{15} & 0 & 0 \\ d_{31} & d_{31} & d_{33} & 0 & 0 & 0 \end{pmatrix}$ (3)	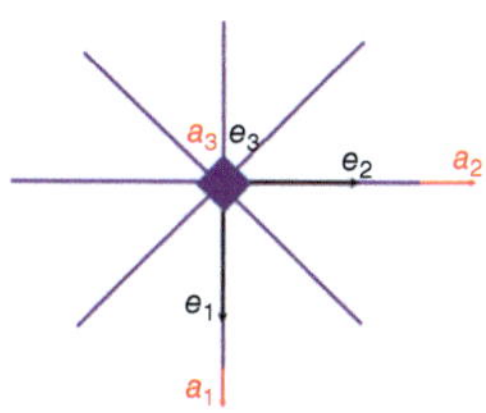

(Continued)

Table 1.4 (Continued)

Crystal system	Crystal class	Piezoelectric tensor (Number of independent coefficients)	Conventional choice of Cartesian coordinate system
	422	$\begin{pmatrix} 0 & 0 & 0 & d_{14} & 0 & 0 \\ 0 & 0 & 0 & 0 & -d_{14} & 0 \\ 0 & 0 & 0 & 0 & 0 & 0 \end{pmatrix}$ (1)	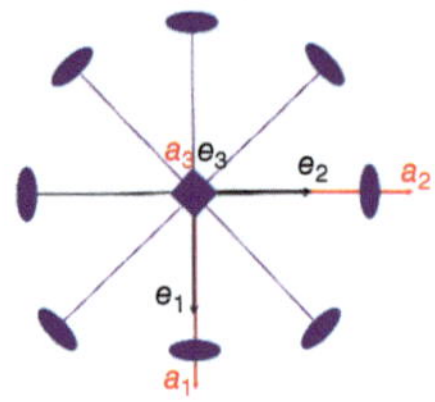
	$\bar{4}2m$	$\begin{pmatrix} 0 & 0 & 0 & d_{14} & 0 & 0 \\ 0 & 0 & 0 & 0 & d_{14} & 0 \\ 0 & 0 & 0 & 0 & 0 & d_{36} \end{pmatrix}$ (2)	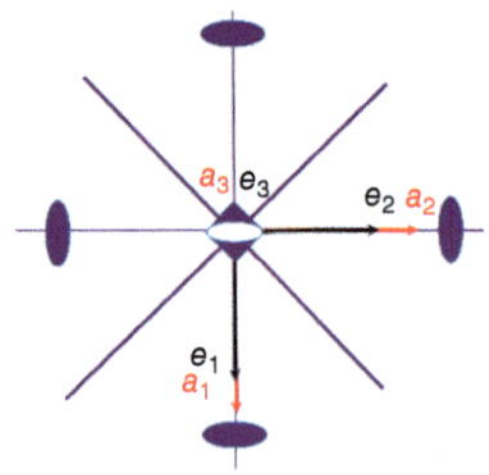
	4	$\begin{pmatrix} 0 & 0 & 0 & d_{14} & d_{15} & 0 \\ 0 & 0 & 0 & d_{15} & -d_{14} & 0 \\ d_{31} & d_{31} & d_{33} & 0 & 0 & 0 \end{pmatrix}$ (4)	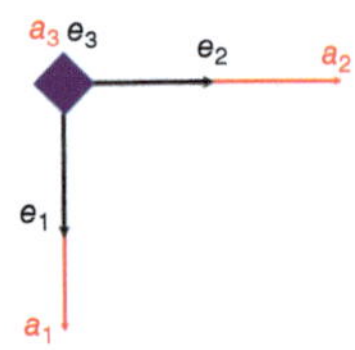
	$\bar{4}$	$\begin{pmatrix} 0 & 0 & 0 & d_{14} & d_{15} & 0 \\ 0 & 0 & 0 & -d_{15} & d_{14} & 0 \\ d_{31} & -d_{31} & 0 & 0 & 0 & d_{36} \end{pmatrix}$ (4)	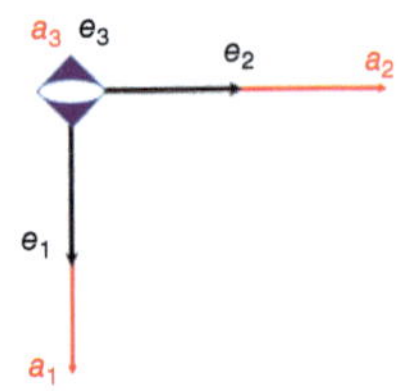
Trigonal	3	$\begin{pmatrix} d_{11} & -d_{11} & 0 & d_{14} & d_{15} & -2d_{22} \\ -d_{22} & d_{22} & 0 & d_{15} & -d_{14} & -2d_{11} \\ d_{31} & d_{31} & d_{33} & 0 & 0 & 0 \end{pmatrix}$ (6)	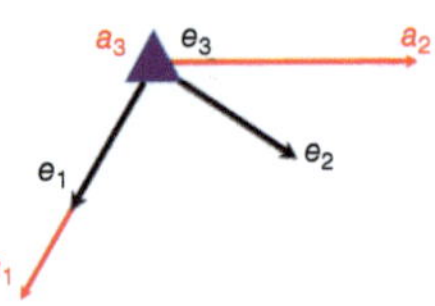
	32	$\begin{pmatrix} d_{11} & -d_{11} & 0 & d_{14} & 0 & 0 \\ 0 & 0 & 0 & 0 & -d_{14} & -2d_{11} \\ 0 & 0 & 0 & 0 & 0 & 0 \end{pmatrix}$ (2)	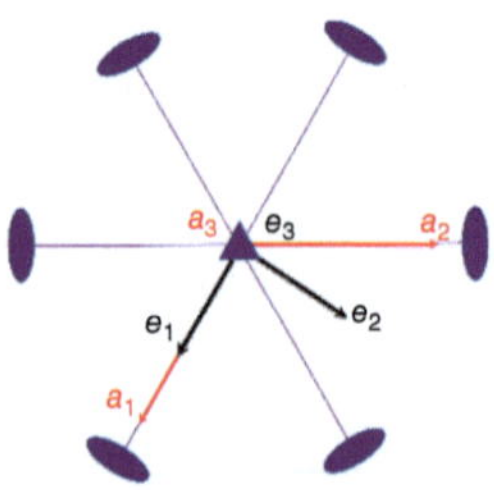

(Continued)

Table 1.4 (Continued)

	$3m$	$\begin{pmatrix} 0 & 0 & 0 & 0 & d_{15} & -2d_{22} \\ -d_{22} & d_{22} & 0 & d_{15} & 0 & 0 \\ d_{31} & d_{31} & d_{33} & 0 & 0 & 0 \end{pmatrix}$	(4)	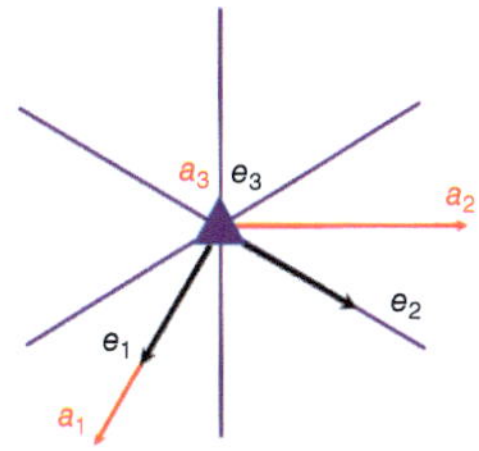
Hexagonal	$6mm$	$\begin{pmatrix} 0 & 0 & 0 & 0 & d_{15} & 0 \\ 0 & 0 & 0 & d_{15} & 0 & 0 \\ d_{31} & d_{31} & d_{33} & 0 & 0 & 0 \end{pmatrix}$	(3)	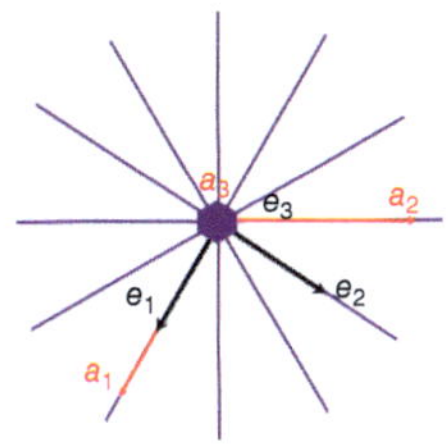
	622	$\begin{pmatrix} 0 & 0 & 0 & d_{14} & 0 & 0 \\ 0 & 0 & 0 & 0 & -d_{14} & 0 \\ 0 & 0 & 0 & 0 & 0 & 0 \end{pmatrix}$	(1)	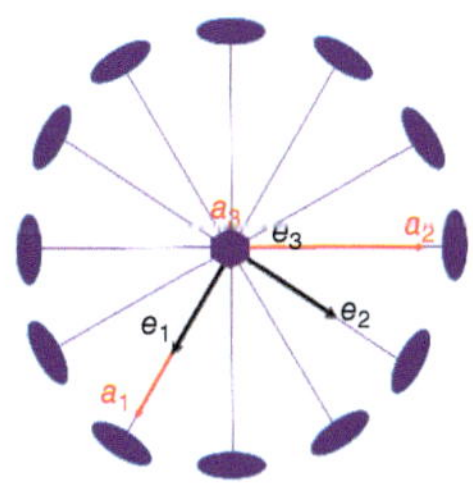
	$\bar{6}m2$	$\begin{pmatrix} d_{11} & -d_{11} & 0 & 0 & 0 & 0 \\ 0 & 0 & 0 & 0 & 0 & -2d_{11} \\ 0 & 0 & 0 & 0 & 0 & 0 \end{pmatrix}$	(1)	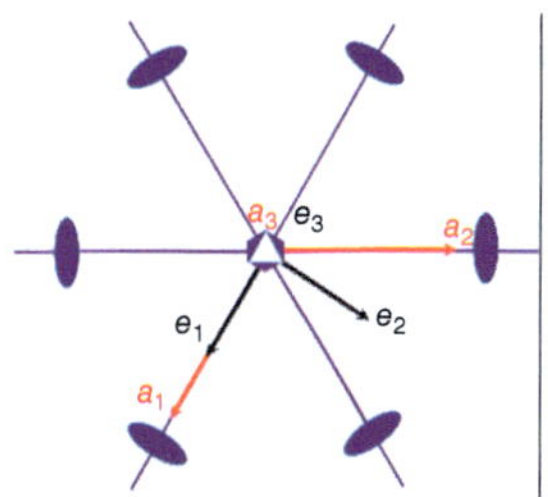
	6	$\begin{pmatrix} 0 & 0 & 0 & d_{14} & d_{15} & 0 \\ 0 & 0 & 0 & d_{15} & -d_{14} & 0 \\ d_{31} & d_{31} & d_{33} & 0 & 0 & 0 \end{pmatrix}$	(4)	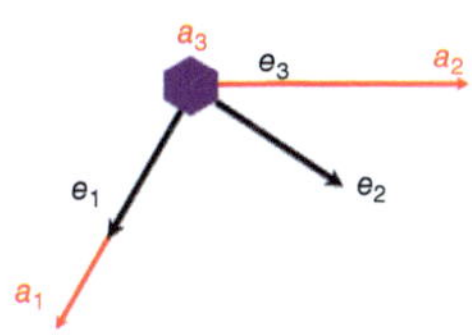
	$\bar{6}$	$\begin{pmatrix} d_{11} & -d_{11} & 0 & 0 & 0 & -2d_{22} \\ -d_{22} & d_{22} & 0 & 0 & 0 & -2d_{11} \\ 0 & 0 & 0 & 0 & 0 & 0 \end{pmatrix}$	(2)	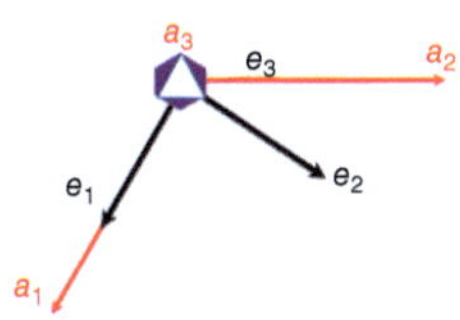
Cubic	$\bar{4}32$ 23	$\begin{pmatrix} 0 & 0 & 0 & d_{14} & 0 & 0 \\ 0 & 0 & 0 & 0 & d_{14} & 0 \\ 0 & 0 & 0 & 0 & 0 & d_{14} \end{pmatrix}$	(1)	The axes of Cartesian coordinate system are parallel to a_1, a_2, a_3

plane for the physical properties. If all the “translation” components of the symmetry operations are removed, then the 230 types of space symmetry groups are reduced to 32 types of point symmetry groups. These types of groups are called crystal classes. 11 of 32 crystal classes, known as Laue classes, and are centrosymmetric, ruling out piezoelectricity straight away. For the remaining 21 non-centrosymmetric crystal classes, the piezoelectricity must be analyzed according to the algorithm outlined previously. Here, we restrict ourselves to the derivation of the piezoelectric coefficients for all possible non-centrosymmetric crystal classes. The interested reader may consult Nye’s dedicated book[12].

1.5.4 The Shapes of Piezoelectric Tensors for Different Crystal Classes

Table 1.4 shows the shapes of tensors of piezoelectric coefficients for 20 piezoelectric crystal classes. The only non-centrosymmetric crystal class where all the piezoelectric coefficients vanish is 432. The first column of Table 1.4 lists the crystal system (crystal system is the symmetry group of the infinite crystal lattice without the unit cell). The second column shows the International Hermann–Mauguin symbol for the crystal classes. The last column adds some information about the assumed orientation of the crystal’s physical coordinate system relative to the crystallographic axes. It involves the drawing of the symmetry diagram so that the orientation of the crystallographic and crystal physical coordinate axes can be related to the symmetry elements of the crystal.

Some additional comments are in place here regarding the choice of the crystal physical Cartesian coordinate system. The following transformation of the Cartesian coordinate system does affect the numerical values of the “independent” piezoelectric coefficients without changing the shape of the piezoelectric tensor itself.

- For the crystal class 1: any change of the coordinate system.
- For the crystal classes 2,3, 4, $\bar{4}$, 6, $\bar{6}$, 4mm, 6mm: rotation around the symmetry axis by an arbitrary angle.
- For the crystal class m: rotation around the normal to the mirror plane.
- For all crystal classes: inversion of the coordinate axes will invert all the piezoelectric coefficients.

References

1 Uchino, K. (1996). *Piezoelectric Actuators and Ultrasonic Motors*. Springer Science & Business Media.

2 Damjanovic, D. (1998). *Curr. Opin. Solid State Mater. Sci.* 3: 469.

3 Uchino, K. (1998). *Smart Mater. Struct.* 7: 273.

4 Trolier-McKinstry, S. and Muralt, P. (2004). *J. Electroceram.* 12: 7.

5 Heywang, W., Lubitz, K., and Wersing, W. (2008). *Piezoelectricity: Evolution and Future of a Technology*. Springer Science & Business Media.

6 Uchino, K. (2017). *Advanced Piezoelectric Materials: Science and Technology*. Woodhead Publisher.

7 von Laue, M. (1912). *Eine Quantitative Prüfung Der Theorie Für Die Interferenz-Erscheinungen Bei Röntgenstrahlen*. Verlag der Kgl. Bayer. Akad. der Wiss.

8 Abrahams, S.C., Bernstein, J.L., and Svensson, C. (1979). *J. Chem. Phys.* 71: 788.

9 Katzir, S. (2007). *The Beginnings of Piezoelectricity: A Study in Mundane Physics*. Springer Science & Business Media.

10 Nye, J.F. and Physics Laboratory J F Nye (1985). *Physical Properties of Crystals: Their Representation by Tensors and Matrices*. Clarendon Press.

11 Bechmann, R. (1958). *Phys. Rev.* 110: 1060.

12 Park, S.-E., Wada, S., Cross, L.E., and Shrout, T.R. (1999). *J. Appl. Phys.* 86: 2746.

13 Niederhauser, J.J., Jaeger, M., Hejazi, M. et al. (2005). *Opt. Commun.* 253: 401.

14 Haertling, G.H. (1999). *J. Am. Ceram. Soc.* 82: 797.

15 Zywitzki, O., Modes, T., Barth, S. et al. (2017). *Surf. Coat. Technol.* 309: 417.

16 Park, S.-E. and Shrout, T.R. (1997). *J. Appl. Phys.* 82: 1804.

17 Li, F., Cabral, M.J., Xu, B. et al. (2019). *Science* 364: 264.

18 Qiu, C., Wang, B., Zhang, N. et al. (2020). *Nature* 577: 350.

19 Landau, L.D. and Lifshitz, E.M. (1959). *Course of Theoretical Physics Vol 7: Theory and Elasticity*. Pergamon Press.

20 Newnham, R.E. (2005). *Properties of Materials: Anisotropy, Symmetry, Structure*. Oxford University Press.

21 Voigt, W. et al. (1928). *Lehrbuch Der Kristallphysik*. Teubner Leipzig.

22 Neumann, F.E. and Meyer, O.E. (1885). *Vorlesungen über die Theorie der Elasticität der festen Körper und des Lichtäthers: gehalten an der Universität Königsberg*. B. G. Teubner.

23 Luger, P. (2014). *Modern X-Ray Analysis on Single Crystals: A Practical Guide*. Walter de Gruyter.

24 Ladd, M. and Palmer, R. (2013). *Structure Determination by X-Ray Crystallography*. Springer.

25 Hahn, T. (2005). *International Tables for Crystallography, Volume A: Space Group Symmetry*. Springer Netherlands.

2

Piezoelectric Effects and Crystallographic Structures

Semën Gorfman[1] and Nan Zhang[2]

[1] *Tel Aviv University, Department of Materials Science and Engineering, Ramat Aviv, Tel Aviv 6997801, Israel*
[2] *Xi'an Jiaotong University, School of Electronic Science and Engineering, Electronic Materials Research Laboratory, Key Laboratory of the Ministry of Education, International Center for Dielectric Research, 28 West Xianning Road, Xi'an 710049, China*

The measurement of the piezoelectric coefficients is challenging, particularly in single crystals of lower symmetry. The achievable electric fields are below 10^7 V/m. For a material with a piezoelectric coefficient of ~10 pm/V, the relative strain is expected to be $<10^{-4}$. For lower electric fields or lower piezoelectric coefficients, the strain might be as small as 10^{-6}. Such small strains can be measured using methods, such as Michelson interferometer, or, the more recently, high-resolution single-crystal X-ray diffraction, as described in this chapter.

2.1 X-ray Diffraction in Single Crystals

2.1.1 Geometry of Single-Crystal X-ray Diffraction Experiment

Figure 2.1 shows the scheme of a high-resolution X-ray diffraction experiment suitable for the determination of piezoelectric coefficients. The X-ray diffraction intensity is measured under an applied external electric field. In order to avoid any irreversible effects created by permanent electric fields, alternating electric fields and synchronization with the X-ray detector are typically used [1–7]. Here, a monochromatic (the wavelength is λ) X-ray beam hits the crystal, and the diffracted X-ray beam is detected either by an open window point detector (e.g. scintillating counter) or a pixel area detector. We assume that the crystal is positioned at the Bragg angle to the primary beam, i.e. a *reciprocal lattice point* $\boldsymbol{B} = h_i\boldsymbol{a}_i^*$ falls exactly on the Ewald sphere. Here, the integer numbers $h_1 = h$, $h_2 = k$, and $h_3 = l$ are the indices of the Bragg reflection. The distribution of X-ray diffraction intensity around the peak (shape of the reflection) is then obtained by "rocking" (rotating by a small

Piezoelectric Materials: From Fundamentals to Emerging Applications, First Edition. Edited by Jiagang Wu.

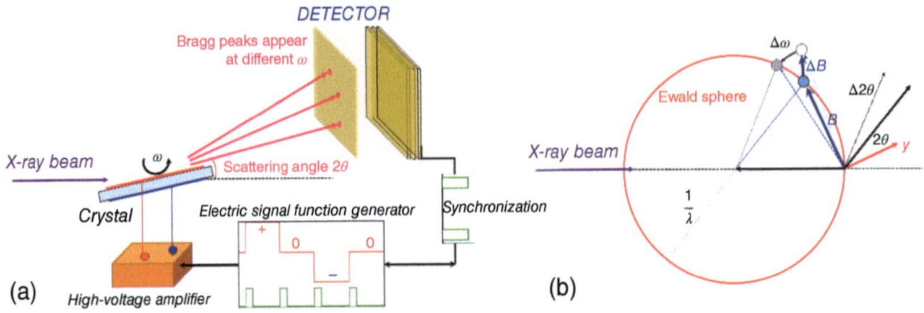

Figure 2.1 (a) The scheme of a single crystal X-ray diffraction experiment for the determination of the piezoelectric coefficients. The experiment is performed under an alternating external electric field, which is synchronized with an X-ray detector, (b) reciprocal space drawing indicating field-induced displacements of the Bragg angle of the reciprocal lattice point from the Ewald sphere. The crystal must be rotated by the angle $\Delta\omega$ to return to the reflecting position, at which the reflection will appear at a slightly different scattering angle $2\theta + \Delta 2\theta$. Source: Semën Gorfman.

angle) a crystal around the current reflecting position, and the intensity is detected and recorded by the X-ray detector [8]. Typically, the width of a rocking curve for high-quality crystal is 0.01°. This width is defined by the crystal's mosaicity, which includes both the size of mosaic blocks and their orientation distribution [9].

If the crystal is piezoelectric, then the external electric field will induce strain. The strain slightly changes the lengths and orientations of direct (and consequently reciprocal) lattice vectors and, accordingly, moves the reciprocal lattice point $\boldsymbol{B}$ by $\Delta\boldsymbol{B}$ away from the Ewald sphere. If the components B_i and ΔB_i (in the crystal physical coordinate system) are measured for several Bragg reflections, then the components of the field-induced strain tensor can be calculated.

2.1.2 Strain and Rotation in Real and Reciprocal Space

The determination of the field-induced strain components, $x_{ij} = d_{kij}E_k$, relies on knowing how such strain affects the geometry of the *reciprocal lattice*. Additionally, we should consider that any strained material may rotate, and the components of the rotation tensor r_{ij} must also be defined. The rotation tensor is antisymmetric, i.e. $r_{ij} = -r_{ji}$ [10, 11].

Let us express the orientation of the direct lattice basis vectors, $\boldsymbol{a}_i$, and reciprocal lattice vectors, $\boldsymbol{a}_j^*$, relative to the crystal physical coordinate system by the matrices $[A]$ and $[B]$, so that:

$$(\boldsymbol{a}_1\ \boldsymbol{a}_2\ \boldsymbol{a}_3) = (\boldsymbol{e}_1\ \boldsymbol{e}_2\ \boldsymbol{e}_3)\begin{bmatrix} A_{11} & A_{12} & A_{13} \\ A_{21} & A_{22} & A_{23} \\ A_{31} & A_{23} & A_{33} \end{bmatrix} \tag{2.1}$$

$$(\boldsymbol{a}_1^*\ \boldsymbol{a}_2^*\ \boldsymbol{a}_3^*) = (\boldsymbol{e}_1\ \boldsymbol{e}_2\ \boldsymbol{e}_3)\begin{bmatrix} B_{11} & B_{12} & B_{13} \\ B_{21} & B_{22} & B_{23} \\ B_{31} & B_{23} & B_{33} \end{bmatrix} \tag{2.2}$$

The columns of the matrices $[A]$ and $[B]$ represent the coordinates of the vectors $\boldsymbol{a}_i$ and $\boldsymbol{a}_j^*$ relative to the crystal physical coordinate system, respectively. If the components of the strain and rotation tensors are expressed in the matrix form as $[x]$ and $[r]$ then:

$$[A(E)] = ([I] + [x] + [r])[A] \tag{2.3}$$

$$[B(E)] = ([I] + [x^*] + [r^*])[B] \tag{2.4}$$

Here, $[I]$ is the unitary matrix, $[x^*]$ and $[r^*]$ are the strain and rotation tensors, which describe the deformation of the reciprocal lattice. The relationship between the direct and reciprocal lattice vectors $\boldsymbol{a}_j\boldsymbol{a}_k^* = \delta_{jk}$ can be rewritten as $[B]^T[A] = [I]$, so that:

$$[B]^T([I] + [x^*]^T + [r^*]^T)([I] + [x] + [r])[A] = [I] \tag{2.5}$$

Opening the brackets, neglecting the second power of strain and rotation (e.g. $[x^*]^T[r] \to [0]$), and considering the fundamental properties of strain $[x]^T = [x]$ and rotation $[r]^T = -[r]$ tensors, we obtain:

$$([x^*] + [x] + [r^*] - [r]) = [0] \tag{2.6}$$

Considering that $[x]$ is symmetric and $[r]$ is antisymmetric (2.6), can be satisfied if and only if

$$[x^*] = -[x]\ [r^*] = [r] \tag{2.7}$$

Thus, the strain of the reciprocal lattice, in comparison to the direct lattice strain, has the same magnitude but opposite sign. The rotations of both lattices are the same. The statement is illustrated in Figure 2.2, which shows direct and reciprocal lattices before and after deformation.

The components of the displacement of a reciprocal lattice point $\boldsymbol{B}^{(n)}$ are then given by the equation:

$$\Delta B_i^{(n)} = -x_{ij}B_j^{(n)} + r_{ij}B_j^{(n)} \tag{2.8}$$

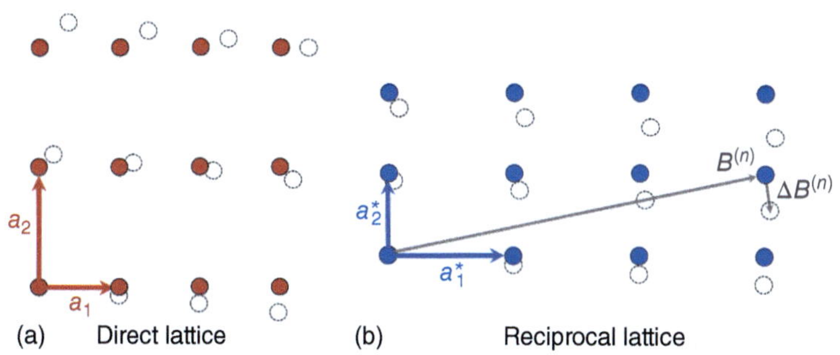

Figure 2.2 Schematic illustration of the relationship between the deformation of the direct (a) and the reciprocal lattice, (b) the filled/open circles show the nodes of both lattices before and after the deformation. Source: Semën Gorfman.

Here, n numbers the reflection. The determination of the strain and rotation components is determined from the system of equations:

$$\begin{cases} \Delta B_1^{(n)} = -x_{11}B_1^{(n)} - x_{12}B_2^{(n)} - x_{13}B_3^{(n)} + r_{12}B_2^{(n)} + r_{13}B_3^{(n)} \\ \Delta B_2^{(n)} = -x_{12}B_1^{(n)} - x_{22}B_2^{(n)} - x_{23}B_3^{(n)} - r_{12}B_1^{(n)} + r_{23}B_3^{(n)} \\ \Delta B_3^{(n)} = -x_{13}B_1^{(n)} - x_{23}B_2^{(n)} - x_{33}B_3^{(n)} - r_{13}B_1^{(n)} - r_{23}B_2^{(n)} \end{cases} \quad (2.9)$$

In the case where all the strain components are present, nine unknown parameters $(x_{11}, x_{22}, x_{33}, x_{23}, x_{13}, x_{12}, r_{23}, r_{13}, r_{12})$ must be determined. Accordingly, the positions $B_i^{(n)}$ and displacements $\Delta B_i^{(n)}$ of at least three reflections must be measured. In practice, when more reflections are accessible, the system of equations is overdetermined and can be solved using the method of least squares.

2.1.3 Angular Displacement of Bragg Peaks

According to Eq. (2.8), the relative $(\Delta B/B)$ displacement of the positions of Bragg peaks in the reciprocal space are $10^{-6} - 10^{-4}$. If the reciprocal lattice point is displaced from the Ewald sphere, the crystal has to be rotated by an angle $\Delta\omega$ to return it to the reflecting position. If so, the reflection will also appear at a slightly different point on the X-ray detector (scattering angle 2θ). For the sake of simplicity, we assume that the axis of crystal rotation, $\boldsymbol{\omega}$, is normal to the plane of primary and diffracted X-ray beams. This case is detailed in [12], where the equation for $\Delta\omega$ was derived as:

$$\Delta\omega = -\frac{\tan\theta x_{ij}B_iB_j}{B^2} - \frac{x_{ij}y_iB_j}{B} + \frac{r_{ij}y_iB_j}{B} \quad (2.10)$$

Here, θ presents half of the scattering angle (2θ) at zero field, B is the length of the reciprocal lattice vector, y_i represents the components of the vector $\boldsymbol{y}$, defined by the equation $\boldsymbol{y} = \frac{[\boldsymbol{B}\times\boldsymbol{\omega}]}{B}$ and shown in Figure 2.1b. The change of the scattering angle 2θ is given by:

$$\Delta 2\theta = -2\frac{\tan\theta x_{ij}B_iB_j}{B^2} \quad (2.11)$$

We can estimate the order of magnitude of the displacement of the Bragg peak by calculating $\tan\theta x_{ij}$. Assuming that $\tan\theta \sim 1$ and $x_{ij} \sim 10^{-6} - 10^{-4}$, the shift in the scattering angle $\Delta 2\theta$, respectively, will be ~0.1 – 10 mdeg.

2.1.4 Example: Determination of Piezoelectric Coefficients in BiB_3O_6

Here we will illustrate the use of high-resolution X-ray diffraction for the determination of the piezoelectric coefficients of a monoclinic BiB_3O_6 crystal. This paragraph is based on the previously published article [12] by the author.

The information related to the structure and symmetry of BiB_3O_6 is summarized in Table 2.1. For this measurement, only crystals of the handedness that is shown in Figure 2.3a were used. For crystals of this type, the piezoelectric coefficient d_{222}

Table 2.1 Space group, lattice parameters, and symmetry allowed components of the piezoelectric tensor for point group 2.

Crystal system (cell setting)	Monoclinic (unique axis a_2)
Space group/point group	$C2/2$
Lattice parameters	$a_1 = 7116(2)\,Å, a_2 = 4.993(2)\,Å$ $a_3 = 6.508(3)\,Å$ $\alpha_1 = 90°, \alpha_2 = 105.62(3)°, \alpha_3 = 90°$
Symmetry allowed piezoelectric coefficients	$\begin{pmatrix} 0 & 0 & 0 & 2d_{123} & 0 & 2d_{112} \\ d_{211} & d_{222} & d_{233} & 0 & 2d_{213} & 0 \\ 0 & 0 & 0 & 2d_{323} & 0 & 2d_{312} \end{pmatrix}$

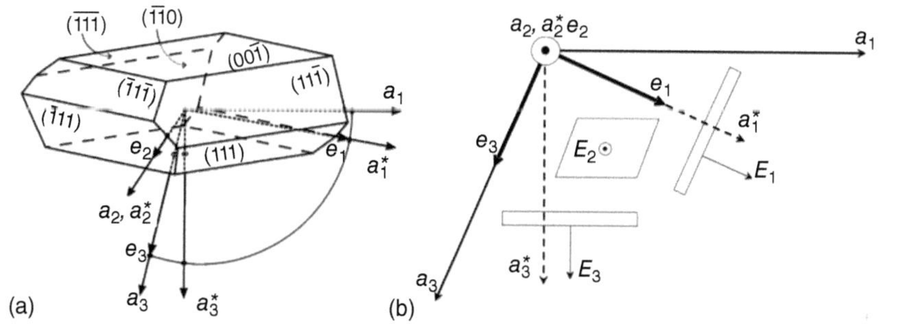

Figure 2.3 (a) Typical morphology of a BiB_3O_6 single crystal, together with the mutual relationship between crystallographic, reciprocal crystallographic, and crystal physical coordinate system, (b) orientation of three plane-shaped samples used for the single crystal X-ray diffraction experiment. Source: Gorfman et al. [12], reproduced with permission of Walter de Gruyter.

is negative. Thin gold contacts were deposited in a vacuum on the sample faces to ensure that the orientation of the electric field was normal to the planes with (100), (010), and (001) Miller indices (see Figure 2.3b). For each sample, the set of rocking curves was collected using an open window point detector. This detector is not position-sensitive; therefore, only the change of the rocking angle $\Delta\omega$ is measured. Such typical electric field-induced shifts of the rocking curve are shown in Figure 2.4.

The shift of the rocking curve for each Bragg reflection can be calculated according to $\Delta\omega = \omega_E - \omega_0$, where ω_E and ω_0 are the mass centers of the diffraction curves measured with and without the electric field.

A short summary of the X-ray data collection is presented in Table 2.2.

The observed shifts of Bragg peaks and model (2.10) were used to determine all the components of the piezoelectric tensor. The overdetermined system of equations was solved by minimizing the sum of squared differences between

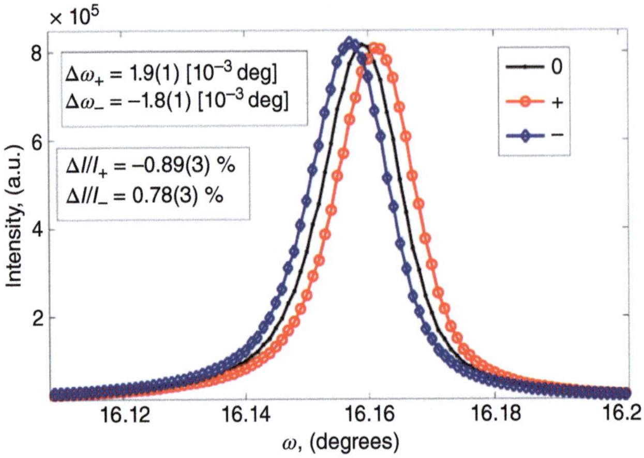

Figure 2.4 Typical modification of a Bragg rocking curve under an applied external electric field with positive and negative polarities. The middle curve corresponds to the average between two zero states. Shifts of the angular positions and relative changes in intensity are extracted from the curves. Source: Semën Gorfman.

Table 2.2 Summary of the experimental conditions used for the data collection.

	Sample 1	Sample 2	Sample 3
Face normal (related to the reciprocal basis)	$\boldsymbol{a}_1^*$	$\boldsymbol{a}_2^*$	$\boldsymbol{a}_3^*$
Face normal (related to the crystal physical coordinate system)	$\boldsymbol{e}_1$	$\boldsymbol{e}_2$	$\boldsymbol{e}_3 \sin\alpha_2 - \boldsymbol{e}_1 \cos\alpha_2$
Piezoelectric constants involved	d_{1ij}	d_{2ij}	$0.96d_{3ij} + 0.27d_{1ij}$
Sample thickness (mm)	0.570	0.533	0.295
Number of measured reflections	56	89	36
Total number of $\Delta\omega$ (including different voltages)	122	228	72

observed and calculated displacements of the Bragg peaks. To account for the electric field-induced strain and rotation, 17 independent parameters for refinement were needed. These include all eight piezoelectric coefficients, which define the strain for all three orientations of the electric field, $x_{ij} = E_k d_{kij}$, and nine components for the rotation tensor (three for each crystal). The piezoelectric coefficients refined from X-ray diffraction data and those determined by Michelson interferometry [13] are listed in Table 2.3.

Table 2.3 The piezoelectric coefficients of BiB_3O_6 (in pm/V) obtained by the X-ray diffraction experiment (second column) and by means of Michelson interferometry [13] (third column).

Piezoelectric coefficient	X-ray diffraction	Michelson interferometer
d_{123}	−6.3(4)	−6.8(6)
d_{112}	−9.2(7)	−8.4(15)
d_{211}	13.7(5)	15.6(6)
d_{222}	−35.8(8)	−39.5(10)
d_{233}	−2.9(2)	−3.8(4)
d_{213}	2.9(2)	5.1(3)
d_{312}	−5.4(3)	−6.2(6)
d_{323}	−14.2(12)	−16.3(20)

Note the lower errors for the coefficients determined from the X-ray experiment.

2.2 Observation of Piezoelectric Effect in Multidomain Ferroelectrics

2.2.1 Ferroelectrics

Ferroelectrics is the class of materials that may exist in two or more energetically equivalent states of spontaneous polarization (e.g. $+P_0$ or $-P_0$). First and foremost, these states are switchable by a sufficiently strong, yet practically achievable, electric field [14, 15]. For the simplest case of *uniaxial* ferroelectrics, the spontaneous polarization is always parallel to the unique crystallographic axis. However, its direction can be inverted under an electric field. The crystallographic structures of these states can be converted from one to another by inversion with respect to some point in the unit cell. The dependence of polarization on the electric field forms a so-called hysteresis loop (Figure 2.5).

The symmetry considerations dictate that the structure of a ferroelectric must belong to one of the 10 polar crystal classes [16], allowing for non-zero components of a polarization vector. All polar crystal classes (1, 2, *m*, *mm*2,4,4*mm*, 3,3*m*, 6,6*mm*) are piezoelectrics, yet not all piezoelectric crystal classes are polar. This means that a piezoelectric material needs not necessarily to be ferroelectric. The most typical example of non-ferroelectric piezoelectric is α-quartz (which belongs to the nonpolar crystal class 32). Nevertheless, ferroelectricity is important for piezoelectricity because the coexistence of these two effects results in a dramatic enhancement of the magnitude of the piezoelectric effect, sometimes by orders of magnitude.

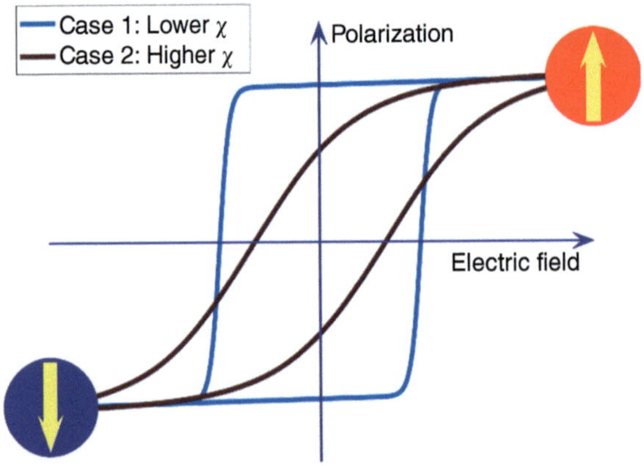

Figure 2.5 Typical polarization–electric field hysteresis loops. Two zero-field/spontaneous polarization states exist, and these states can be switched by an electric field. Depending on the crystallographic structure, stoichiometry, and microstructure of the material, polarization inversion can occur at the coercive fields or spread over broader electric field intervals. Accordingly, this results in "sharper" and "blunter" hysteresis loops (case 1 and case 2 in the figure). They correspond to the cases of lower and high zero-field susceptibility χ. Source: Semën Gorfman.

2.2.2 Piezoelectric Effect in Polycrystalline Ferroelectric Ceramics

Most market-dominating functional piezoelectric materials are ferroelectrics [17]. Many of them are hard to grow as single crystals but easy to prepare as polycrystalline ceramics [18, 19]. For example, $PbZr_{1-x}Ti_xO_3$ (PZT) ferroelectric ceramics dominate the industry of piezoelectric devices entirely, though attempts to replace PZT with similar lead-free materials are undertaken [20]. Piezoelectricity in ferroelectric ceramics has been the subject of numerous books [21–23] and review papers [24–26]. Therefore, we will only cover this subject briefly.

We will start by inspecting the symmetry of piezoelectric effect in polycrystalline ceramics. Any non-textured polycrystalline material where all possible orientations of polycrystalline grains are present equally is macroscopically isotropic and therefore not piezoelectric. In some cases, the presence of texture may break this isotropy sufficiently to make a material piezoelectric. This is exactly the case for the virgin (non-textured) and poled (textured) states of ferroelectric ceramics (Figure 2.6). Imagine a disk/pellet-shaped device so that an external electric field is applied normal to the pellet faces. Assume that it is heated above the Curie temperature (the temperature at which spontaneous polarization vanishes [14]) and slowly cooled under an electric field. In this way, the initially polycrystalline ferroelectric device is "made" piezoelectric [18].

Given that the direction of spontaneous polarization is fixed relative to the crystallographic coordinate system of each grain, the orientation of spontaneous polarization (global, i.e. relative to the pellet) alters from one grain to another (Figure 2.6). The process of poling ensures that the projection of the spontaneous

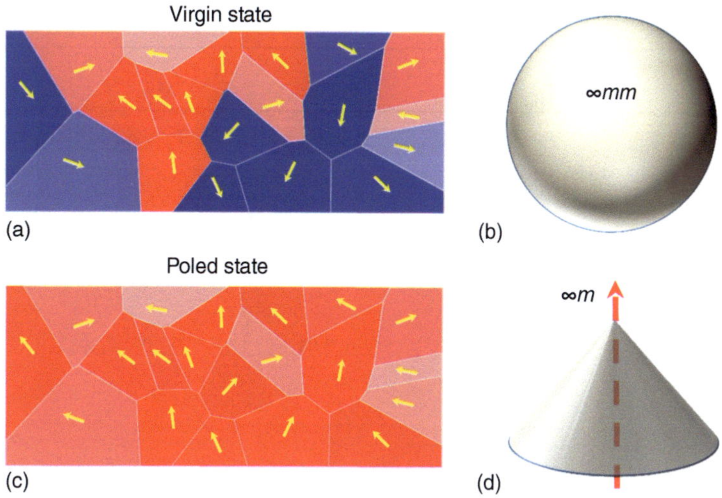

Figure 2.6 Schematic illustration of the appearance of piezoelectricity in polycrystalline ferroelectric (a) un-poled virgin state: the arrow in each polycrystalline grain indicates the direction of spontaneous polarization in each grain. For a large number of grains, the net polarization is zero (b), and shows that the macroscopic symmetry of such an un-poled state possesses a type of macroscopic symmetry group of a sphere (∞mm), (c) shows the same model as in (a) for the polycrystalline material, but after poling (cooling under electric field) in the vertical direction, (d) shows the resulting macroscopic symmetry, which is the symmetry of a cone (∞m). Source: Semën Gorfman.

polarization vector in the direction of the applied electric field becomes positive in every grain. This process changes the macroscopic type of point symmetry from ∞mm (a sphere) to ∞m (a cone). Both point groups are non-crystallographic and are not listed in Table 1.4 as such. However, we can still use Eq. (1.19) for the derivation of independent piezoelectric coefficients.

Let us define the Cartesian coordinate system with e_3 parallel to the poling direction/cone axis. It is easy to show that the Voigt matrix of piezoelectric coefficients for the crystal class ∞m is similar to that of the crystal classes $6mm$, $4mm$:

$$[d] = \begin{pmatrix} 0 & 0 & 0 & 0 & d_{15} & 0 \\ 0 & 0 & 0 & d_{15} & 0 & 0 \\ d_{31} & d_{31} & d_{33} & 0 & 0 & 0 \end{pmatrix} \tag{2.12}$$

This means that piezoelectric properties of any poled ferroelectric ceramic can be described by three independent piezoelectric coefficients $d_{31}(=d_{32})$, d_{33}, $d_{15}(=d_{24})$.

Due to the wide applicability of poled ferroelectric ceramics, these piezoelectric coefficients are most frequently reported in the literature, with the notation "d_{33}" often used (mistakenly) as a "synonym" for a piezoelectric coefficient in general. It is worth looking at the geometrical meaning of these coefficients. Table 2.4 shows schematic drawings, including applied mechanical stress with resulting polarization as well as an electric field with resulting mechanical deformations.

Table 2.4 Geometrical meaning of the d_{31}, d_{33}, d_{15} piezoelectric coefficients.

Coefficient	d_{31}	d_{33}	d_{15}
Equations for direct piezoelectric effect	$P_3 = d_{31}\sigma_{11}$	$P_3 = d_{33}\sigma_{33}$	$P_1 = d_{15}\sigma_{13}$
Schematic drawing	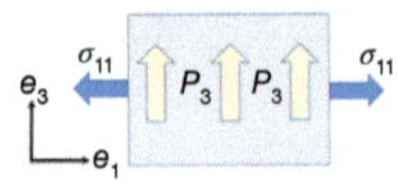	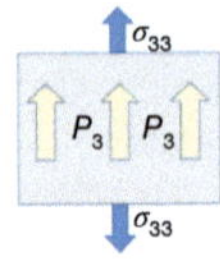	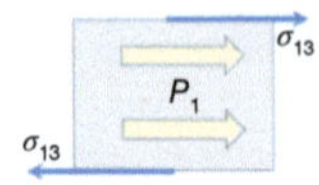
Equations for converse piezoelectric effect	$x_{11} = d_{31}E_3$	$x_{33} = d_{33}E_3$	$x_{13} = \frac{1}{2}d_{15}E_1$
Schematic drawings	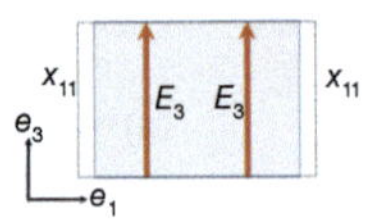	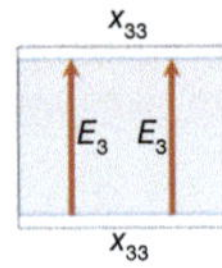	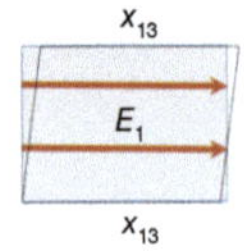

2.2.3 Quadratic Electrostriction and Piezoelectricity

The previous paragraph considered only the symmetry-related aspects of piezoelectricity in polycrystalline ferroelectrics. In this section, we will discuss the possible origin of enhanced values of these coefficients.

Piezoelectricity in ferroelectrics originates from the coupling of polarization and strain. Indeed, electric polarization is mainly associated with the displacement of atoms (ions) from their original positions in the unit cell. Such displacements will also change the lattice parameters to values that minimize the free energy of the resulting periodic crystal. The possible distortion of the unit cell is schematically shown for the perovskite structure (e.g. $BaTiO_3$) in Figure 2.7.

A strong coupling between polarization and strain makes ferroelectrics different from most ferromagnetics, where magnetization is decoupled from strain, so that magnetization may change in a rigid geometrical structure of the crystal. However, this is inaccurate for a comparatively narrow, albeit quite significant, class of multiferroic materials [27, 28]. In ferroelectrics, the polarization couples to the strain via a quadratic electrostriction law [29, 30] that defines the dependence of the strain on the polarization in the following form:

$$x = QP^2 \tag{2.13}$$

Here, Q is the electrostriction coefficient. The same equation in tensor form is

$$x_{ij} = Q_{ijkl}P_kP_l \tag{2.14}$$

Here, Q_{ijkl} are the components of the fourth-rank tensor of electrostriction coefficients. Figure 2.8 outlines the electric field dependence of strain. These curves (known as "butterfly-shaped" strain-field loops) are obtained by squaring those in

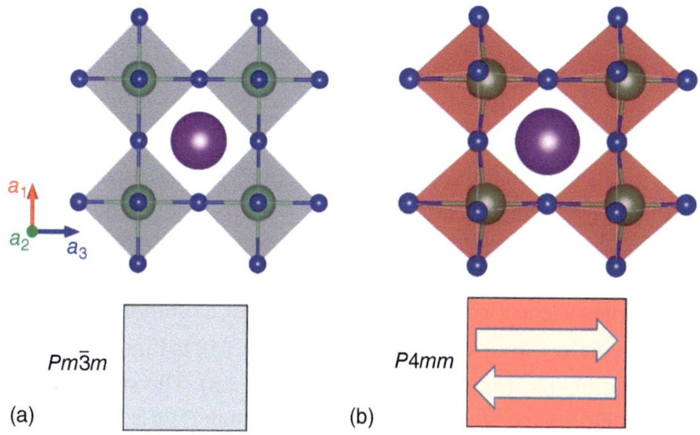

Figure 2.7 Schematic drawing of the perovskite structure in (a) non-ferroelectric $Pm\bar{3}m$ and (b) ferroelectric $P4mm$ phase with switchable spontaneous polarization along the tetragonal $\boldsymbol{c}$ ($\boldsymbol{a}_3$) axis/fourfold symmetry axis. This polarization is associated with the displacement of atoms from their original position, which in turn modifies the unit cell's lengths and thus introduces mechanical strain relative to the non-ferroelectric phase. Source: Semën Gorfman.

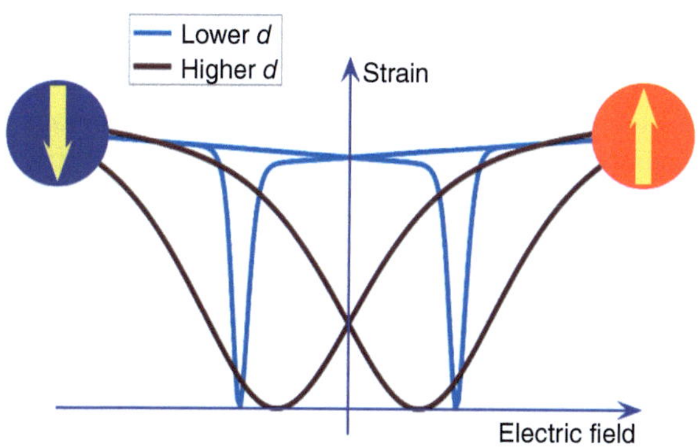

Figure 2.8 Typical strain–electric field hysteresis loops, which are obtained from polarization–electric field loops, assuming the quadratic electrostriction law for the dependence between strain and polarization. Similar to Figure 2.5, two cases, "sharp" and "blunt" switching processes, are considered. The second case results in higher zero-field piezoelectric constants, d. Source: Semën Gorfman.

Figure 2.5. The effective piezoelectric coefficients[1] – the slope of the $x(E)$ curve at the specific value (e.g. $E = 0$) of the electric field – are introduced as:

$$d = \frac{\partial x}{\partial E} = 2QP\frac{\partial P}{\partial E} = 2\epsilon_0 QP\chi \tag{2.15}$$

1 We use the word "effective" for the piezoelectric coefficient in order to acknowledge the fact that the value of this coefficient is not constant but, instead, depends on the position in the strain-field loop/external electric field.

Here, $\epsilon_0 = 8.85 \cdot 10^{-12}\ \frac{\text{F}}{\text{m}}$, χ is the dielectric susceptibility. The same equation in the tensor form is:

$$d_{kij} = \frac{\partial x_{ij}}{\partial E_k} = 2\epsilon_0 Q_{ijmn} P_m \chi_{nk} \tag{2.16}$$

As it follows from (2.16), the magnitude of the "effective" piezoelectric coefficient in ferroelectrics is proportional to the spontaneous polarization and dielectric susceptibility. Therefore, the enhancement of a piezoelectric coefficient in ferroelectrics is associated with a high spontaneous polarization and high susceptibility. For example, a zero-field piezoelectric coefficient, $d = \frac{\partial x}{\partial E}(E = 0)$, as demonstrated in Figure 2.8 is higher when dielectric susceptibility $\chi = \frac{1}{\epsilon_0}\frac{\partial P}{\partial E}(E = 0)$ is higher as well (compared to the polarization–electric field curves in Figure 2.5). The strategy for the enhancement of piezoelectric coefficients in ferroelectric materials is to activate the polarization inversion mechanisms at a lower electric field and thus to "smear" those usually sharp edges of the hysteresis loop.

2.2.4 Extrinsic and Domain-Wall Motion Mechanisms of Piezoelectricity in Ferroelectrics

In ferroelectrics, piezoelectricity can also be driven by "extrinsic"/"mesoscopic" mechanisms [31]. These mechanisms are different from the "intrinsic"/"atomistic" ones discussed in the previous paragraph. The most common type of such "extrinsic" mechanism is the so-called "domain-wall motion." There are many indications of the dominating role of domain-wall motion in "enhancing" the piezoelectric effect in ferroelectrics [32–35]. Domain-wall motion can contribute to the piezoelectric effect if domains are ferroelectric and ferroelastic at the same time [23]. Switching between two domain states involves the simultaneous switching between spontaneous polarization and spontaneous strain. Such domains commonly appear in perovskites (ABX_3), where thermally driven phase transitions between higher symmetry (e.g. cubic $Pm\bar{3}m$) and lower symmetry (e.g. tetragonal $P4mm$) phases occur, as shown in Figure 2.7. The lower symmetry phase is ferroelectric and strained (relative to the paraelectric phase). Assuming that the "cubic" lattice parameter of the high-symmetry phase is a_0, and "tetragonal" lattice parameters in the low-symmetry phase are a and c, three different domain variants (Table 2.5) will appear.

Here in Table 2.5, $x_a = \frac{a-a_0}{a_0}$ and $x_c = \frac{c-a_0}{a_0}$.

Consider the case where the microstructure contains domains with their spontaneous polarization being developed along $\boldsymbol{e}_1$ and $\boldsymbol{e}_2$ axes (Domains 1 and 2 in Table 2.5), and an electric field is applied parallel to $\boldsymbol{e}_1$. Because of the energy of interaction between the electric field and the polarization equals $-\boldsymbol{P} \cdot \boldsymbol{E}\, V$ (V is the interaction volume) [36], the application of an electric field along $\boldsymbol{e}_1$ will cause the first domain to grow, while the application of an electric field along $\boldsymbol{e}_2$ will cause this domain to shrink. The average strain is given by:

$$[x] = \eta(E)[x^{(1)}] + (1 - \eta(E))[x^{(2)}] \tag{2.17}$$

Table 2.5 Polarization and strain in three ferroelectric/ferroelastic domains, formed during the phase transitions between higher-symmetry $Pm\overline{3}m$ and tetragonal $P4mm$ in the materials with the structure of a perovskite (e.g. $BaTiO_3$).

	Domain 1	**Domain 2**	**Domain 3**
Polarization	$\boldsymbol{P} = P\boldsymbol{e}_1$	$\boldsymbol{P} = P\boldsymbol{e}_2$	$\boldsymbol{P} = P\boldsymbol{e}_3$
Strain	$\begin{pmatrix} x_c & 0 & 0 \\ 0 & x_a & 0 \\ 0 & 0 & x_a \end{pmatrix}$	$\begin{pmatrix} x_a & 0 & 0 \\ 0 & x_c & 0 \\ 0 & 0 & x_a \end{pmatrix}$	$\begin{pmatrix} x_a & 0 & 0 \\ 0 & x_a & 0 \\ 0 & 0 & x_c \end{pmatrix}$
Drawing	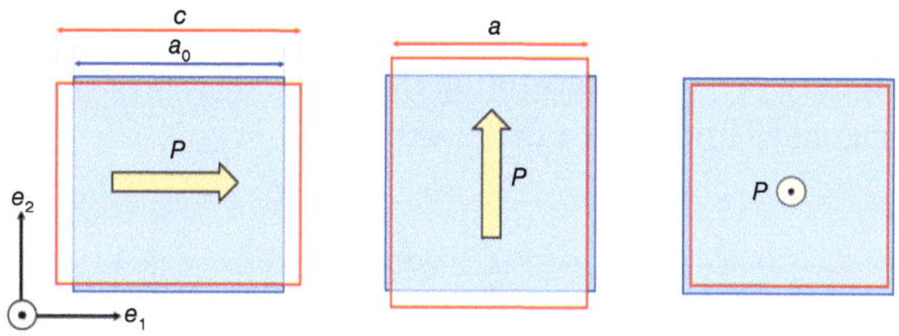		

The axes of the Cartesian coordinate system $\boldsymbol{e}_i$ are parallel to the basis vectors of the conventional crystallographic coordinate system in the cubic phase $\boldsymbol{a}_i$. The last row of the table shows schematic drawings of the shape of the unit cell (before and after the phase transition) along with the direction of the polarization vector.

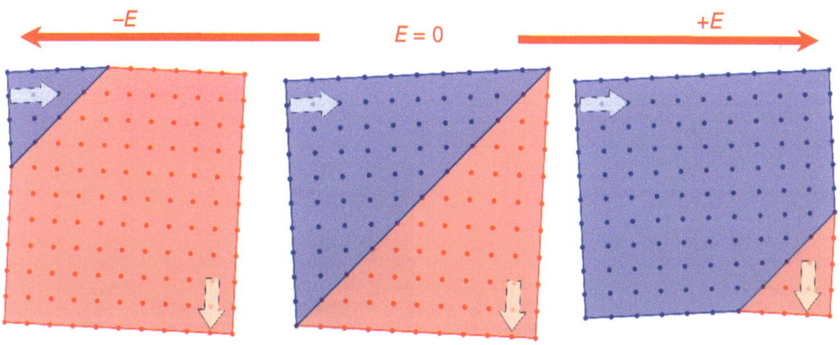

Figure 2.9 Electric-field-induced deformation, mediated by domain-wall motion. The image shows two domains (Domain 1 and Domain 2) of tetragonal symmetry, separated by the domain-wall parallel to the lattice planes with the Miller indices (101). The spontaneous polarization is parallel to the tetragonal *c*-axis of each domain, as depicted by the thick arrow inside the domain. Domain-wall motion changes the volume fraction of domains, thus minimizing the energy of interaction between the electric field (shown by the arrow on the top) and spontaneously polarized domains. Source: Semёn Gorfman.

Here, $\eta(E)$ and $(1 - \eta(E))$ are the volume fractions of Domains 1 and 2. Figure 2.9 illustrates the following statement: a domain-wall between these domains moves, hence changes the shape of the material in a flexible and controllable way. The ferroelectric nature of these domains enables their control by an external electric field. Their ferroelastic nature provides routes to deform the material through domain-wall motion and tailor the material's functionality.

2.2.5 Piezoelectricity at the Morphotropic Phase Boundary

The electromechanical coupling of perovskite-based solid solutions deserves special attention. The most typical example of such solid solutions is the perovskite-based PZT, where the *A* positions are occupied by Pb, while the *B* positions are shared between Zr and Ti. The structure, symmetry, and crystallographic direction of spontaneous polarization depend on the temperature T and the composition parameter x. Figure 2.10 shows the x–T phase diagram (reported by Jaffe for the first time [37]): the Morphotropic Phase Boundary (MPB) is the sharp and nearly vertical boundary ($x \sim 0.48$), separating the rhombohedral Zr-rich phase ($x < 0.48$) and the tetragonal Ti-rich phase ($x > 0.48$).

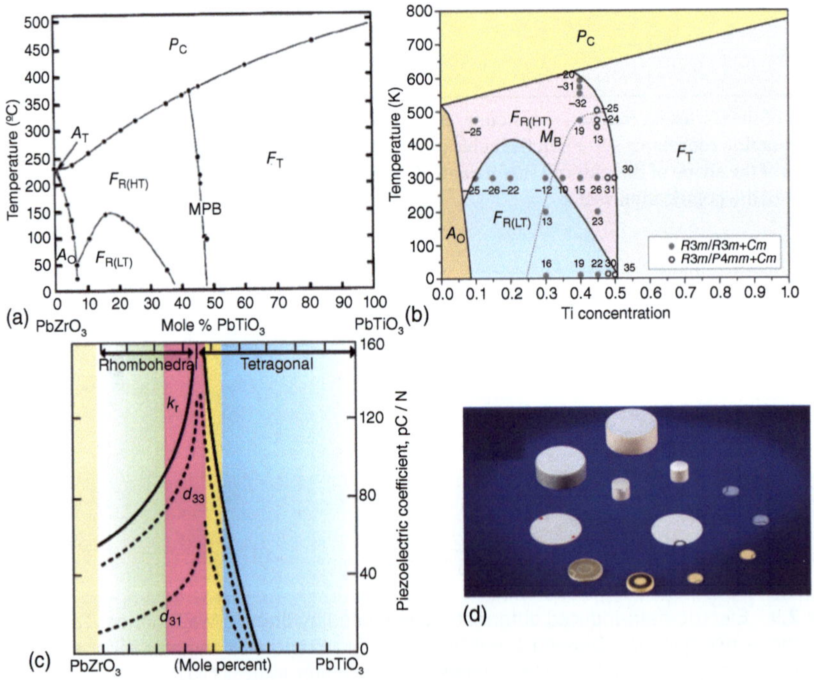

Figure 2.10 (a) The phase diagram of PZT, originally introduced for the first time by Jaffe. Source: Jaffe et al. [37], Reproduced with permission of Elsevier. (b) The most recent phase diagram of PZT Source: Adapted from Zhang et al. [38], which includes lower symmetry (e.g. *Cm*) and coexisting phases. (c) The enhancement of the piezoelectric coefficients across the Morphotropic Phase Boundary. Source: Zhang et al. [39], reproduced with permission of the International Union of Crystallography. (d) The photographs of devices made of PZT. Source: CeramTec GmbH / https://www.ceramtec-group.com/en/?utm_source=ceramtec.com &utm_medium=redirects&utm_campaign=/http://www.ceramtec.com/ / last accessed under May 05, 2023.

The MPB is a long-standing puzzle for crystallographers: the space symmetry groups describing the structures on both sides of the MPB (e.g. *R*3*m*/*P*4*mm*) are not group–subgroup related to each other, and therefore, a sharp boundary should not exist. The MPB is noteworthy for materials scientists: piezoelectric coefficients at the MPB are higher by two orders of magnitude than when they are away from the MPB (may reach up to 400 pC/N). Piezoelectricity in PZT has been the topic of much attention [38, 40, 41]. Understanding the mechanisms of piezoelectricity enhancement at the MPB from a structural point of view is the key to replicate these properties in other (e.g. lead-free) materials [20].

One particular breakthrough in the understanding of PZT properties was the discovery of a bridging monoclinic phase [42–44]. In the series of high-resolution synchrotron X-ray powder diffraction experiments, Noheda and coworkers showed that the symmetry of PZT near the MPB is monoclinic (e.g. *Cm*), as opposed to rhombohedral (e.g. *R*3*m*) or tetragonal (e.g. *P*4*mm*). This opened the possibility for the mechanisms related to the rotation of the polarization vector in the monoclinic mirror plane [45–48], as well as creating even more complicated and mobile domain patterns [49].

References

1 Gorfman, S., Schmidt, O., and Tsirelson, V. (2013). *ZAAC* 639: 1953.
2 Gorfman, S. (2014). *Crystallogr. Rev.* 20: 210.
3 Fujimoto, I. (1978). *Phys. Rev. Lett.* 40: 941.
4 Fujimoto, I. (1980). *Jpn. J. Appl. Phys.* 19: L345.
5 Guillot, R., Allé, P., Fertey, P. et al. (2002). *J. Appl. Crystallogr.* 35: 360.
6 Vergentev, T., Bronwald, I., Chernyshov, D. et al. (2016). *J. Appl. Crystallogr.* 49: 1501.
7 Choe, H., Heidbrink, S., Ziolkowski, M. et al. (2017). *J. Appl. Crystallogr.* 50: 975.
8 He, B.B. (2018). *Two-Dimensional X-Ray Diffraction*. Hoboken, NJ: Wiley.
9 Warren, B.E. (1990). *X-Ray Diffraction*. Courier Corporation.
10 Nye, J.F. (1985). *Physical Properties of Crystals: Their Representation by Tensors and Matrices*. Oxford University Press.
11 Newnham, R.E. (2005). *Properties of Materials: Anisotropy, Symmetry, Structure*. Oxford University Press.
12 Gorfman, S., Schmidt, O., Pietsch, U., and Becker, P. (2007). *Z. Kristallogr.* 222: 396.
13 Haussühl, S., Bohatý, L., and Becker, P. (2006). *Appl. Phys. A* 82: 495.
14 Lines, M.E. and Glass, A.M. (1979). *Principles and Applications of Ferroelectrics and Related Materials*. OUP Oxford.
15 Scott, J.F. (2007). *Science* 315: 954.
16 Whatmore, R. (2017). *Handbook of Electronic and Photonic Materials*. Ferroelectric materials, Springer, Springer International Publishing.
17 Trolier-McKinstry, S., Zhang, S., Bell, A.J., and Tan, X. (2018). *Annu. Rev. Mater. Res.* 48 (48): 191.

18 Uchino, K. (2006). *Materials Science and Technology* (ed. R.W. Cahn, P. Haasen, and E.J. Kramer), 123. Weinheim: Wiley-VCH Verlag GmbH & Co. KGaA.

19 Haertling, G.H. (1999). *J. Am. Ceram. Soc.* 82: 797.

20 Rödel, J., Jo, W., Seifert, K.T.P. et al. (2009). *J. Am. Ceram. Soc.* 92: 1153.

21 Heywang, W., Lubitz, K., and Wersing, W. (2008). *Piezoelectricity: Evolution and Future of a Technology*. Springer Science & Business Media.

22 APC International and Ltd (2011). *Piezoelectric Ceramics: Principles and Applications*. APC International.

23 Tagantsev, A., Eric Cross, L., and Fousek, J. (2011). *Domains in Ferroic Crystals and Thin Films*. Springer Science & Business Media.

24 Damjanovic, D. (1999). *Rep. Prog. Phys.* 61: 1267.

25 Hao, J., Li, W., Zhai, J., and Chen, H. (2019). *Mater. Sci. Eng. R Rep.* 135: 1.

26 Panda, P.K. and Sahoo, B. (2015). *Ferroelectrics* 474: 128.

27 Spaldin, N.A., Cheong, S.-W., and Ramesh, R. (2010). *Phys. Today* 63: 38.

28 Fiebig, M., Lottermoser, T., Meier, D., and Trassin, M. (2016). *Nat. Rev. Mater.* 1: 16046.

29 Sundar, V. and Newnham, R.E. (1992). *Ferroelectrics* 135: 431.

30 Li, F., Jin, L., Xu, Z., and Zhang, S. (2014). *Appl. Phys. Rev.* 1: 011103.

31 Damjanovic, D. (2005). *J. Am. Ceram. Soc.* 88: 2663.

32 Jones, J.L., Hoffman, M., Daniels, J.E., and Studer, A.J. (2006). *Appl. Phys. Lett.* 89: 092901.

33 Zhang, N., Gorfman, S., Choe, H. et al. (2018). *J. Appl. Crystallogr.* 51: 1396.

34 Pramanick, A., Prewitt, A.D., and Forrester, J.S. (2012). *Crit. Rev. Solid State Mater. Sci.* 37: 243.

35 Gorfman, S., Choe, H., Zhang, G. et al. (2020). *J. Appl. Crystallogr.* 53: 1039.

36 Jackson, J.D. (2007). *Classical Electrodynamics*. Wiley.

37 Jaffe, B., Cook, W.R., and Jaffe, H. (1971). *Piezoelectric Ceramics* (ed. B. Jaffe, W.R. Cook, and H. Jaffe), 135–183. Academic Press.

38 Zhang, N., Yokota, H., Glazer, A.M. et al. (2014). *Nat. Commun.* 5: 5231.

39 Zhang, N., Yokota, H., Glazer, A.M. et al. (2018). *IUCrJ* 5: 73.

40 Randall, C.A., Kim, N., Kucera, J.P., and Cao, W. (1998). *J. Am. Ceram. Soc.* 81: 677.

41 Zhang, N., Yokota, H., Glazer, A.M., and Thomas, P.A. (2011). *Acta Crystallogr. B* 67: 386.

42 Noheda, B., Cox, D.E., Shirane, G. et al. (1999). *Appl. Phys. Lett.* 74: 2059.

43 Noheda, B., Cox, D.E., Shirane, G. et al. (2000). *Phys. Rev. B: Condens. Matter Mater. Phys.* 63: 014103.

44 Noheda, B., Gonzalo, J.A., Cross, L.E. et al. (2000). *Phys. Rev. B: Condens. Matter Mater. Phys.* 61: 8687.

45 Damjanovic, D. (2010). *Appl. Phys. Lett.* 97: 062906.

46 Choe, H., Bieker, J., Zhang, N. et al. (2018). *IUCrJ* 5: 417.

47 Fu, H. and Cohen, R.E. (2000). *Nature* 403: 281.

48 Noheda, B., Cox, D.E., Shirane, G. et al. (2001). *Phys. Rev. Lett.* 86: 3891.

49 Bokov, A.A., Long, X., and Ye, Z.-G. (2010). *Phys. Rev. B: Condens. Matter Mater. Phys.* 81: 172103.

3

Lead-Based Piezoelectric Materials

Fei Li

Xi'an Jiaotong University, School of Electronic and Information Engineering, State Key Laboratory for Mechanical Behavior of Materials, Electronic Materials Research Lab, Key Lab of Education Ministry/International Center for Dielectric Research, No. 12, Xianning West Road, Xi'an 710049, China

3.1 Introduction

As a conversion medium between mechanical and electrical energies, piezoelectric materials, are widely used in sensing, energy exchange, energy storage, refrigeration, catalysis, photovoltaics, flexible devices, and other fields. Researchers favored these because of their unique electromechanical conversion function. Ferroelectrics with perovskite structure, such as $BaTiO_3$ and lead zirconate titanate (PZT), are the most studied and widely used piezoelectric materials. Among these, lead-based materials, represented by PZT, have dominated the market for decades due to their exceptional performances. With the rapid development of piezoelectric devices, the research field of piezoelectric materials is also keeping pace. Therefore, the exploration of new and high-performance ferroelectric ceramic materials and related devices has become a research hotspot in the field of advanced functional ceramics.

Due to the limited space of the topic, this chapter only covers some contents of recent advances in lead-based piezoelectric materials and their applications, and some omissions are inevitable. The ferroelectric materials, such as $LiNbO_3$, potassium sodium niobate, bismuth layered structure, and tungsten–bronze structure, are not introduced. Interested readers may refer to Chapter 4 for lead-free piezoelectric materials.

3.2 Brief History of Lead-Based Piezoelectric Materials

3.2.1 Lead Zirconate Titanate, PZT

The first qualitative leap of the piezoelectric effect came from the discovery of barium titanate ($BaTiO_3$) in the 1940s (Figure 3.1). Its piezoelectric coefficient was about 190 pC/N, much higher than that of quartz (2 pC/N) [1]. $BaTiO_3$ is also the first ferroelectric material with a perovskite structure. The ferroelectric material is

Piezoelectric Materials: From Fundamentals to Emerging Applications, First Edition. Edited by Jiagang Wu.

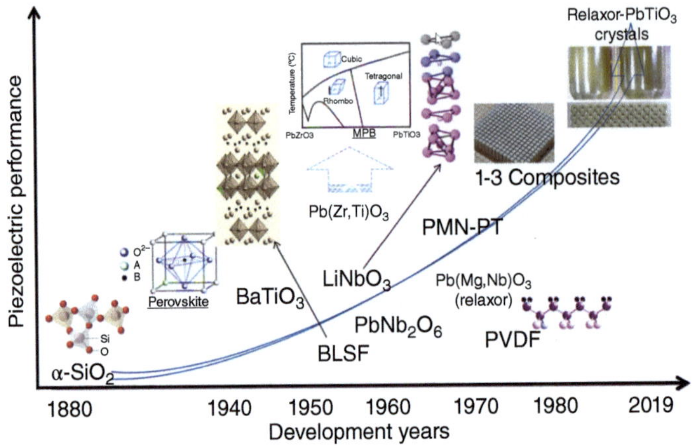

Figure 3.1 Development of piezoelectric materials with milestones/breakthroughs.

a kind of material with spontaneous polarization, and the polarization vector can be reversed by an electric field [2]. Since then, piezoelectric materials have officially entered the era of ferroelectricity.

As a piezoelectric material for transducers, barium titanate was not utilized for a long period before it was replaced by the well-known lead zirconate titanate ($PbZrO_3$–$PbTiO_3$) ceramics. In the 1950s, Jaffe found better piezoelectric properties ($d_{33} \sim 300$ pC/N) and better temperature stability than barium titanate in PZT solid solutions [3]. Without a long period of basic research, PZT ceramics have been industrially produced and have become a widely used piezoelectric material in various industries. To date, PZT components used in the world are still from the 1960s or 1970s (Figure 3.2).

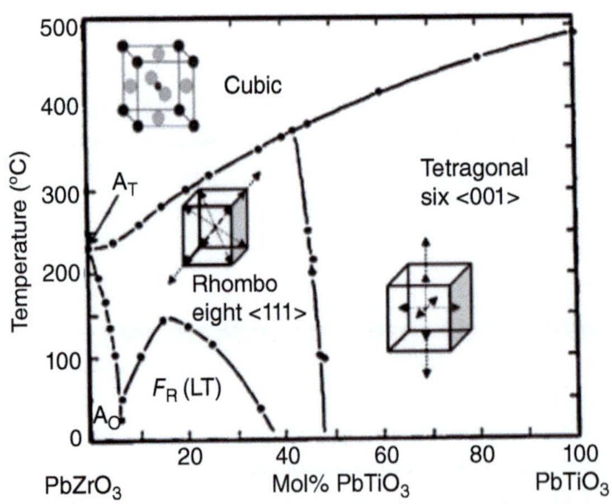

Figure 3.2 Morphotropic phase boundary (MPB) in the $Pb(Zr,Ti)O_3$ system. Source: Jaffe et al. [4]. Reproduced with permission of Elsevier.

In terms of academic research, the emergence of PZT has provided two research directions in the field of ferroelectric materials, which are still very active today. One is the phase boundary (morphotropic phase boundary, MPB). In PZT solid solutions, $PbTiO_3$ (PT) shows a tetragonal ferroelectric phase, and its piezoelectric coefficient d_{33} is less than 100 pC/N, whereas $PbZrO_3$ (PZ) is antiferroelectric and lacks piezoelectric activity. However, when PZ is added to PT, and the molar ratio of PZ to PT reaches 53 : 47, the crystal structure changes from the tetragonal phase to the rhombohedral phase; that is, the direction of spontaneous polarization changes from ⟨100⟩ direction to ⟨111⟩ direction. More surprisingly, the piezoelectric coefficient d_{33} of PZT can reach about 300 pC/N near this MPB component. Although the tetragonal and rhombohedra ferroelectric phases have essential differences in the direction of spontaneous polarization, their cell parameters are very close to those of the cubic phase, so they are called morphotropic phases. In the PZT phase diagram, the boundary between these two phases is called the "morphotropic phase boundary." Even 70 years after the emergence of PZT, finding and constructing MPB is still the preferred method to improve the piezoelectric properties of ferroelectric materials.

The second is property modification via dopants. The doping of trace elements in PZT can induce completely different properties. From the point of view of the application, doping can be divided into two types, "soft" and "hard." "Soft" doping can improve the piezoelectric coefficient of PZT ceramics (d_{33} can reach 500–800 pC/N); the main elements include A-site dopants (e.g. Ba^{2+}, Sr^{2+}, and lanthanide rare earth) and B-site dopants (e.g. Nb^{5+} and Ta^{5+}). "Hard" doping can reduce the dielectric loss and mechanical loss in PZT ceramics by an order of magnitude; for example, the dielectric loss can be reduced from 1% to 1‰, to meet the application requirements of some high-power emission-type transducers. Hard doping usually uses low-valence elements (e.g. $Fe^{2+/3+}$ and $Mn^{2+/3+}$) to replace Ti^{4+} or Zr^{4+} in the B site.

To design piezoelectric materials with better performance, researchers have carried out a lot of research work to explore the doping modification mechanism of PZT ceramics. At present, most of the explanations of doping mechanisms are from the perspective of domain-wall motion. In short, "hard" doping can make the ferroelectric domain structure of PZT ceramics more stable, which is not easily changed under the action of the external field, resulting in the reduction of the piezoelectric effect and the loss related to the domain-wall motion. The "soft" doping results in the enhancement of the domain activity of PZT ceramics, which easily disturbed by external fields, and thus exhibit higher piezoelectric properties.

3.2.2 Lead-based Relaxor Ferroelectrics

The perovskite structure has a high tolerance for element substitution. Materials researchers have carried out a lot of doping attempts and obtained a large number of lead-based perovskite ferroelectric ceramics, among which a class of materials is known as ferroelectric relaxors [5–7]. The core characteristics of relaxor ferroelectrics are as follows: (i) heterogeneity in crystal structure at the nanoscale (as shown in Figure 3.3); (ii) polar nano-regions appear in relaxor ferroelectrics due to their heterogeneous structure at the nanoscale, leading to strong dielectric relaxation

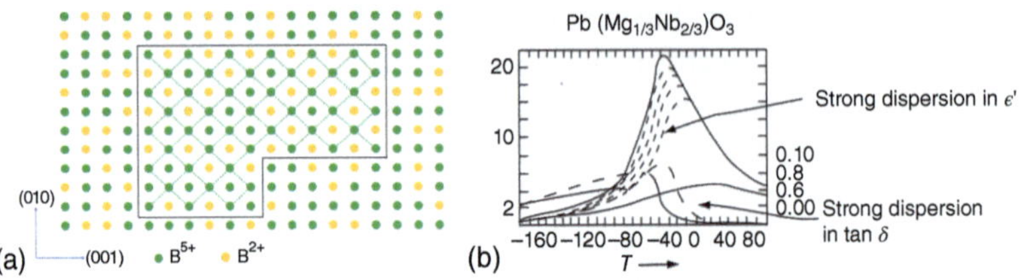

Figure 3.3 (a) Microstructural characteristic of relaxor ferroelectrics (Nanoscale ordered and disordered structure). Source: Li [6]. Reproduced with permission of Science China Press. (b) Dielectric dispersion in PMN as a function of temperature and frequency. Source: Cross [5]. Reproduced with permission of Taylor & Francis.

behaviors in such materials. The most classic ferroelectric relaxors are those based on PZT, where Zr is fully replaced by two different elements in a certain proportion (e.g. Mg^{2+}:Nb^{5+} = 1 : 2, In^{3+}:Nb^{5+} = 1 : 1) to form PMN–PT or PIN–PT solid solutions.

Relaxor ferroelectrics have been a key research topic since the 1960s and have been widely used in electro-optical modulators, capacitors, and low-hysteresis actuators. For example, as a deformable mirror actuator, they were used in the Hubble Telescope in the United States in the 1990s to successfully improve the telescope resolution. However, the classical relaxor ferroelectric ceramics, such as PMN–PT ceramics, do not show obvious advantages in piezoelectric properties and, therefore, have not attracted enough attention.

What is worth mentioning is that the importance of relaxor ferroelectrics in piezoelectric applications was not widely noticed until the gradual maturity of relaxor ferroelectric single crystal growth technology in the 1990s. During 1993–1997, Park and Shrout [8] grew high-quality PZN–PT relaxor ferroelectric single crystals with a size of 20 mm × 20 mm using a high-temperature flux method. A record-high piezoelectric coefficient of ∼2500 pC/N, an electromechanical coupling coefficient of ∼94%, along with a strain of ∼1.7% were obtained. This research marked a significant leap in the performance of piezoelectric materials at that time. Thus, it is honored as "a revolutionary breakthrough in the field of piezoelectric materials in the past 50 years," which is of revolutionary significance for the development of high-performance piezoelectric sensors, transducers, and other piezoelectric devices [9].

3.3 Recent Progress on Lead-Based Piezoelectric Materials

3.3.1 Local Structure Heterogeneity to Boost Piezoelectricity

As for the local structure heterogeneity in lead-based piezoelectrics, it is important to mention the origin of the high piezoelectric response of relaxor-PT ferroelectric single crystals (PMN–PT and PZN–PT) (Figure 3.4). Compared with classical ferroelectrics, the most distinct microstructural feature of relaxor–PTs (compositions around MPB) is the presence of nanoscale heterogeneous polar regions. Based

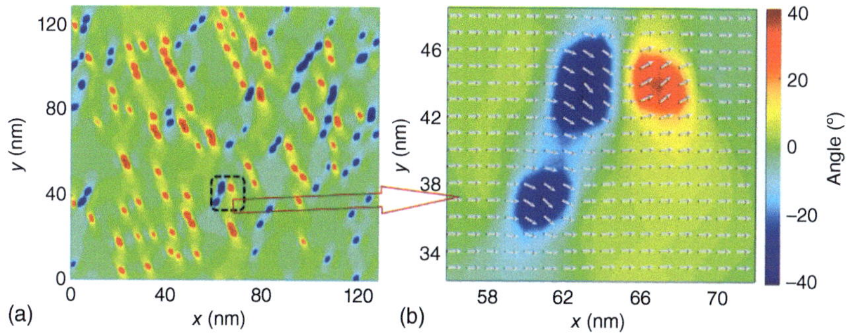

Figure 3.4 (a) Phase-field-simulated microstructures for a [100]-poled PNR-ferroelectric composite at 100 K. (b) The enlarged square area in (a), where the polar vectors of every grid are shown. The *x* and *y* axes represent the [100] and [010] crystallographic directions, respectively. The unit for *x* and *y* axes is nanometer. The color bar denotes the angle between the polar vector of the grids and the [100] direction. Source: Adapted from Li et al. [10].

on a structure–property relationship, such a special polar state is thus believed to be responsible for the enhanced piezoelectric properties of relaxor–PTs. Through phenomenological theories and phase-field simulations, Li et al. [10] reported that the relaxor members in relaxor ferroelectrics can generate structural heterogeneity at the nanoscale, resulting in additional energies such as electrostatic and elastic energies associated with heterogeneous interfaces. In addition to the long-range ferroelectric phase transition, these energies can further flatten the energy pattern, which is responsible for the outstanding piezoelectricity in relaxor–PT piezoelectrics (Figure 3.5).

Inspired by this feature in relaxor–PTs, it is expected to achieve high piezoelectricity if the long-range ordering of dipoles, inherently existing in ferroelectrics, is disrupted on the nanoscale. For ceramics and single crystals, element doping is a relatively simple method that may change the local structure. Li et al. [11, 12] introduced local structure heterogeneity via Sm doping in $Pb(Mg_{1/3}Nb_{2/3})O_3$–$PbTiO_3$ (Sm–PMN–PT) piezoelectrics to manipulate the interface energy and flattening of the thermodynamic energy profile further improved the piezoelectric properties of the materials. High piezoelectric coefficients up to 1500 and 4000 pC/N were achieved in piezoelectric ceramics and single crystals, respectively. This work provides a new idea and example for the design of high-performance piezoelectric materials, which may be used in the design and synthesis of many high-end materials.

Subsequently, this strategy to enhance the piezoelectricity by deliberately introducing "local structural heterogeneity" has been extended to other systems, such as PMN–PZT and KNN, and this approach has been proven to be effective in enhancing piezoelectric response in these materials.

3.3.2 Texture Technology (Templated Grain Growth)

The piezoelectric properties of relaxor ferroelectric single crystals (d_{33}: 1500–2000 pC/N; Strain: 1.7%) and electromechanical coupling coefficient (k_{33}: 90–92%) are

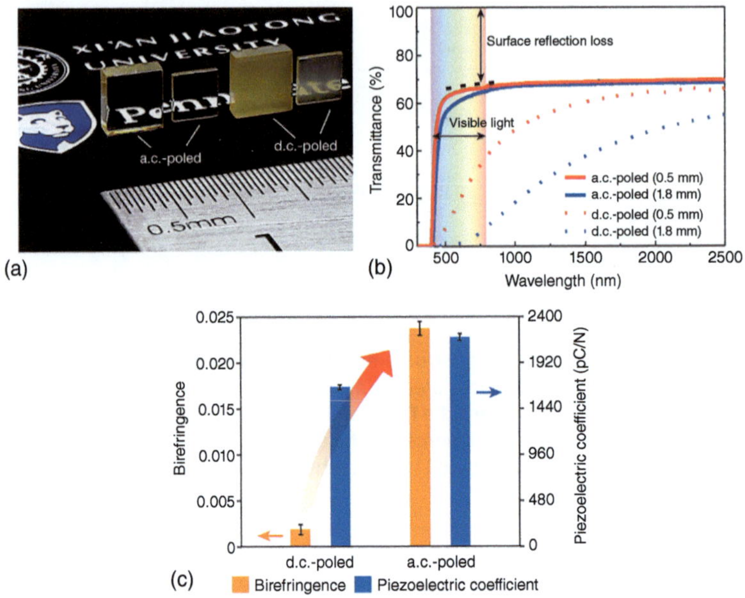

Figure 3.5 Comparison of the properties of a.c.-poled and conventional d.c.-poled [001]-oriented PMN-28PT crystals: (a) photo of AC-poled and DC-poled PMN-28PT crystals. The thicknesses of the crystals are 0.5 and 1.8 mm, respectively, (b) light transmittance of the AC-poled and DC-poled PMN-28PT crystals, (c) piezoelectric coefficient d_{33}, and birefringence of the AC-poled and DC-poled PMN-28PT crystals. Source: Qiu et al. [11]. Reproduced with permission of Springer.

significantly higher than those of ceramics with the same composition and phase transition temperature. From the perspective of practical application, however, there are still many problems in the application of relaxor ferroelectric single crystals in many piezoelectric devices, such as high production cost, lower phase transition temperatures, and difficulties in heterotypic or co-typic preparations. In recent years, scholars have paid much attention to the study of textured piezoelectric ceramics to obtain piezoelectric properties similar to relaxor ferroelectric single crystals and avoid the above problems. The textured piezoelectric ceramics are designed to orientate the ceramic grains to obtain the piezoelectric and electromechanical coupling coefficients that closely resemble single crystals.

The orientation of grains in polycrystalline ceramic materials is considered to be randomly distributed. However, in the process of molding and sintering, the application of certain special conditions such as heat, force, electricity, and magnetism will guide the grain to align or grow along some directions to obtain the preferred orientation of the grain, which is called texture [13]. The templated grain growth (TGG) technology is a common method to fabricate textured ceramics, which depends on the reaction and preferential growth of directionally arranged grains in fine matrix powders to achieve the directionally arranged grains (shown in Figure 3.6). Till now, TGG has still been the most commonly used technique for fabricating textured piezoelectric ceramics due to its simple processing, high-quality textured productions, and ease of industrialization.

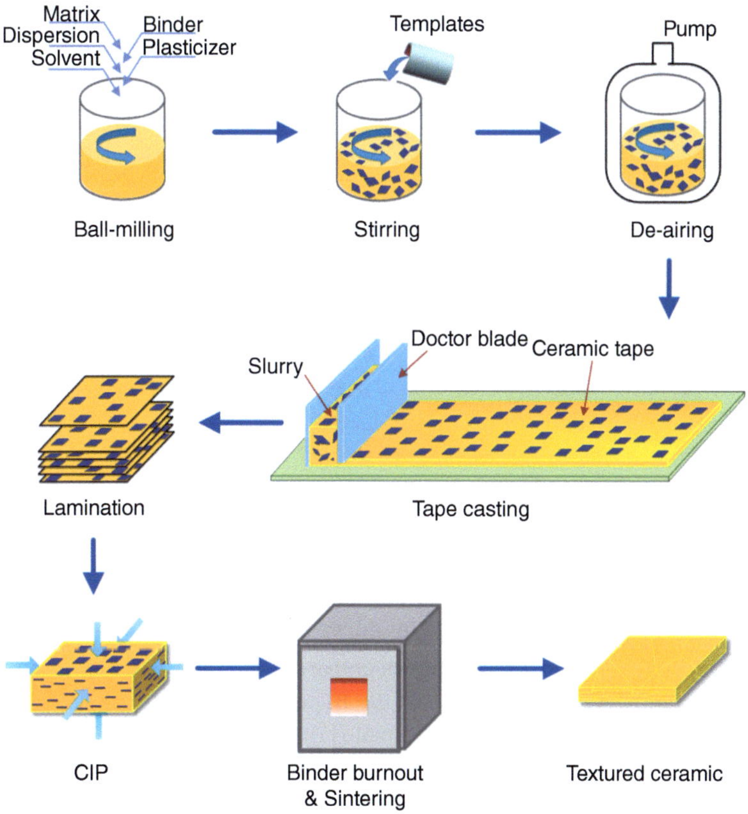

Figure 3.6 The sketch map of templated grain growth progress.

The core of TGG technology is the template, which should meet the following requirements: (i) stable chemical and thermal properties; (ii) a high length-to-thickness ratio (~10) to facilitate directional arrangement; (iii) a similar crystal structure to the matrix and low lattice mismatch (<10%).

Compared with lead-free textured piezoceramics, there are few available templates for lead-based textured piezoceramics, such as $BaTiO_3$ (BT) [14–16], $SrTiO_3$ (ST) [17–19], $0.4(Na_{0.5}Bi_{0.5})TiO_3$–$0.6PbTiO_3$ (BNT–0.6PT) [20], and $PbTiO_3$ (PT) [21]. BNT–0.6PT and PT templates have been reported in very limited literature and are limited to specific systems. The most commonly used templates for lead-based textured piezoelectric ceramics are BT and ST.

PMN–PT is the first lead-based piezoceramics to be textured. The physical properties of the textured PMN–PT ceramics are listed in Table 3.1. In 2001, Sabolsky et al. prepared PMN–35PT textured ceramics with a texture degree of 82% [15]. The strain could reach 0.3% under an electric field of 45 kV/cm, which was twice that of random ceramics of the same composition. In 2005, Kwon et al. [18] and Brosnan et al. [24] used the ST template to prepare PMN–32.5PT textured ceramics. Although the mechanical properties of the ceramics were optimized to some extent, and the strain was 0.3% (under 50 kV/cm electric field), the texture degree was only 70%. In 2012, Yan et al. obtained high-quality PMN–32.5PT textured ceramics with

Table 3.1 Electromechanical properties of some textured PMN–PT ceramics.

Sample	Template content	Template size (μm)	F_{001} (%)	d_{33} (pC/N)	ε_{33}	T_{rt} (°C)	T_c (°C)
PMN–32PT [15]	BT 5%	~100	82	~711 45 kV/cm	2900	~60	170
PMN–32.5PT [18]	ST 3%	5–15	73	~990 10 kV/cm	3500	/	148
PMN–32.5PT [18]	ST 5%	5–15	65	~1200 10 kV/cm	3700	—	120
PMN–32.5PT [22]	BT 5%	~10	90	630	2750	80	160
PMN–32.5PT [23]	BT 1%	5–10	98	1000	2591	90	160
PMN–25PT [20]	NBT–PT 7%	5–10	92	855	3490	60	129

98% using a high-quality BT template [23], and its d_{33} was 1000 pC/N, which was twice that of random ceramics of the same composition. With the increase in BT template content, however, both d_{33} and ε_{33} decreased continuously due to the low d_{33} and ε_{33} of the BT template. Compared with the ST (SrTiO3) template, the Curie temperature of PMN–32.5PT textured ceramics prepared with the BT template did not change significantly, which indicates that the BT template is more suitable for texturing lead-based ceramics than the ST template [18, 22].

After decades of development, the piezoelectric properties of ⟨001⟩ oriented PMN–PT textured ceramics have greatly improved, but the phase transition temperatures (T_{rt} and T_c) of PMN–PT ceramic systems are relatively low, which limits the temperature stability of such textured piezoelectric ceramics. To overcome the abovementioned issue, scholars have introduced a third end member containing In^{3+}, Sc^{3+}, and Zr^{4+} to improve phase transition temperatures. In addition, the corresponding textured ceramics are listed in Table 3.2 with the physical properties.

Chang et al. prepared 0.28PIN–0.40PMN–0.32PT textured ceramics using 5% BT templates [25]. The d_{33} coefficient and coercive field were 927 pC/N and 8.8 kV/cm, respectively. Moreover, T_{rt} and T_c are 110 and 200 °C, respectively, which are

Table 3.2 Electromechanical properties of some ternary lead-based textured ceramics.

Sample	Template content	F_{001} (%)	d_{33} (pC/N)	ε_{33}	E_c (kV/cm)	T_{rt} (°C)	T_c (°C)
PIN–PMN–PT [25]	BT 5%	97	927	2400	8.8	~110	200
PIN–PMN–PT [26]	BT 7%	95	780	2850	10.6	—	225
Mn–PMN–PZT [27]	BT 3%	93	720	3700	—	~120	210
PMN–PYN–PT [28]	BT 3%	99	1100	2230	7.6	~100	205
PIN–PSN–PT [29]	BT 3%	99	850	2120	6.8	~170	261
PIN–PSN–PT [29]	BT 5%	99	1100	2310	7.5	~110	248

improved compared with the PMN–PT system. Yan et al. fabricated Mn-doped 0.4PMN–0.25PZ–0.35PT ceramics with 3% BT templates [27]. The d_{33} coefficient ~720 pC/N and the k_{31} coefficient ~53% were much higher than that of ordinary ceramics of the same composition. Recently, Yang et al. used the BT template to prepare PIN–PSN–PT texture ceramics with a high texture degree of ~99% [29]. When the template content is 5%, the d_{33} coefficient and k_{33} coefficient of the texturing ceramics are as high as 1100 pC/N and 90%, respectively, which are comparable to those of the single crystal.

3.3.3 Solid-state Crystal Growth (SSCG)

Solid-state crystal growth (SSCG) has been investigated as an alternative method for fabricating single crystals. During SSCG, a small single-crystal seed diffuses to combine into a high-density polycrystal, as shown in Figure 3.7. After prolonged annealing, usually at temperatures between 100 and 200 °C below the melting point, the fine grains in the matrix are consumed and become a large single crystal. This process is called solid-state single-crystal growth because it does not involve the complete melting of the main components. Therefore, this method is suitable for high-temperature melting, volatile components, and the growth of non-congruent fused materials. Compared with other single-crystal growth processes, the crystal growth rate of the SSCG process is relatively low, ranging from 50 to 200 μm/h depending on the material system.

Solid-state conversion of single crystals has recently been proven to be a very successful way to produce piezoelectric single crystals for commercial usage. For example, single crystals such as $Pb(Mg_{1/3}Nb_{2/3})O_3$–$PbTiO_3$ (PMN–PT) and $Pb(Mg_{1/3}Nb_{2/3})O_3$–$Pb(Zr,Ti)O_3$ (PMN–PZT) are now produced by the SSCG method.

The possibility of growing a single crystalline PMN–PT material by the SSCG method was recognized many years ago. In 1998, Li et al. [32] used a method of embedding a $PbTiO_3$ (PT) single crystal into polycrystalline $Pb(Mg_{1/3}Nb_{2/3})O_3$ (PMN) to grow PMN–PT single crystals. Khan et al. [33] reported a solid-state growth of the PMN–35 mol% PT single crystal using the same method. Recently, the growth of undoped and Mn-doped 71PMN–29PT high-quality single crystals

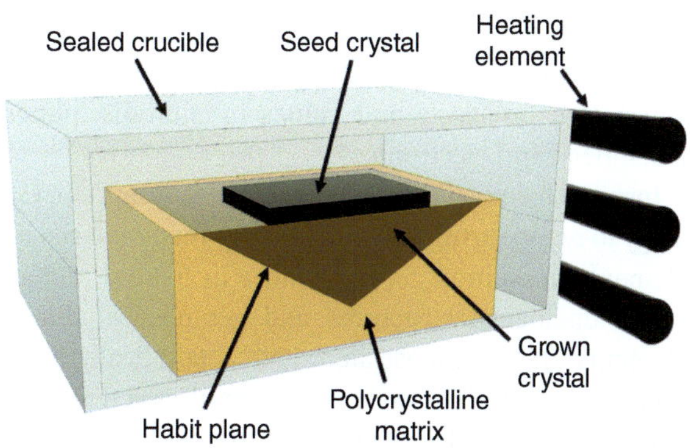

Figure 3.7 Cross-section of the solid-state crystal growth (SSCG) technique. Source: Kabakov et al. [30]. Reproduced with permission of Royal Society of Chemistry.

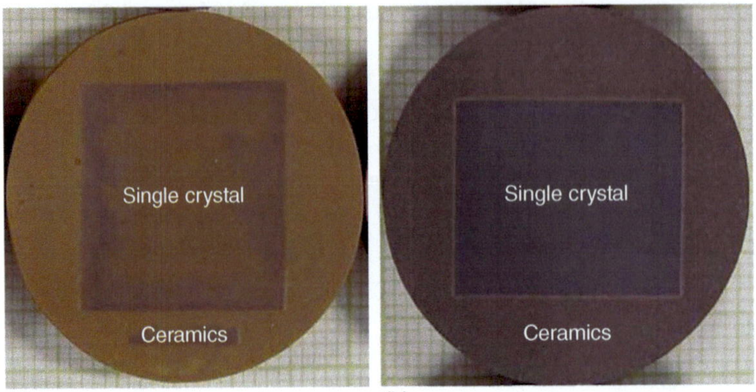

(a) (b)

Figure 3.8 Polished surfaces of 71PMN−29PT single crystals grown by SSCG method: (a) undoped and (b) Mn-doped. Source: Oh et al. [31]. Reproduced with permission from Korean Ceramic Society.

using the SSCG method was reported [31], in which excellent piezoelectric and electromechanical properties of both single crystals were demonstrated. Such single crystals were manufactured by Ceracomp Co., Ltd. from South Korea (Figure 3.8), which has become well-known for its production of high-quality piezoelectric single crystals via the SSCG method. Company Ceracomp Co. also introduced a method for obtaining PMN–PT single crystal ceramics by seeding with a $BaTiO_3$.

Zhang et al. [34] demonstrated single crystal growth of PMN–PZT using the SSCG technique. Further investigation on PMN–PZT single crystals obtained by the SSCG method was extended to doping of manganese (Mn), iron (Fe), or even indium (In). Until now, the SSCG technique was shown to be the only method to produce large relaxor–PZT single crystals.

3.3.4 Piezoelectric Metamaterials

The natural piezoelectric ceramic has only five nonzero piezoelectric strain constants. The design ideas of piezoelectric devices can be greatly expanded by artificially exciting piezoelectric modes that are necessary for engineering applications but not present in natural piezoelectric ceramics. This new way of thinking is of great significance in scientific research and also has important engineering application prospects and values in multidisciplinary fields, such as materials, machinery, and electronics.

Inspired by quasi-symmetry breaking, Yang et al. [35] have introduced a methodology to realize artificial anisotropy by metamaterial design to excite all the nonzero elements in contrast to zero values in natural materials. By elaborately programming topological structures and geometrical dimensions of the PZT-5H unit elements, they demonstrated, both theoretically and experimentally, that tunable nonzero or ultrahigh values of overall effective piezoelectric coefficients can be obtained. This work breaks the traditional cognition that there are only five nonzero piezoelectric coefficients d_{ij} in piezoelectric ceramic materials for more than 70 years, and brings a new design idea for the future development of piezoelectric devices.

3.4 Recent Progress on the Applications of Lead-Based Piezoelectrics

Lead-based piezoelectrics (ceramics, single crystals, and films), due to their excellent dielectric, piezoelectric, electro-optic, and pyroelectric properties, have been widely used in the fields of medical ultrasonic diagnosis, hydroacoustic transducer, precision drive, and electro-optic modulation. In this section, the recent developments in applications of lead-based piezoelectrics based on the materials introduced previously were summarized.

In the aspect of the underwater acoustic transducers, using high-performance PIN–PMN–PT single crystal, the un-prestressed triplet transducer has been developed. The resonant frequency of the single crystal transducer is 3 kHz lower than that of the PZT transducer of the same size, whereas the transmitting voltage response is 3 dB higher, and the free-field receiving sensitivity is 6 dB higher [36]. Based on the high transverse piezoelectric coefficient d_{32} and low dielectric loss of PIN–PMN–PT single crystal, a new miniaturized low-frequency and low-noise vector hydrophone was developed. The sound pressure sensitivity of the hydrophone without amplification can reach −212.9 dB@100 Hz and the overall equivalent noise pressure level of the hydrophone can reach 52.8 dB@100 Hz, which is significantly lower than the marine environmental noise in the operating band of 60–1000 Hz [37].

In terms of the medical ultrasonic transducers, a two-dimensional transducer array for ultrasonic therapy was developed using Sm-doped PMN–PT ceramics, whose output sound pressure is twice that of a PZT ceramic array of the same size [38]. At present, animal experiments on this ultrasonic transducer have been completed, and the expected therapeutic effect has been achieved [39]. A high-frequency ultrasonic transducer with a center frequency of 40 MHz was developed using Sm-doped PMN–PT ceramics. Compared with the PZT ceramic transducer (bandwidth <40%), the transducer has a wider working bandwidth (80%) and significantly reduced insertion loss, showing application potential in the field of high-frequency medical ultrasound [40]. An ultrasonic transducer for deep brain stimulation was developed using Sm-doped PMN–PT single crystal [41]. Compared with the traditional PZT piezoelectric ceramic transducers, the energy conversion efficiency of the transducer was greatly improved, showing potential applications in the fields of deep brain stimulation and analgesia.

In terms of the electro-optic modulator, an electro-optical Q-switch was developed using transparent PIN–PMN–PT relaxor ferroelectric single crystal, which is comparable to the electro-optical switching of commercial potassium dihydrogen deuterated phosphate (DKDP) and $LiNbO_3$ crystals in terms of output optical energy and optical pulse width; however, the device size and the driving voltage

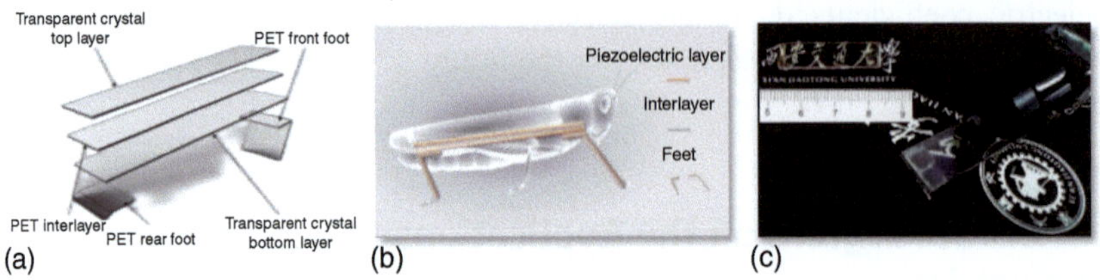

Figure 3.9 Schematic picture for transparent millirobot (a) splitting assembly structure, (b) schematic structure of the bio-inspired millirobot with a grasshopper, (c) prototype millirobot with a wide-angle camera (The inset shows the photo for the transparent millirobot). Source: From Gao et al. [43]. Reproduced with permission from AIP Publishing LLC.

are greatly reduced, which shows that transparent relaxor ferroelectric single crystal can achieve miniaturization, portability, and low power consumption in electro-optical devices [42].

In recent years, the design of bio-inspired transparent millirobots has gained significant attention, and many transparent millirobots have been proposed. However, a critical drawback of these transparent millirobots is the high driving voltage, which hinders their minimization and untethered travel. In the present work, based on the structure and motion of grasshoppers, we designed a transparent millirobot driven by low driving voltage with [011]-oriented $Pb(In_{1/3}Nb_{2/3})O_3$–$Pb(Mg_{1/3}Nb_{2/3})O_3$–$PbTiO_3$ (PIN–PMN–PT) ferroelectric crystal (Figure 3.9). It shows the traveling speed of up to 276.5 mm/s at 100 V and the minimum driving voltage for initial motion as low as 3 V, which is more than 2 orders of magnitude smaller than that of state-of-the-art transparent millirobots based on dielectric elastomers. The robustness of the millirobot is also demonstrated by its mechanical loading stability, as it can carry 78 g mass (97.5 times its weight), and cryogenic tolerance, as it can move at −150 °C.

Adaptive lenses (ALENS) are a smart optical system for controlling focal length by various input stimuli. Traditional piezoelectric ALENS are prepared using piezoceramics with transparent liquid. However, the drawbacks of this design include the small output displacements and the unstable optical axis due to the low piezoelectric coefficients of ceramics, as well as the evaporation and leakage of the liquid. To overcome the abovementioned issues, we proposed a radial extension-arching structured piezoelectric ALENS based on $Pb(In_{1/2}Nb_{1/2})O_3$–$Pb(Mg_{1/3}Nb_{2/3})O_3$–$PbTiO_3$ (PIMNT) relaxor ferroelectric crystals (PIMNT–ALENS) (Figure 3.10). In the design, the transparency liquid was replaced by polydimethylsiloxane (PDMS) film. Our newly designed ALENS possesses the advantages of fast response (~10^2 μs) and high focal sensitivity (8.5 cm/V), far outperforming the conventional ALENS based on piezoceramics actuators (~10^3 μs, ~10^{-1} cm/V) and dielectric elastomer actuators (DEA) (~10^5 μs and ~10^{-2} cm/V).

In terms of piezoelectric motors and actuators, a novel shear-bend coupling electromechanical model was proposed based on high-performance piezoelectric materials, which led to the development of the smallest ultrasonic piezoelectric motor, which solved the problems of extended travel, large thrust, and high

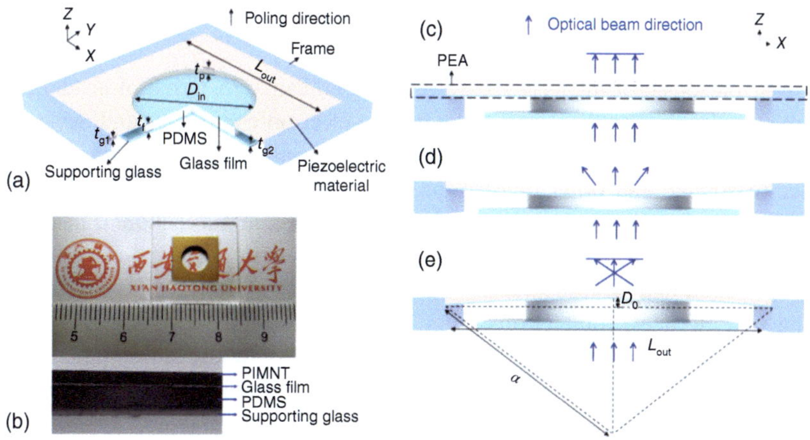

Figure 3.10 Diagram of the proposed PIMNT–ALENS: (a) the schematic of PIMNT–ALENS, (b) a prototype of PIMNT–ALENS. (The bottom figure is the cross section of PIMNT–ALENS.) The cross section of PIMNT–ALENS, (c) at an unactuated state, (d) at a negative voltage, and (e) at a positive voltage. Source: From Qiao et al. [44]. Reproduced with permission from AIP Publishing LLC.

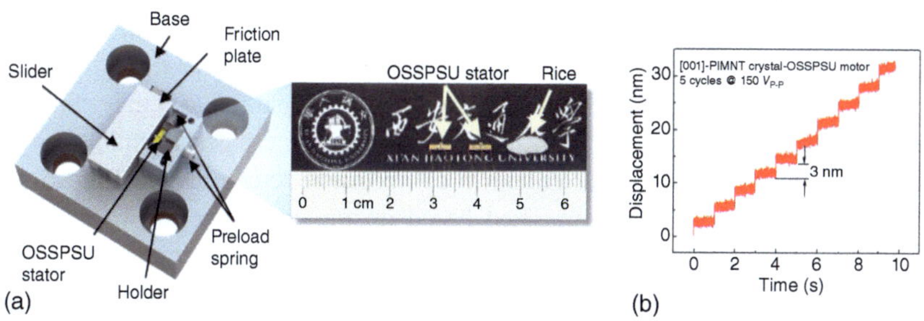

Figure 3.11 Schematic and nanostep motion characteristics of the proposed miniature linear OSSPSU motor: (a) basic design of the miniature linear ultrasonic OSSPSU motor, (b) the minimum step displacement of the OSSPSU motor. Source: From Liu et al. [45]. Reproduced from Springer Nature/CC BY 4.0.

accuracy of the lens zoom image stabilization system in tight spaces. On the other hand, to meet the demand for high precision and a wide range of motion of piezoelectric micro-actuators for image stabilization technology, a piezoelectric actuator using relaxation ferroelectric single crystal was fabricated. A miniaturized high-precision ultrasonic motor was developed (Figure 3.11). The size of the motor stator is much smaller than a grain of rice. This motor has both high displacement resolution (3 nm) and high velocity per unit volume ($4.66\,s^{-1}\,mm^{-2}$), and its displacement resolution is 2 orders of magnitude higher than the traditional ultrasonic motor.

3.5 Challenges and Prospects

Lead-based piezoelectric materials have great significance in terms of academic research and practical applications in the field of electronic materials and devices. Since the early 1970s, these materials have had excellent dielectric, piezoelectric, electro-optical, and other physical properties for broadband high-sensitivity medical ultrasonic transducers, underwater acoustic communication transducers, high-performance piezoelectric sensors, piezoelectric drivers, ultra-small electro-optic modulators, and other devices.

In the past decades, joint efforts from the scientific and technological communities around the world have led to breakthroughs in the development of lead-based piezoelectrics. The microstructure and macroscopic physical properties of lead-based piezoelectrics have been characterized and analyzed in a more comprehensive and detailed way, and the theoretical analysis of the relationship between the properties and the structure has been further deepened. These studies have greatly promoted the performance optimization of piezoelectrics and the development of related electronic devices. However, with the abovementioned research progress, it is necessary to point out that there are still many important scientific and technical problems that need to be solved.

(1) The first is the precise design of local heterogeneity. Regulating the local structure of ferroelectric materials is an effective method to improve the piezoelectric properties. The question then is: How to precisely design local heterogeneity? Chemical doping is a relatively simple method that may change the local structure. However, with so many elements in the periodic table, how to choose the proper one? For chemical composition screening, the following points for reference are summarized based on previous reports: (i) obvious differences between the doped cation and the substituted one in radius, valence state, and electronic structure; (ii) avoiding generating oxygen vacancies; and (iii) limited damage to the long-range ordering of dipoles.
(2) SSCG method is useful for the growth of single-crystal materials with high melting temperatures, volatile components, and incongruent melting. It realizes the transformation from ceramic to single crystal. However, SSCG method is also faced with some issues, including crystal defects, small crystal size, low quality, and difficulty in controlling growth rate. The growth mechanism is still unclear. The correlation between crystal structure, lattice defects, growth conditions, and crystal shape should be understood to guide single-crystal growth via SSCG method. The hope is that SSCG method can be used to prepare more crystals and grow high-quality crystals with large sizes, so as to expand its application range.
(3) The exploration of novel piezoelectric metamaterials. Based on the piezoelectric topological order structure, piezoelectric metamaterials can realize all 18 nonzero piezoelectric strain coefficient matrices and even enhance some piezoelectric coefficients. It offers an alternative possibility of designing new piezoelectric devices via sequentially architected piezoelectric elements and

the resulting synergistic stress–strain effects. Based on the research status of piezoelectric metamaterials, the challenges are summarized as follows: (i) The theory of dynamic homogenization of nonhomogeneous materials with microstructures has not been thoroughly developed. So most of the existing research on the properties of piezoelectric superstructures can only rely on numerical simulation. (ii) 3D printing technology provides an important means for the preparation of personalized and refined microstructures; however, there are few types of research on piezoelectric metamaterials. (iii) The existing piezoelectric materials still have shortcomings in their properties, which may become an obstacle to the further popularization and application of piezoelectric metamaterials. The development of piezoelectric materials with better properties will provide more diverse choices for the design of piezoelectric metamaterials.

References

1 Von Hippel, A., Breckenridge, R., Chesley, F., and Tisza, L. (1946). High dielectric constant ceramics. *Ind. Eng. Chem.* 38 (11): 1097–1109.

2 Zhang, S., Li, F., Jiang, X. et al. (2015). Advantages and challenges of relaxor-$PbTiO_3$ ferroelectric crystals for electroacoustic transducers – a review. *Prog. Mater Sci.* 68: 1–66.

3 Jaffe, B., Roth, R.S., and Marzullo, S. (1954). Piezoelectric properties of lead zirconate-lead titanate solid-solution ceramics. *J. Appl. Phys.* 25 (6): 809–810.

4 Jaffe, B., Cook, W.R., and Jaffe, H. (1971). Piezoelectric ceramics. In: *Piezoelectric Ceramics* (ed. B. Jaffe), 92. Academic Press.

5 Cross, L.E. (1987). Relaxor ferroelectrics. *Ferroelectrics* 76 (1): 241–267.

6 Li, F. (2022). Exploration of the mechanisms enabling high performance in relaxor ferroelectric crystals. *Chin. Sci. Bull.* 67: 3872–3881.

7 Li, F., Zhang, S., Damjanovic, D. et al. (2018). Local structural heterogeneity and electromechanical responses of ferroelectrics: learning from relaxor ferroelectrics. *Adv. Funct. Mater.* 28 (37): 1801504.

8 Park, S.-E. and Shrout, T.R. (1997). Ultrahigh strain and piezoelectric behavior in relaxor based ferroelectric single crystals. *J. Appl. Phys.* 82 (4): 1804–1811.

9 Service, R.F. (1997). Shape-changing crystals get shiftier. *Science* 275 (5308): 1878–1878.

10 Li, F., Zhang, S., Yang, T. et al. (2016). The origin of ultrahigh piezoelectricity in relaxor-ferroelectric solid solution crystals. *Nat. Commun.* 7 (1): 1–9.

11 Qiu, C., Wang, B., Zhang, N. et al. (2020). Transparent ferroelectric crystals with ultrahigh piezoelectricity. *Nature* 577: 350–354.

12 Li, F., Cabral, M.J., Xu, B. et al. (2019). Giant piezoelectricity of Sm-doped $Pb(Mg_{1/3}Nb_{2/3})O_3$-$PbTiO_3$ single crystals. *Science* 364: 264–268.

13 Fei, L. (2022). Exploration of the mechanisms enabling high performance in relaxor ferroelectric crystals. *Chin. Sci. Bull.* 67 (32): 3873–3881.

14 Remeika, J. and Jackson, W.M. (1954). A method for growing barium titanate single crystals. *JACS* 76 (3): 940–941.

15 Sabolsky, E., James, A., Kwon, S. et al. (2001). Piezoelectric properties of ⟨001⟩ textured $Pb(Mg_{1/3}Nb_{2/3})O_3$–$PbTiO_3$ ceramics. *Appl. Phys. Lett.* 78 (17): 2551–2553.

16 Liu, D., Yan, Y., and Zhou, H. (2007). Synthesis of micron-scale platelet $BaTiO_3$. *J. Am. Ceram. Soc.* 90 (4): 1323–1326.

17 Watari, K., Brahmaroutu, B., Messing, G.L. et al. (2000). Epitaxial growth of anisotropically shaped, single-crystal particles of cubic $SrTiO_3$. *J. Mater. Res.* 15 (4): 846–849.

18 Kwon, S., Sabolsky, E.M., Messing, G.L., and Trolier-McKinstry, S. (2005). High strain, ⟨001⟩ textured $0.675Pb(Mg_{1/3}Nb_{2/3})O_3$–$0.325PbTiO_3$ ceramics: templated grain growth and piezoelectric properties. *J. Am. Ceram. Soc.* 88 (2): 312–317.

19 Wu, J., Chang, Y., Lv, W. et al. (2018). Topochemical transformation of single crystalline $SrTiO_3$ microplatelets from $Bi_4Ti_3O_{12}$ precursors and their orientation-dependent surface piezoelectricity. *CrystEngComm* 20 (22): 3084–3095.

20 Poterala, S.F., Trolier-McKinstry, S., Meyer, R.J., and Messing, G.L. (2011). Processing, texture quality, and piezoelectric properties of $\langle 001\rangle_C$ textured $(1-x)Pb(Mg_{1/3}Nb_{2/3})TiO_3-xPbTiO_3$ ceramics. *J. Appl. Phys.* 110 (1): 014105.

21 Yan, Y., Zhou, J.E., Maurya, D. et al. (2016). Giant piezoelectric voltage coefficient in grain-oriented modified $PbTiO_3$ material. *Nat. Commun.* 7: 13089.

22 Yan, Y., Cho, K.-H., and Priya, S. (2011). Templated Grain Growth of ⟨001⟩-textured $0.675Pb(Mg_{1/3}Nb_{2/3})O_3$–$0.325PbTiO_3$ piezoelectric ceramics for magnetic field sensors. *J. Am. Ceram. Soc.* 94 (6): 1784–1793.

23 Yan, Y., Wang, Y.U., and Priya, S. (2012). Electromechanical behavior of [001]-textured $Pb(Mg_{1/3}Nb_{2/3})O_3$–$PbTiO_3$ ceramics. *Appl. Phys. Lett.* 100 (19): 192905.

24 Brosnan, K.H., Messing, G.L., Meyer, R.J., and Vaudin, M.D. (2006). Texture measurements in ⟨001⟩ fiber-oriented PMN-PT. *J. Am. Ceram. Soc.* 89 (6): 1965–1971.

25 Chang, Y., Watson, B., Fanton, M. et al. (2017). Enhanced texture evolution and piezoelectric properties in CuO-doped $Pb(In_{1/2}Nb_{1/2})O_3$–$Pb(Mg_{1/3}Nb_{2/3})O_3$-$PbTiO_3$ grain-oriented ceramics. *Appl. Phys. Lett.* 111 (23): 232901.

26 Wei, D.-d., Yuan, Q.-b., Zhang, G.-q., and Wang, H. (2015). Templated grain growth and piezoelectric properties of ⟨001⟩-textured PIN–PMN–PT ceramics. *J. Mater. Res.* 30 (14): 2144–2150.

27 Yan, Y., Cho, K.-H., and Priya, S. (2012). Piezoelectric properties and temperature stability of Mn-doped $Pb(Mg_{1/3}Nb_{2/3})$–$PbZrO_3$–$PbTiO_3$ textured ceramics. *Appl. Phys. Lett.* 100 (13): 132908.

28 Chang, Y., Wu, J., Liu, Z. et al. (2020). Grain-oriented ferroelectric ceramics with single-crystal-like piezoelectric properties and low texture temperature. *ACS Appl. Mater. Interfaces* 12 (34): 38415–38424.

29 Yang, S., Li, J., Liu, Y. et al. (2021). Textured ferroelectric ceramics with high electromechanical coupling factors over a broad temperature range. *Nat. Commun.* 12 (1): 1–10.

30 Kabakov, P., Dean, C., Kurusingal, V. et al. (2020). Solid-state crystal growth of lead-free ferroelectrics. *J. Mater. Chem. C* 8 (23): 7606–7649.

31 Oh, H.-T., Joo, H.-J., Kim, M.-C., and Lee, H.-Y. (2018). Effect of Mn on dielectric and piezoelectric properties of 71PMN-29PT [$71Pb(Mg_{1/3}Nb_{2/3})O_3$–$29PbTiO_3$] single crystals and polycrystalline ceramics. *J. Korean Ceram. Soc.* 55 (2): 166–173.

32 Li, T., Scotch, A.M., Chan, H.M. et al. (2005). Single crystals of $Pb(Mg_{1/3}Nb_{2/3})O_3$-35 mol% $PbTiO_3$ from polycrystalline precursors. *J. Am. Ceram. Soc.* 81 (1): 244–248.

33 Khan, A., Meschke, F.A., Li, T. et al. (2004). Growth of $Pb(Mg_{1/3}Nb_{2/3})O_3$-35 mol% $PbTiO_3$ single crystals from (111) substrates by seeded polycrystal conversion. *J. Am. Ceram. Soc.* 82 (11): 2958–2962.

34 Zhang, S., Lee, S.-M., Kim, D.-H. et al. (2007). Electromechanical properties of PMN–PZT piezoelectric single crystals near morphotropic phase boundary compositions. *J. Am. Ceram. Soc.* 90 (12): 3859–3862.

35 Yang, J., Li, Z., Xin, X. et al. (2019). Designing electromechanical metamaterial with full nonzero piezoelectric coefficients. *Sci. Adv.* 5 (11): eaax1782.

36 Tian, F., Liu, Y., Ma, R. et al. (2021). Properties of PMN-PT single crystal piezoelectric material and its application in underwater acoustic transducer. *Appl. Acoust.* 175: 107827.

37 Li, Z., Yang, S., Wang, S. et al. (2020). Development of a new kind low frequency low-noise vector sensor. *Sens. Actuators, A* 301: 111743.

38 Zhang, Z., Su, M., Li, F. et al. (2020). New Sm-PMN-PT ceramic-based 2-D array for low-intensity ultrasound therapy application. *IEEE Trans. Ultrason. Ferroelectr. Freq. Control* 67 (10): 2085–2094.

39 Zhang, Z., Liu, R., Li, G. et al. (2021). A dual-mode 2D matrix array for ultrasound image-guided noninvasive therapy. *IEEE Trans. Biomed. Eng.* 68 (12): 3482–3490.

40 Zhang, Z., Li, F., Chen, R. et al. (2018). High-performance ultrasound needle transducer based on modified PMN-PT ceramic with ultrahigh clamped dielectric permittivity. *IEEE Trans. Ultrason. Ferroelectr. Freq. Control* 65 (2): 223–230.

41 Zhang, T., Liang, H., Wang, Z. et al. (2022). Piezoelectric ultrasound energy–harvesting device for deep brain stimulation and analgesia applications. *Sci. Adv.* 8 (15): eabk0159.

42 Liu, X., Tan, P., Ma, X. et al. (2022). Ferroelectric crystals with giant electro-optic property enabling ultracompact Q-switches. *Science* 376 (6591): 371–377.

43 Gao, X., Qiao, L., Qiu, C. et al. (2022). A robust, low-voltage driven millirobot based on transparent ferroelectric crystals. *Appl. Phys. Lett.* 120 (3): 032902.

44 Qiao, L., Gao, X., Jin, H. et al. (2022). Ultrahigh focal sensitivity in a relaxor ferroelectric crystal-based piezoelectric adaptive lens. *Appl. Phys. Lett.* 121 (8): 082902.

45 Liu, J., Gao, X., Jin, H. et al. (2022). Miniaturized electromechanical devices with multi-vibration modes achieved by orderly stacked structure with piezoelectric strain units. *Nat. Commun.* 13 (1): 6567.

4

Lead-free Perovskite Piezoelectric Materials – Part One

Jiagang Wu

Sichuan University, College of Materials Science and Engineering, No. 24 South Section 1, Yihuan Road 610065, Chengdu, China

4.1 Potassium Sodium Niobate (KNN)

Potassium sodium niobate {(K, Na)NbO_3, KNN}, a solid solution of ferroelectric $KNbO_3$ and antiferroelectric $NaNbO_3$ (Figure 4.1), was first synthesized in 1954 by Shirane et al. [2], whereas its piezoelectric properties were first reported four years later by Egerton [3]. Pristine KNN ceramics show relatively moderate piezoelectricity (e.g. d_{33} = 80–100 pC/N) but a high Curie temperature (T_c = 420 °C), rendering it a promising lead-free piezoelectric candidate for replacing lead-based counterparts. Although KNN was reported in 1954, it did not receive enough attention until 2004 because of the inferior piezoelectric properties compared to lead-based ones. In 2004, Saito et al. reported high piezoelectric properties in textured KNN-based piezoceramics comparable to commercial soft PZT-4 ceramics [4], which ignited the passion for investigating KNN-based materials. In particular, due to the demand for sustainable development, high-performance lead-free piezoelectric ceramics are widely studied to hopefully replace lead-based counterparts. In addition, the establishment of related regulations or laws, for example, the Restriction of Hazardous Substances Directive (RoHS) released by the EU, accelerates the development of lead-free piezoelectric materials [5]. Currently, KNN-based piezoelectric materials are investigated in the forms of ceramics, thin films, and single crystals. In this section, we will briefly introduce the latest advances in KNN-based piezoelectric materials. Readers are suggested to refer to previous reviews or books for more fundamental descriptions of KNN-based piezoelectric materials [1, 6].

4.1.1 KNN-Based Ceramics

4.1.1.1 Synthesis of KNN Ceramics

Although KNN ceramics are eco-friendly, it is difficult to synthesize ideal KNN ceramics due to the volatilization of alkali metal (i.e. K/Na) and the easy formation of other secondary phases [1]. The volatilization of K/Na results in the deviation

Piezoelectric Materials: From Fundamentals to Emerging Applications, First Edition. Edited by Jiagang Wu.

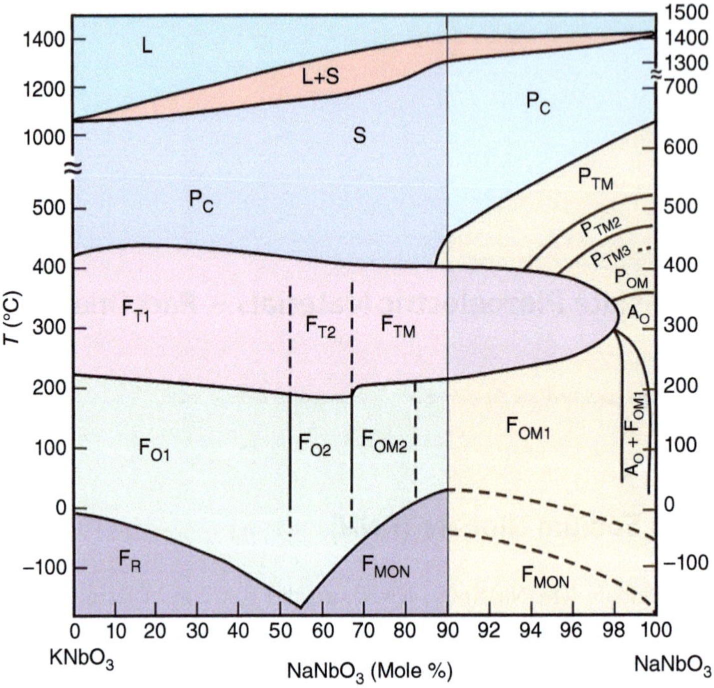

Figure 4.1 Phase diagram of (K, Na)NbO_3 solid solution. Source: Reproduced with permission from Li et al. [1]; © 2013, Wiley.

from the stoichiometric ratio, and the different reaction ratios between alkali carbonates (i.e. K_2CO_3 and Na_2CO_3) and niobium oxide (Nb_2O_5) induce the secondary phases [7]. Thus, some methods have been proposed to improve the sinterability of KNN ceramics, such as adding sintering aids (e.g. ZnO, MnO_2, Fe_2O_3, CuO, and others) and utilizing advanced sintering technologies (e.g. hot-pressed sintering, microwave sintering, and spark plasma sintering) [6, 8]. Sintering aids in effectively reducing the sintering temperature and thus decreases the volatilization of K/Na, and advanced sintering technologies also reduce the temperature and dwelling time of the sintering process. Except for the difficulty in synthesizing ideal KNN ceramics, recent reports show that the polymorph of Nb_2O_5 also significantly affects the phase structure and microstructure [9, 10]. By using two different polymorphs of Nb_2O_5 (i.e. the monoclinic M and the orthorhombic O), the calcined KNN powders (i.e. KNN-O and KNN-M) show different X-ray diffraction (XRD) patterns and microstructure. A pure KNN perovskite phase and a relatively homogeneous microstructure were obtained in calcined KNN-O powders, whereas a mixed $KNbO_3$-like and $NaNbO_3$-like perovskite and a heterogeneous microstructure were observed in calcined KNN-M powders [9]. Interestingly, increasing the calcination temperature also intensifies the microstructural heterogeneity of calcined KNN-O powders due to the presence of a liquid phase at higher temperatures. Such a phenomenon is attributed to the different kinetics of forming $KNbO_3$-like and $NaNbO_3$-like particles at a certain temperature. As a result, although the piezoelectric coefficient (d_{33}) is very similar among these KNN ceramics containing different Nb_2O_5 precursors, KNN ceramics fabricated using orthorhombic Nb_2O_5 show better dielectric properties. Therefore, before starting to fabricate KNN-based materials, choosing the appropriate Nb_2O_5 precursor is also important for guaranteeing the expected results.

4.1.1.2 Piezoelectric Properties

The piezoelectric property (such as most studied small-signal piezoelectric coefficient d_{33}), which measures the ability of interconverting mechanical and electrical energies, is one of the most important factors to evaluate the practicability of piezoelectric materials [11]. Compared with commercial PZT-based ceramics, pristine KNN ceramics show much inferior d_{33} values, which is the main obstacle to KNN ceramics' development during 1954–2003. In 2004, a milestone was achieved by Saito et al., who reported high piezoelectric properties in textured KNN-based piezoceramics comparable to commercial soft PZT-4 ceramics [4]. Although the high cost and complexity of texturing are not practical for the industry, their proposed orthorhombic-tetragonal (O–T) phase boundary has inspired others to pursue high d_{33} values by focusing on the phase boundary engineering. Meanwhile, other supplementary approaches have also been proposed to further improve d_{33} values, such as adding sintering aids, texturing, poling conditions, two-step sintering, and optimizing the sintering process [12–21]. In this section, we first summarize the development of the phase boundary and its effect on d_{33} values, then briefly introduce the texturing method, and finally conclude the physical mechanisms of high d_{33} values.

Phase Boundary Engineering High piezoelectric properties of lead-based piezoceramics mainly originate from the morphotropic phase boundary (MPB) that separates the rhombohedral (R) and the tetragonal (T) phases and has the R–T phase coexistence [22]. The R–T phase coexistence shows a relatively flattened Gibbs free energy profile due to the comparable energies of the R and T phases, consequently resulting in the enhanced dielectric and piezoelectric response [11]. Analogous to $KNbO_3$ and barium titanate ($BaTiO_3$) ceramics, pristine KNN ceramics undergo two ferroelectric–ferroelectric and one ferroelectric-parameter phase transitions with temperature, that is, rhombohedral (R) $\xrightarrow{-123^\circ C}$ orthorhombic (O) $\xrightarrow{220^\circ C}$ tetragonal (T) $\xrightarrow{420^\circ C}$ cubic (C) (Figure 4.2) [11, 25]. It is easy to see that a single room-temperature O phase accounts for the inferior dielectric and piezoelectric response of pristine KNN ceramics. Inspired by the R–T phase coexistence found in lead-based piezoceramics, researchers turned to constructing a room-temperature multiphase coexistence in KNN ceramics. The chemical modification is one of the most useful methods to shift the ferroelectric–ferroelectric phase transition temperature points. Although KNN ceramics were first reported in 1954, its chemical modification started in 2003. In 2004, Saito et al. co-doped Li^+, Ta^{5+}, and Sb^{5+} into KNN ceramics and constructed an O–T phase boundary near room temperature in $(K_{0.44}Na_{0.52}Li_{0.04})(Nb_{0.84}Ta_{0.10}Sb_{0.06})O_3$ (abbreviated

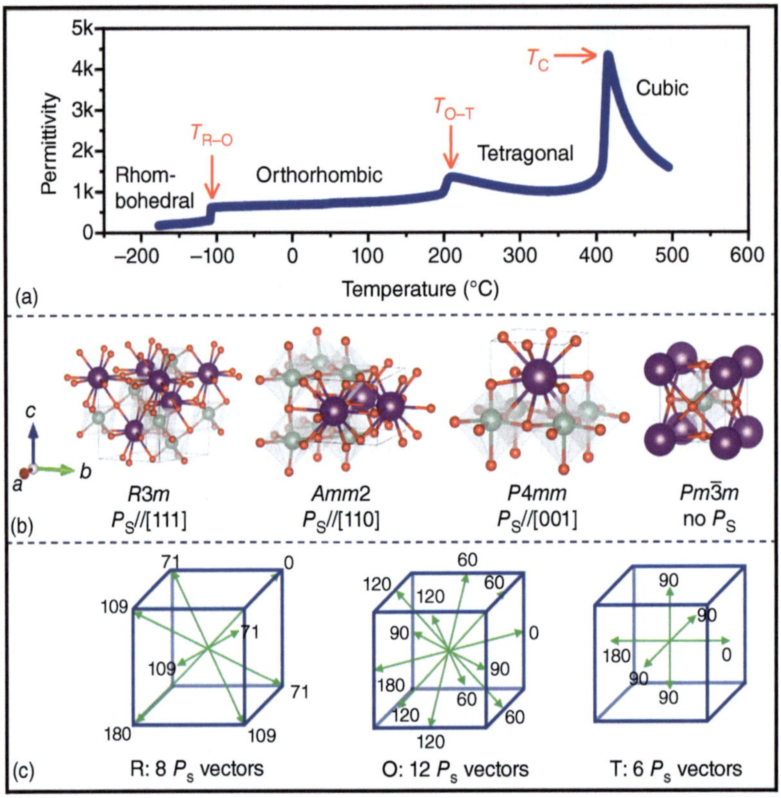

Figure 4.2 (a) Temperature-dependent relative permittivity of (K, Na)NbO_3 ceramics, (b) lattice structural schematic diagrams for R, O, T, and C phases, (c) schematic diagrams of spontaneous polarization (P_s) for R, O, and T phases. Source: Adapted from Lv et al. [23, 24].

to LF4) ceramics [4]. They reported d_{33} values of 300 and 416 pC/N in textured and non-textured states, respectively. In particular, textured LF4 (LF4T) shows excellent piezoelectric properties and temperature stability comparable to those of commercial PZT-4. Although the high cost and complexity of texturing are not practical for the industry, their proposed O–T phase boundary has inspired others to pursue high d_{33} values by focusing on the O–T phase boundary. Then, researchers developed many chemical dopants that can effectively shift the O–T phase transition temperature to near room temperature, for example, Li and some Li-containing species (e.g. $LiSbO_3$ and $LiTaO_3$) [1, 6]. Although these chemical dopants can construct an O–T phase boundary near room temperature, the most reported d_{33} values were less than 300 pC/N, that is, for non-textured LF4, and a few reported d_{33} values were within 300–400 pC/N that were still lower than LF4T and inferior to those of lead-based piezoceramics (Figure 4.3a,c). Later research reported that doping Sb, $BiScO_3$, and $BaZrO_3$ can shift the rhombohedral–orthorhombic phase transition temperature (T_{R-O}) to room temperature, although the obtained d_{33} values were still much lower than 300 pC/N [27–29]. Although these efforts failed to further improve d_{33} values to higher than those of non-textured and textured LF4

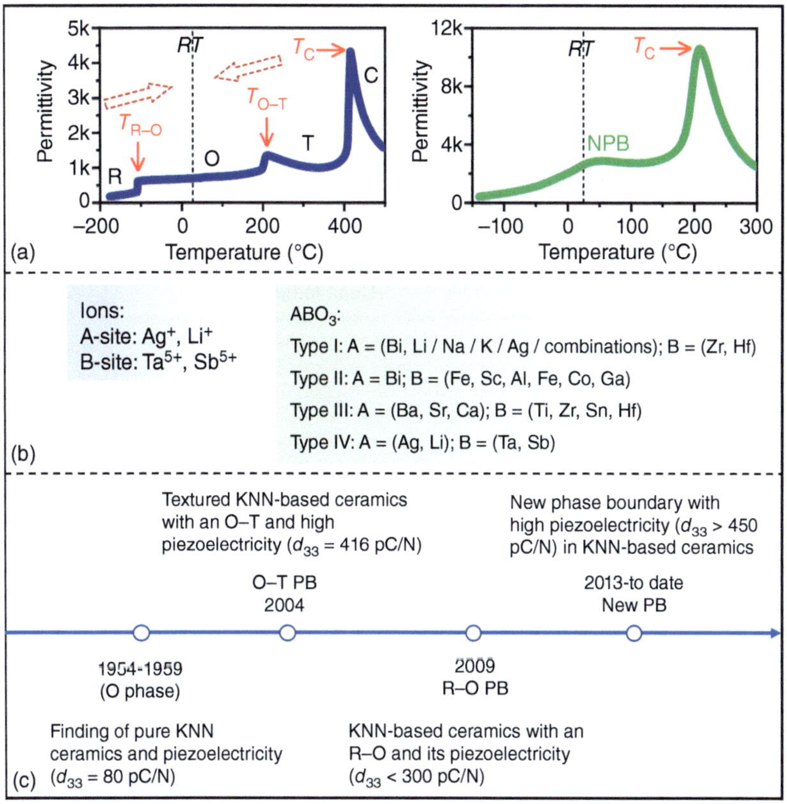

Figure 4.3 (a) Phase transition sequences of pristine KNN ceramics and strategies of constructing room-temperature phase boundaries, (b) chemical dopants for shifting $T_{R\text{-}O}$ and $T_{O\text{-}T}$, (c) development of phase boundaries of KNN ceramics. "PB" is the abbreviation of phase boundary. Source: (a) Reproduced with permission from Lv et al. [23]; © 2020, Royal Society of Chemistry; (b) Reproduced with permission from Lv et al. [26]; © 2022, American Chemical Society.

ceramics, researchers accumulated rich experience in screening chemical dopants that can regulate the phase transition temperature (Figure 4.3b).

Since researchers have developed chemical dopants that can decrease (increase) $T_{O\text{-}T}$ ($T_{R\text{-}O}$), one may wonder if it possible to simultaneously shift $T_{R\text{-}O}$ and $T_{O\text{-}T}$ to room temperature and construct a phase boundary akin to the MPB of the Pb(Zr, Ti)O_3 system (Figure 4.3a). Although some attempts were made to improve the piezoelectric constant of KNN-based ceramics, their d_{33} is still much inferior to the PZT-based ceramics [30, 31]. In 2013–2014, Wu et al. proposed a new chemical dopant $Bi_{0.5}M_{0.5}ZrO_3$ (M = Na, Li, K, Ag, and their combinations) where Zr^{4+} can increase $T_{R\text{-}O}$ and $(Bi_{0.5}M_{0.5})^{2+}$ can decrease $T_{O\text{-}T}$, and then designed a new formula of $(1-x)(K_{1-y}Na_y)(Nb_{1-z}Sb_z)O_3\text{–}xBi_{0.5}(Na_{1-w}K_w)_{0.5}ZrO_3$ [32]. Through elaborately tailoring the values of x, y, z, and w, they constructed a room-temperature R–T phase boundary and then achieved a d_{33} value as high as 490 pC/N, superior to those of LF4T and comparable to those of commercial lead-based piezoceramics.

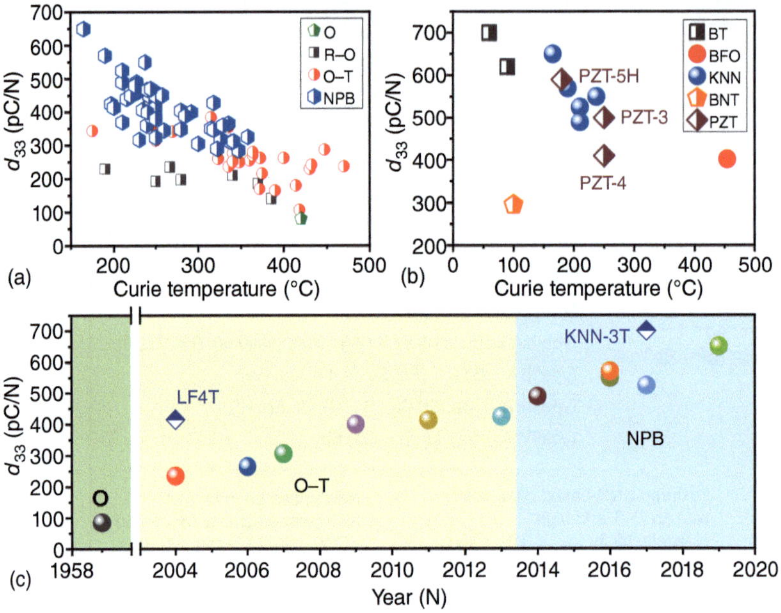

Figure 4.4 (a) d_{33} versus T_c among KNN-based piezoceramics with different phase boundaries, (b) d_{33} versus T_c among different piezoceramics, (c) historical evolution of KNN-based piezoceramics' d_{33} values. Source: Reproduced with permission from Ref. [23]; © 2020, Royal Society of Chemistry.

Thereafter, they designed different formulas and continuously improved d_{33} values of KNN-based ceramics to 490–650 pC/N [33–36]. More importantly, their designed formulas were successfully verified by other groups [26]. As a result, they termed such a phase boundary constructed by simultaneously shifting $T_{R\text{-}O}$ and $T_{O\text{-}T}$ to near room temperature and greatly promoting the d_{33} as the new phase boundary (NPB) and clarified the difference between the NPB and other phase boundaries (i.e. R–O and O–T) [23]. Figure 4.4 shows the historical evolution of KNN-based ceramics' d_{33} values. Therefore, the phase boundary engineering achieved by the necessary chemical dopants has been proved to be one of the most promising approaches to improve the piezoelectric properties of KNN-based ceramics.

Texturing Except for the phase boundary engineering, the texturing is another important method to improve the piezoelectricity of piezoceramics and also applies to KNN ceramics. As we know, the grains and the ferroelectric domains within grains of piezoceramics prepared by the conventional solid-phase sintering method are randomly distributed. Although applying a high electric field can re-orient ferroelectric domains along a specific direction, the coupling effects between grains and the back-switching of re-oriented domains significantly reduce the degree of the orientation of ferroelectric domains, which is the main reason why most randomly distributed piezoceramics show inferior piezoelectric properties compared to their well-oriented single crystal counterparts. However, fabricating single crystals is usually high-cost and time-consuming. Thus, researchers developed

Table 4.1 Representative piezoelectric systems and their corresponding templets.

Material systems	Templates	References
(K, Na)NbO_3-based	$NaNbO_3$	[38]
$BaTiO_3$-based	$BaTiO_3$	[39]
$(Bi_{0.5}Na_{0.5})TiO_3$-based	$Bi_4Ti_3O_{12}$	[40]
	$NaNbO_3$	
	$Bi_{4.5}Na_{0.5}Ti_4O_{15}$	
	$Bi_{2.5}Na_{3.5}Nb_5O_{18}$	
$Pb(Mg_{1/3}Nb_{2/3})O_3$-$PbTiO_3$-based	$BaTiO_3$	[41]
$PbTiO_3$	$PbTiO_3$	[42]

Table 4.2 Piezoelectric properties of representative textured and non-textured KNN-based ceramics.

Composition	Orientation/degree	d_{33} (pC/N)	References
$(K_{0.44}Na_{0.52}Li_{0.04})(Nb_{0.86}Ta_{0.10}Sb_{0.04})O_3$	<001>/91% Random	416 300	[4]
$(K_{0.463}Na_{0.537})_{0.946}Li_{0.054}NbO_3$	<001>/81% Random	254 168	[43]
$0.97(K_{0.5}Na_{0.5})_{0.98}Ag_{0.02}(Nb_{0.96}Sb_{0.04})O_3$–$0.03(Bi_{0.5}Na_{0.5})ZrO_3$	<001>/97.6% Random	550 290	[44]
$0.96(K_{0.5}Na_{0.5})(Nb_{0.965}Sb_{0.035})$–$0.01CaZrO_3$–$0.03(Bi_{0.5}K_{0.5})HfO_3$	<001>/98% Random	700 300	[38]

the texturing method to synthesize piezoceramics with grains oriented along one specific direction, as described in Chapter 9 in detail. The core of the texturing method is to use templates with one specifically oriented direction as seeds that induce other raw particles to grow along this direction during the sintering process [37]. Considering the compatibility between templates and raw particles, choosing an appropriate template is key to fabricating high-quality textured piezoceramics, as listed in Table 4.1. Since the phase boundary engineering can effectively improve piezoelectric properties, the researcher thus fabricated textured KNN-based ceramics containing the phase boundary with the purpose of further enhancing piezoelectric properties, as listed in Table 4.2. Of particular importance is that Zhai et al. obtained an ultra-high d_{33} of 700 pC/N in <001>-textured $0.96(K_{0.5}Na_{0.5})(Nb_{0.965}Sb_{0.035})$–$0.01CaZrO_3$–$0.03(Bi_{0.5}K_{0.5})HfO_3$ (96KNNS–1CZ–3BKH), which is superior to those of other textured and non-textured KNN-based ceramics [38]. Indeed, the textured KNN-based ceramics show much better piezoelectric properties than their non-textured counterparts. However, it should be pointed out that the texturing method is also expensive and complicated and may not be suitable for large-scale industrial applications.

Physical Mechanisms The physical mechanisms are discussed to explain the high piezoelectric properties of the phase-boundary-featured KNN-based ceramics, particularly for the NPB-featured ones. Generally, for perovskite ferroelectric ceramics, its piezoelectricity can be expressed as

$$d_{33} = 2{\cdot}Q{\cdot}\varepsilon_r{\cdot}P_r \tag{4.1}$$

where Q, ε_r, and P_r are the electrostrictive coefficient, relative dielectric permittivity, and remanent polarization [45]. On the other hand, piezoelectricity also originates from intrinsic and extrinsic contributions [46]. The intrinsic contribution mainly comes from the lattice distortion associated with the phase structure, and the extrinsic one originates from the domain switching and domain-wall motion [46]. In fact, both statements are intimately correlated because both ε_r and P_r are closely related to the phase structure and the ferroelectric domain. So, we first summarized the phase structure and ferroelectric domains and then discussed the physical mechanisms of the enhanced piezoelectricity through phase boundary engineering.

Since the phase boundary engineering aimed to shift the phase transition temperature to the room temperature, a room-temperature multiphase coexistence is highly anticipated. The incipient methods to identify the phase structure rely on the XRD patterns and temperature-dependent dielectric properties. However, the highly overlapped diffractions always puzzle researchers' eyes due to the similar lattice parameters among the R, O, and T phases. Besides, the chemical dopants significantly suppress the dielectric anomalies of inter-ferroelectric phase transitions (particularly for the R–O), making it difficult to accurately obtain the phase transition temperature points. Considering the distinct difference of the P_s vectors among R, O, and T phases (Figure 4.5a,b), the scanning transmission electron microscopy (STEM), working in the modes of the high angle annular dark field mode (STEM–HAADF) and the convergent beam electron diffraction (STEM–CBED), thus has been proposed to detect the phase structure. By means of STEM–CBED and STEM–HAADF, Wu et al. found the NPB possessed an R–T or an R–O–T multiphase coexistence, and these multiphases coexist in the nanoscale (Figure 4.5c–i) [33–35].

As the extrinsic contribution, the ferroelectric domains significantly affect the piezoelectric properties of piezoceramics, for which the observation is thus of particular importance. Currently, three methods are used to observe the ferroelectric domains of KNN-based ceramics, that is, chemical etching, piezoresponse force microscopy (PFM), and transmission electron microscopy (TEM) [48]. The chemical etching can observe the ferroelectric domains for a large-scale area but also destructs the tiny domains such as ultra-fine nanodomains due to the etching process [16]. Therefore, the PFM and the TEM are the most widely used methods [49]. Through these three methods, researchers conducted observations of the ferroelectric domains in different KNN-based ceramics, as summarized in Figure 4.8. Pristine KNN (or KNN ceramics with an O phase) exhibit the regular domain configurations with a zig-zag shape and a micron size (Figure 4.6a) [50, 53]. These regular domains hinge on the symmetry of the O phase. On the contrary, the O–T phase boundary shows complicated domain configurations with a nano size (Figure 4.6b,c) [51, 52]. In particular, Zuo et al. observed the striped domains,

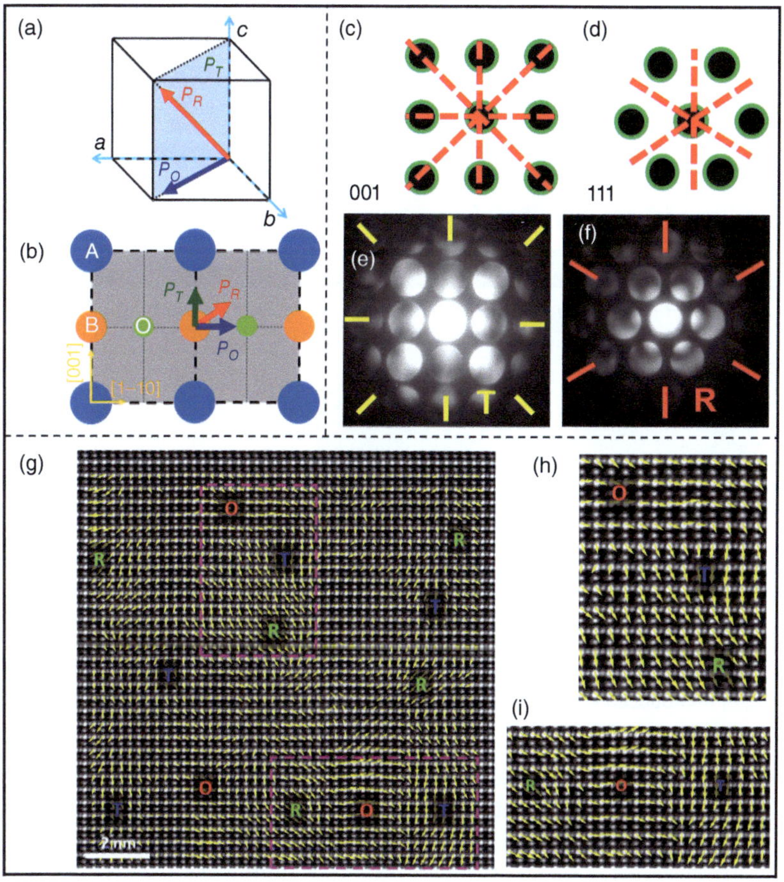

Figure 4.5 (a) Schematic diagram of the P_s vectors of R, O, and T phases in one unit cell, (b) schematic diagram of the P_s vectors of R, O, and T phases in one unit cell viewed along [110] direction; simulated CBED pattern for (c) a T phase along [001] and for (d) a R phase along [111] direction; measured CBED patterns along (e) [001] and (f) [111] directions for KNNS–BNKH ceramics; (g) measured STEM–HAADF image for KNNS–BNKZ–AS ceramics; (h,i) enlarged visions of the areas in (g) marked with the dashed magenta boxes. The dashed lines in (c,d) indicate the mirrors, along which the left and right sides are mirror symmetry. Source: (a,b) Reproduced with permission from Zhao et al. [47]; © 2018, American Chemical Society; (c,d) Reproduced with permission from Lv et al. [23]; © 2020, Royal of Society Chemistry; (e,f) Reproduced with permission from Zheng et al. [35]; © 2017, Royal of Society Chemistry; (g–i) Reproduced with permission from Tao et al. [36]; © 2019, American Chemical Society.

comprising alternate and hierarchical nanodomains with the size of 20 and 50 nm, in the O–T phase boundary (Figure 4.6b) [51]. In the same way, the NPB also exhibits similar striped domains comprising hierarchical nanodomains (Figure 4.6d–f) [35]. Of particular interest is that the NPB-featured KNN-based ceramics, with a more complex composition design, even show ultra-fine nanodomains with a scale of ~2 nm (Figure 4.6g) [34]. These ultra-fine nanodomains were reported to be heterogeneous in strain (Figure 4.6h) [34]. These special nanodomains were also

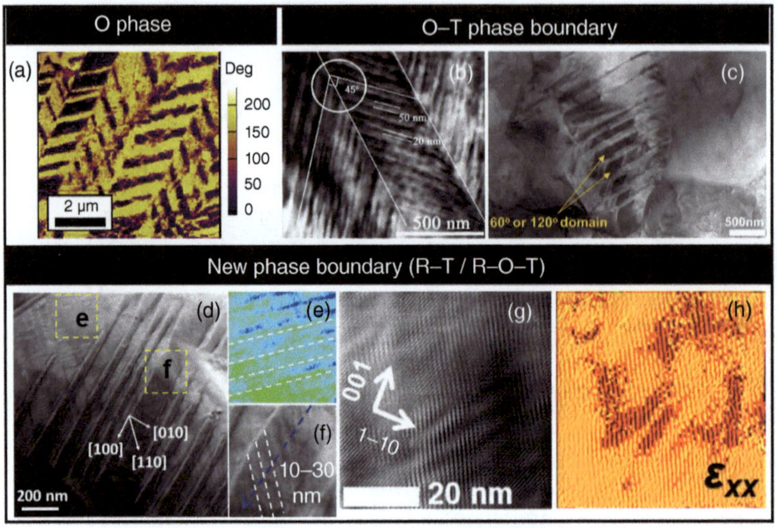

Figure 4.6 Domain configurations of KNN-based ceramics with (a) the O phase, (b–c) the O–T phase boundary, and (d–h) the NPB. Source: (a) Reproduced with permission from Ref. [50]; © 2019, AIP Publishing; (b) Reproduced with permission from Ref. [51]; © 2011, AIP Publishing; (c) Reproduced with permission from Ref. [52]; © 2018, Royal Society of Chemistry; (d–f) Reproduced with permission from Ref. [35]; © 2017, Royal Society of Chemistry; (g,h) Reproduced with permission from Ref. [34]; © 2016, Wiley.

widely reported in lead-based piezoelectric materials and were believed to be responsible for ultra-high piezoelectricity [54].

Here, we discussed the physical mechanisms of the enhanced piezoelectric properties of KNN-based ceramics. As discussed above, shifting the phase transition temperature to near room temperature results in the multiphase coexistence, particularly for the NPB. According to the phenomenological theory, these coexisting phases possess comparable free energy, thus constructing a relatively flattened energy barrier among different phases, as imaged by the phase-field simulations (Figure 4.7a) [25, 36]. Such a flattened energy barrier allows an easy polarization rotation among different phases (Figure 4.7b) and consequently endows piezoceramics with a fast response to the external stimuli, including the electric field and the strain. As a result, the phase-boundary-engineered KNN-based ceramics show much higher dielectric permittivity near room temperature compared to the pristine ones (Figure 4.7d).

On the other hand, the nanoscale multiphase coexistence also induces the nanoscale domains and the polar nanoregions (PNRs). Nanoscale domains are prone to switching under electric fields and stress and play an important role in enhancing piezoelectricity of piezoceramics [24]. In particular, it has been reported in KNN-based ceramics that a smaller domain size corresponds to a higher d_{33} value [24]. The PNRs, originating from the nanoscale fluctuation of the P_s induced by the chemical dopants, coexist with the long-range ferroelectric ordering (LRFO) (Figure 4.7c) and are responsible for the macroscopically observed relaxor-featured

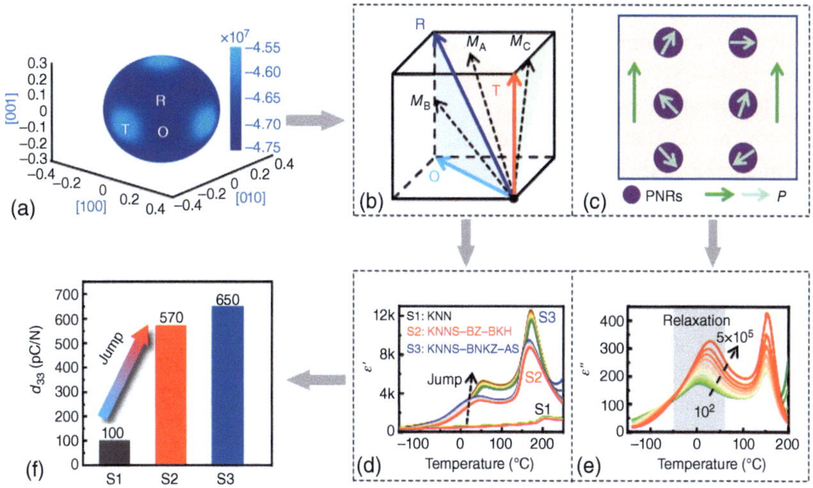

Figure 4.7 (a) Gibbs free energy profile for an R–O–T phase coexistence, (b) schematic diagram of the P_s vectors of R, O, T, and M phases in one unit cell, (c) schematic diagram of relaxor-featured multiphase coexistence, (d) temperature-dependent real part of permittivity (ε') of three representative KNN-based ceramics, (e) temperature- and frequency-dependent imaginary part of permittivity (ε'') of KNNS–BZ–BKH ceramics, (f) comparison of d_{33} values among three representative KNN-based ceramics. Source: (a–c) Reproduced with permission from Tao et al. [36]; © 2019, American Chemical Society; (e) Reproduced with permission from Lv et al. [55]; © 2022, Elsevier.

multiphase coexistence (Figure 4.7e) [55]. These PNRs can further promote the domain switching of nanodomains due to the fast response to the electric field and thus benefit the piezoelectricity [56, 57]. The easy switching of nanodomains and PNRs guarantees relatively high P_r values, even with a substantial content of chemical dopants. Consequently, according to Eq. (4.1), d_{33} values are significantly improved due to the enhanced $\varepsilon_{r,}$ and the remaining high P_r values (Figure 4.7f). In addition, recent work shows that the dopants-induced T phase and high-density nanoscale heterostructures with low-angle polar vectors and the electric-induced phase transition can also explain the enhanced piezoelectricity [57, 58]. Note that the recently uncovered physical mechanisms are not in conflict with Figure 4.7 but are actually further analysis.

Structural Heterogeneity Except for abovementioned methods, the structural heterogeneity has also been proposed to enhance the electric-induced strain property of KNN-based ceramics. Under the guidance of the phase-field simulation, Wang et al. found that a core–shell structure, comprising a ferroelectric core and paraelectric shell, may substantially improve the electric-induced strain due to the enhanced domain-wall motion [59]. Thus, they prepared the core–shell structure by taking the example of $(Na_{0.49}K_{0.49}Li_{0.02})(Nb_{0.8}Ta_{0.2})O_3$–$x$$CaZrO_3$ (KNNLT–xCZ) ceramics and using the two-step preparation process. As a result, the core–shell structured KNNLT-3.5CZ sample shows much higher electric-induced strain and temperature stability compared to the structurally homogeneous one. Therefore, although the two-step preparation process may not be practical for industry, it should be an alternative way to enhance the strain property of KNN-based ceramics.

4.1.1.3 Temperature Stability

Apart from the piezoelectric properties, the temperature stability is also another important factor that decides the practical applications of KNN-based ceramics. Being different from the MPB established in the Pb(Zr, Ti)O_3 (PZT) piezoelectric system, the phase boundary of KNN-based ceramics is simultaneously affected by the composition and the temperature (termed as the polymorphic phase boundary, PPB), whereas the MPB is only decided by the composition (when $T < T_c$), as compared in Figure 4.8. Therefore, PZT-based ceramics show stable piezoelectric properties varying with temperature, whereas the significantly reduced piezoelectric properties with temperature have been widely observed in KNN-based ceramics and have become one of the most challenging issues for its practical applications.

Currently, three measurement methods are utilized to evaluate the temperature stability of KNN-based ceramics. The first one measures *in situ* d_{33} values varying with temperature, the second one is measuring *in situ* electric-induced strain varying with temperature, and the third one is measuring d_{33} values after annealing poled-samples at different temperatures. The detailed description is referred to the previous review [23]. Clearly, the first one can reflect the real temperature stability of piezoceramics when servicing in a device, whereas the third one only reflects the thermal stability. The second one is debatable because the most reported electric-induced strain was obtained under high electric fields, which is impossible or should be avoided when servicing as actuators. Indeed, some researchers found that increasing the amplitude of the applied electric field enhanced the temperature stability of electric-induced strain [60]. To improve the temperature stability of KNN-based ceramics, many approaches, including ceramic composites, texturing, regulating the phase ratio of multiphase coexistence, changing the phase transition temperature, and structural heterogeneity, have been proposed [48]. Here we mainly introduced the ceramic composites and other approaches are referred to in the previous summary [48].

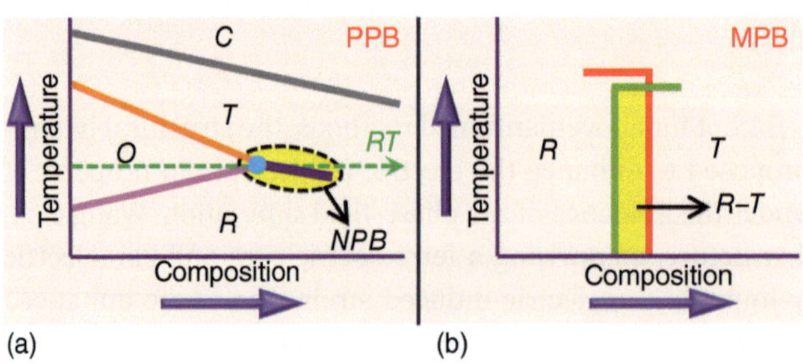

Figure 4.8 Intrinsic traits of the polymorphic phase boundary (PPB) and morphotropic phase boundary (MPB): (a) general phase diagram for constructing the NPB, (b) phase diagram of a typical MPB. Source: Reproduced with permission from Lv et al. [23]; © 2020, Royal Society of Chemistry.

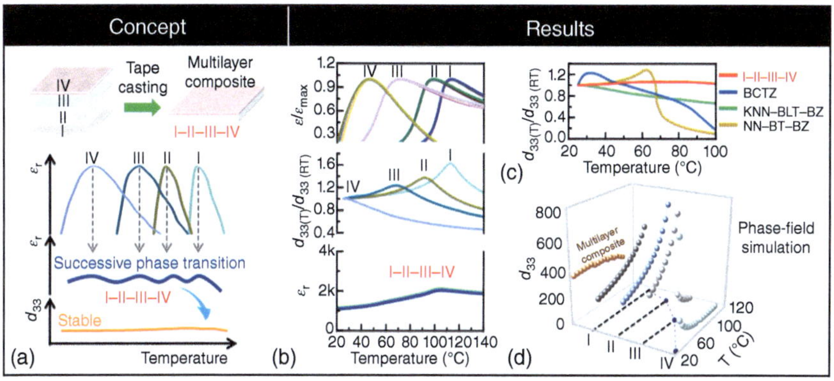

Figure 4.9 (a) Schematic illustration of fabricating a multilayer composite by using KNN-based ceramics with different T_{O-T} values, (b) dielectric and piezoelectric properties of the multilayer composite and its ingredients, (c) comparison of temperature-dependent d_{33} among representative lead-free piezoceramics, (d) phase-field simulation for temperature-dependent d_{33} of the multilayer composite and its ingredient. Source: Reproduced with permission from Zheng et al. [61]; © 2022, John Wiley & Sons.

Considering the intrinsic PPB behavior of KNN-based ceramics and the enhanced dielectric and piezoelectric response near inter-ferroelectric phase transition temperature, Wu et al. proposed to construct a ceramic composite that comprised four compositions with different but successive T_{O-T}s, and expected to obtain the stable dielectric and piezoelectric properties over a relatively wide temperature range, as shown in Figure 4.9a [61, 62]. Then, they prepared the compositionally graded KNN-based multilayer composite by tape casting and obtained the relatively stable dielectric permittivity over a temperature range of 20–140 °C (Figure 4.9b). Of particular importance is that the ceramic composite showed a stable d_{33} value of 330 pC/N within 20–100 °C, much superior to other lead-free piezoceramics (Figure 4.9c). The phase-field simulation indicated that the enhanced temperature stability was attributed to the synergistic contribution of the continuous phase transition, the strain gradient, and the complementary effect of each constituent layer (Figure 4.9d). Note that Li et al. also reported the enhanced temperature stability in the layered KNN-based ceramic composite very soon after Wu et al. [63]. Therefore, the composite ceramics using different compositions is highly promising for resolving the temperature stability of KNN-based other lead-free piezoceramics.

4.1.1.4 Other Emerging Phenomena

Except for the rapid development in the traditional directions (e.g. the dielectric, piezoelectric, ferroelectric properties, and the physical mechanisms), recent years have also witnessed the development of KNN-based materials in other fields, such as piezo-catalysis and biomedical treatment [64, 65]. As we know, piezoelectric materials can generate the net surface potential difference under the mechanical force. Thus, researchers proposed to use the net surface potential difference to conduct the electrolysis of water, which can generate the hydroxyl radicals ($\cdot OH^-$) and

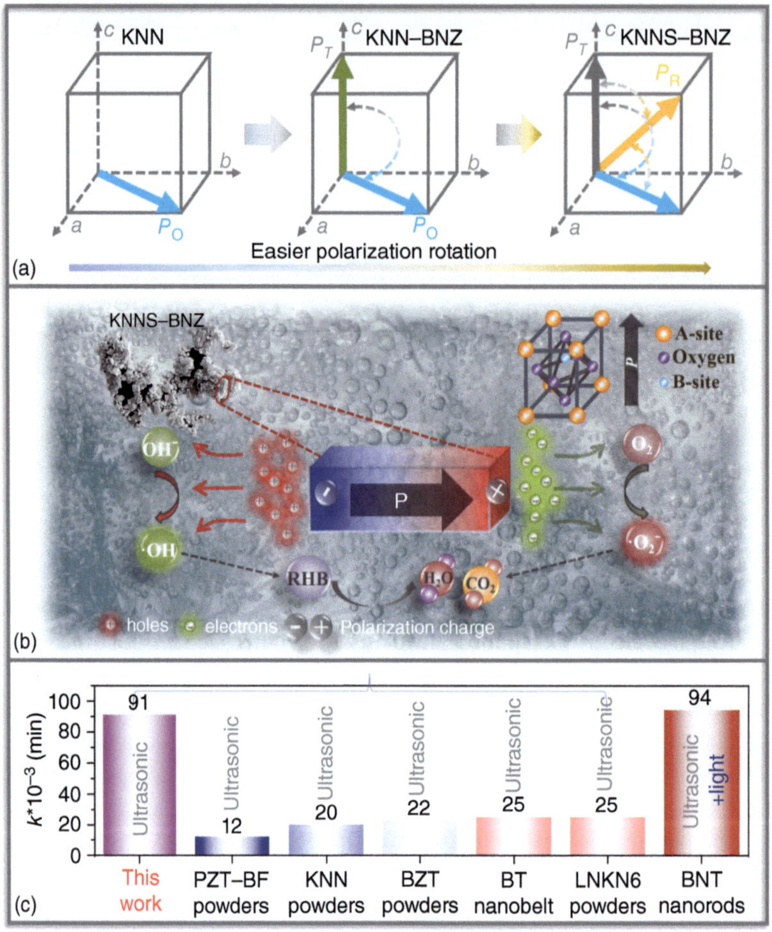

Figure 4.10 (a) Schematic diagram of polarization rotation among Ps vectors of different crystallographic symmetries for KNN, KNN–BNZ, and KNNS–BNZ materials, (b) in the section of the catalytic process, a schematic diagram illustrating the generation of electrons and holes and their transferring for piezo-catalysis, (c) comparison of *k* values of piezo-catalytic activities using different perovskite-structured piezoelectric materials. Source: Reproduced with permission from Sun et al. [64]; © 2022, Elsevier.

superoxide radicals ($\cdot O_2^-$) that exactly can degrade the organic pollutants, including the rhodamine B (RhB), the methylene blue (MB), and the methyl orange (MO) (Figure 4.10b) [64]. Based on the consideration of the phase boundary engineering, Wu et al. proposed to apply KNN-based materials to the piezo-catalysis and modulate the performance by regulating the polarization rotation (Figure 4.10a) [64]. Thus, they prepared three typical KNN-based materials, that is, KNN, KNN–BNZ, and KNNS–BNZ, which show the incremental polarization rotation (Figure 4.10a). As a result, a high reaction rate constant (or degradation rate *k*) of 91×10^{-3}/minutes was obtained for KNNS–BNKZ samples, superior to those of KNN, KNN–BNZ, and other lead-free piezoelectric materials (Figure 4.10c). Thus, KNN-based piezoelectric materials are promising for sewage treatment. In addition, KNN-based piezoceramics were also reported to be anti-microbial due to their piezoelectricity and benign biocompatibility, promising for biomedical treatment [59]. KNN-based piezoceramics were also used for the ultrasound transducers, which is specially discussed in Chapter 15. Therefore, except for being used in traditional electronic devices (e.g. sensors and actuators), KNN-based materials are also promising for other fields.

4.1.2 KNN-based Thin Films

Compared to the rapid development of KNN-based piezoceramics, KNN-based thin films develop slowly due to the difficulty in fabricating high-quality thin films. The difficulty hinges on the volatilization of alkali metal (K/Na) and the complicated stress and thickness effects, which stop the researchers from obtaining the expected composition and phase structure. Currently, the sol–gel, RF magnetron sputtering, and pulsed laser deposition are used to fabricate KNN-based thin films. Considering the volatilization of K/Na, the sol–gel method is the most used one. Yao et al. combined the phase boundary engineering and the texturing to prepare [001]-textured $0.95(K_{0.48}Na_{0.52})(Nb_{0.95}Sb_{0.05})O_3$–$0.05Bi_{0.5}(Na_{0.82}K_{0.18})_{0.5}ZrO_3$ ([001]-KNNS–BNKZ) thin films that possessed an R–T phase coexistence [66, 67]. They obtained a large effective piezoelectric strain coefficient (d_{33}^{eff}) up to 250 pm/V in [001]-KNNS–BNKZ, superior to those of other KNN-based thin films and other piezoelectric thin films (Figure 4.11). Considering the integration and miniaturization of micro-electromechanical systems, developing high-performance KNN-based thin films should be given more attention in the future.

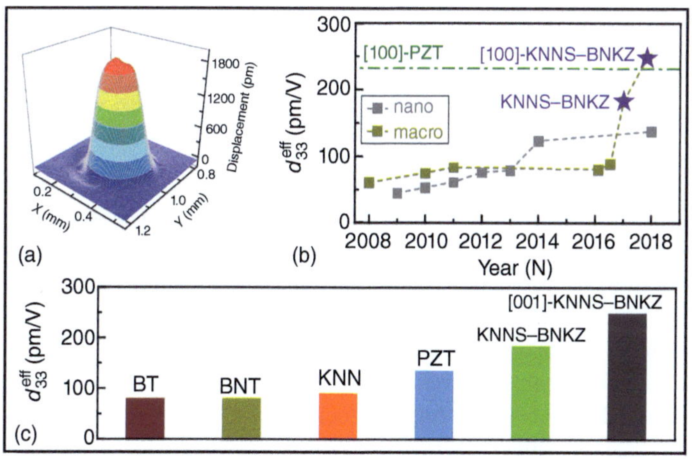

Figure 4.11 (a) Three-dimensional (3D) drawing of the piezoelectric dilatation for <100>-oriented KNNS–BNKZ thin films, tested under 5 V at 1.5 kHz, (b) historical evolutions of (*d*eff 33) values of KNN-based thin films, (c) *d*eff 33 values of representative piezoelectric thin films. Source: (a,b) Reproduced with permission from Ref. [66]; © 2019, Wiley; (c) Reproduced with permission from Ref. [67]; © 2017, Wiley.

Table 4.3 Piezoelectric properties of KNN-based single crystals with different compositions.

Compositions	Orientation	d_{33} (pC/N)	T_c (°C)	References
$(K_{0.5}Na_{0.5})NbO_3$	001	160	423	[68]
$0.5\%MnO_2$–$(K_{0.5}Na_{0.5})NbO_3$	001	270	416	[68]
$0.95(K_{0.5}Na_{0.5})NbO_3$–$0.05LiNbO_3$	001	405	426	[69]
$K_{0.95}Li_{0.05}Ta_{0.61}Nb_{0.39}O_3$	—	431	15	[70]
$K_{0.40}Na_{0.60}Ta_{0.41}Nb_{0.59}O_3$	100	560	201	[71]
$(K_{0.287}Na_{0.691}Li_{0.022})(Nb_{0.870}Ta_{0.063}Sb_{0.067})O_3$	001	732	279	[72]
$(99.6K_{0.5}Na_{0.5}NbO_3$–$0.4LiBiO_3)$-$0.375\%MnO_2$	002	1050	~400	[73]

4.1.3 KNN-based Single Crystals

Compared to piezoceramics, their single crystal counterparts can further improve piezoelectric properties due to the high degree of orientation, even though the high-cost of single crystals. Currently, several methods are used to fabricate KNN-based single crystal, such as the seed-free solid-state crystal growth (SFSSCG) method, the Bridgman method, top-seeded solution growth method (TSSG), and the flux method. However, unlike the easy phase boundary engineering in KNN-based piezoceramics, it is impossible to freely choose chemical dopants when growing KNN-based single crystals due to the difficulty in obtaining a good chemical homogeneous. Thus, KNN-based single crystals can embrace a few chemical dopants, such as Li^+, Sb^{5+}, and Ta^{5+}. Table 4.3 lists the piezoelectric properties of different KNN-based single crystals. Note that an ultra-high d_{33} value of 1050 pC/N was obtained in $(99.6K_{0.5}Na_{0.5}NbO_3$–$0.4LiBiO_3)$–0.375% MnO_2 single crystals prepared by the SFSSCG method [73]. Considering the low-cost of the SFSSCG method, it is thus expected this method may be practical for industrial applications.

In addition, some researchers also found that KNN-based single crystals showed an ultra-high large-signal piezoelectric coefficient ($d_{33}^* = S_{max}/E_{max}$, where S_{max} and E_{max} are the maximum strain and electric field). Hu et al. fabricated $(K_xNa_{1-x})NbO_3$ single crystals with an engineered compositional gradient, in which they reported an ultra-high d_{33}^* of 9000 pm/V, superior to the highest value reported in state-of-the-art lead-based single crystals (~2500 pm/V) [74]. More important, the temperature dependence of the ultra-high d_{33}^* is also superior to that of representative relaxor-$PbTiO_3$ single crystals. Therefore, KNN-based single crystals could be used for some special fields that require high piezoelectricity.

4.2 Barium Titanate

Barium titanate ($BaTiO_3$, BT) is a typical ferroelectric and piezoelectric material with an ABO_3 perovskite structure, in which Ba^{2+} occupies the A site and Ti^{4+} occupies

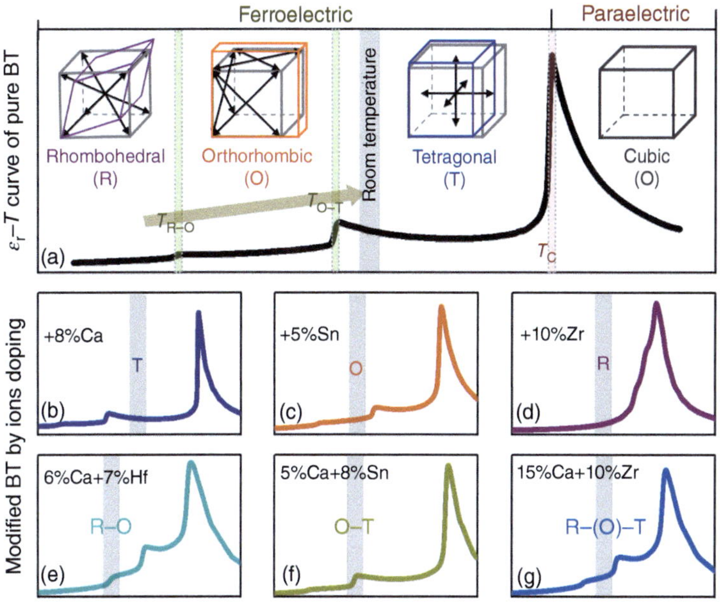

Figure 4.12 (a) Temperature-dependent phase structure and phase transition of pure BT, (b–g) the possible phase structure can be obtained in BT-based ceramics at room temperature through chemical modification.

the B site. When the temperature drops from high temperature or rises from low temperature, the phase structure of BT will undergo three transitions; thus, BT has four substructures (four pure phase structures). When the phase transition occurs, the dielectric properties of BT will change significantly, as shown in Figure 4.12. Over ~120 °C, BT presents the cubic (C) paraelectric structure without ferroelectricity. When the temperature decreases to below ~120 °C, the phase structure of BT will transform from paraelectric to ferroelectric phases. Therefore, the T_C of BT is near 120 °C. When the temperature decreases from T_C to ~0 °C, the tetragonal (T) phase presents in BT. From 0 to near −90 °C, the orthorhombic (O) phase presents in BT, and the rhombohedral (R) phase presents in BT below −90 °C [48, 75–77]. The phase transition temperature between the O and T phases and the R and O phases are usually called $T_{O\text{-}T}$ and $T_{R\text{-}O}$, respectively. The R–O, O–T, and T–C phase transitions are temperature-dependent in BT material, known as the polymorphic phase transition (PPT). For polycrystalline BT ceramic, the phase transition is gradual due to the random distribution of grain orientation and internal stress between different grains. Therefore, BT ceramic is actually in a state of two-phase coexistence near the phase transition temperature.

Figure 4.12a shows that the different phase structure can be obtained in BT at room temperature through shifting $T_{R\text{-}O}$ and $T_{O\text{-}T}$. According to the different values of $T_{R\text{-}O}$ and $T_{O\text{-}T}$, the room-temperature phase structure can be pure T, O, or R phase when $T_{R\text{-}O}$ and $T_{O\text{-}T}$ are away from room temperature. The O–T or R–O phase boundary (phase coexistence) will present at room temperature when $T_{O\text{-}T}$

Table 4.4 Influence of typical doping ions and additives on the phase transition temperature of BT.

Doping ions or additives	Occupying site	T_{R-O}	T_{O-T}	T_C
Li^+, Na^+, K^+	A site	→	→	→
Ca^{2+}		↓	↓	→
Sr^{2+}, Mg^{2+}		—	—	↓
Bi^{3+}, Al^{3+}, La^{3+}, Ce^{3+}, etc.		—	—	—
Zr^{4+}, Sn^{4+}, Hf^{4+}	B site	↑	↑	↓
Ce^{4+}		↑	↑	→
Fe^{3+}, Ni^{3+}, Nb^{5+}, Ta^{5+} etc.		↑	↑	↓
$CaZrO_3$	Additives (A and B sites)	→	→	→
$CaSnO_3$		↑	↑	↓
$CaHfO_3$		→	→	→
Bi-based compounds, e.g., $Bi(Mg_{2/3}Nb_{1/3})O_3$		↑	↑	↓

↓, decreasing; ↑, increasing; →, little change.

or T_{R-O} is shifted to room temperature. If T_{R-O} and T_{O-T} simultaneously move to room temperature and are close to each other, the possible R–O–T even R–T phase boundary may be obtained, as displayed in Figure 4.12b–g. Besides, there is a special case that T_{R-O}, T_{O-T}, and T_C are all close to a narrow temperature scope and even totally get together, forming the quasi-quadruple point with convergent R, O, T, and C phases [78, 79].

The most used method to shift T_{R-O} and/or T_{O-T} is the chemical modification, including A- and/or B-site cations doping/substitution, adding other perovskite-type compounds, etc. [48, 79, 80]. The normally employed doping cations and perovskite-type additives are displayed in Table 4.4. The ion doping or substitution strategy generally uses other isovalent or aliovalent cations with similar radii to replace partially A- and/or B-site cations of BT (Ba^{2+} and Ti^{4+}, respectively). Until now, it is found that chemical modification always leads to the simultaneous movement of T_{R-O} and T_{O-T}, and T_{R-O} always moves faster than T_{O-T}. Except for the pure T phase, other possible phase structures can only be achieved when T_{R-O} and T_{O-T} are shifted to high temperatures based on the phase transition temperature of pure BT ceramic. In this process, the T_C of BT always moves to a low temperature after chemical modification [78].

The general isovalent cations include A-site Ca^{2+}, and B-site Zr^{4+}, Hf^{4+}, and Sn^{4+}. Ca^{2+} doping at A site shows little effect on T_C but can shift T_{R-O} and T_{O-T} to low temperature by the similar moving speed [81, 82]. Zr^{4+}, Hf^{4+}, and/or Sn^{4+} doping at the B site have quick shifting effects to high temperature on T_{R-O} and T_{O-T}, and the moving speed of T_{R-O} is always faster than T_{O-T} [83, 84]. However, the T_C of BT will decrease when these ions are doped into the B site. Until now, most high-piezoelectric BT-based ceramics are obtained via the abovementioned isovalent ions doping. Other isovalent ions, such as Sr^{2+}, Mg^{2+}, and Ce^{4+}, can also shift $T_{R\text{-}O}$ and $T_{O\text{-}T}$ while being difficult to achieve high piezoelectricity in BT because they bring about diffusive phase transition in BT or deteriorate the sinterability of BT ceramic. For aliovalent ion doping, the $T_{R\text{-}O}$, $T_{O\text{-}T}$, and/or T_C can also be regulated. However, with the addition of aliovalent ions into the BT lattice, the cation defects and large lattice distortion are inevitable due to the charge difference and quite different ion radii. Therefore, aliovalent ions doping usually negatively affects the ferroelectric polarization and long-range ferroelectric orders of BT, resulting in the diffusion of ferroelectric–ferroelectric or/and ferroelectric–paraelectric phase transitions, even relaxor phase transition [80]. Similar regulating effects can be obtained after adding the perovskite-type compounds consisting of above discussed isovalent or aliovalent cations into BT ceramic.

The piezoelectric properties of BT ceramic can be improved by forming the phase boundary at room temperature via A-site Ca^{2+} and/or B-site $Zr^{4+}/Hf^{4+}/Sn^{4+}$ doping. After changing the doping ions and content, the different phase structures will show, as presented in Figure 4.12b–g and Table 4.5. For example, the T phase can be obtained in pure BT ceramic or Ca^{2+} doping BT-based ceramics [76, 85]; the R phase can be obtained in 10% Zr^{4+}/Hf^{4+} doping BT-based ceramics [86, 87]; the O phase can be obtained in 4–5% Sn^{4+} doping BT-based ceramics [88, 98]; R–O phase boundary can be obtained in 6% Ca^{2+} and 7% Hf^{4+} (or 7.5% Zr^{4+}) co-doping BT-based ceramics [89]; O–T phase boundary can be obtained in 5% Ca^{2+} and 8% Sn^{4+} co-doping BT-based ceramics [90]; R–O–T phase boundary can be obtained in 15% Ca^{2+} and 10% Zr^{4+}/Hf^{4+} co-doping BT-based ceramics [86, 96]; R–O–T–C four phases convergence can be obtained in 11% Sn^{4+} doping BT-based ceramics [97]. As presented in Figure 4.13 and Table 4.5, for pure BT or BT-based ceramics with T, O, or R phases, the piezoelectric coefficient d_{33} is less than ~240 pC/N. For BT-based ceramics with R–O phase boundary, the d_{33} can be increased to ~350–360 pC/N. For BT-based ceramics with O–T phase boundary, the d_{33} can be further improved to ~500–570 pC/N. For BT-based ceramics with R–T or R–O–T phase boundary, the d_{33} can be improved to ~530–700 pC/N. For BT-based ceramics with convergent R, O, T, and C phases, the d_{33} can be increased to >1100 pC/N [47, 76, 85–98]. It is worth noting that the piezoelectric coefficient of BT-based ceramics with phase boundary is much larger than those with single phase, and the piezoelectric coefficient increases in the order of the phase boundaries: R–O < O–T < R–O–T. The improved dielectric constant ε_r and relatively high remnant polarization P_r in phase boundary are beneficial for the improvement of piezoelectricity according to the equation $d_{33} \sim P_r \cdot \varepsilon_r$ [45, 99]. In addition, based on the reported results until now, the $T_{R\text{-}O}$ and $T_{O\text{-}T}$ of BT-based ceramics will merge into T_C peak before $T_{R\text{-}O}$ catches up with $T_{O\text{-}T}$ since the T_C of BT is relatively low and always shifts to low temperature when $T_{R\text{-}O}$ and $T_{O\text{-}T}$ are elevated [78]. Therefore, it is difficult to obtain a real R–T phase boundary with the absence of the O phase in BT-based ceramics. The reported R–T phase boundary in BT-based ceramics has been confirmed to be the R–O–T phase boundary in subsequent studies [100–105].

Some key breakthroughs exist in high-piezoelectric BT-based ceramics through phase boundary formation via chemical modification. In 2009, a breakthrough

Table 4.5 Comparison of room-temperature phase structure, electrical properties, and Curie temperature for BT-based piezoceramics via chemical modification.

System	Phase structure	d_{33} (pC/N)	ε_r	P_r (μC/cm^2)	T_C (°C)	References
$BaTiO_3$	T	190	1700	12.87	120	[76]
$(Ba_{0.9}Ca_{0.1})TiO_3$	T	170	900	8.24	130	[85]
$Ba(Ti_{0.9}Hf_{0.1})O_3$	R	185	1450	12.23	82	[86]
$Ba(Ti_{0.9}Zr_{0.1})O_3$	R	200	1500	10.94	83	[87]
$Ba(Ti_{0.96}Sn_{0.04})O_3$	O	240	1900	12.83	95	[88]
$(Ba_{0.94}Ca_{0.06})(Ti_{0.93}Hf_{0.07})O_3$	R–O	355	2050	13.52	98	[89]
$(Ba_{0.94}Ca_{0.06})(Ti_{0.925}Zr_{0.075})O_3$	R–O	360	1800	14.54	98	[89]
$BaTiO_3$–$0.1CaTiO_3$–$0.075(BaZr_{0.5}Hf_{0.5})O_3$	O–T	500	2800	12.97	102	[85]
$(Ba_{0.95}Ca_{0.05})(Ti_{0.92}Sn_{0.08})O_3$	O–T	568	4800	10.65	62	[90]
$Ba(Ti_{0.84}Hf_{0.16})O_3$–$0.48(Ba_{0.7}Ca_{0.3})TiO_3$	R–T	410	2800	7.44	102	[91]
$Ba(Ti_{0.88}Sn_{0.12})O_3$–$0.3(Ba_{0.7}Ca_{0.3})O_3$	R–T	530	3800	10.75	75	[92]
$Ba(Ti_{0.8}Hf_{0.2})O_3$–$0.5(Ba_{0.7}Ca_{0.3})TiO_3$	R–T	550	3500	11.20	80	[93]
$Ba(Ti_{0.8}Zr_{0.2})O_3$–$0.5(Ba_{0.7}Ca_{0.3})TiO_3$	R–T	620	3000	14.80	90	[94]
$(Ba_{0.85}Ca_{0.15})(Ti_{0.9}Hf_{0.1})O_3$	R–O–T	540	4200	11.0	83	[86]
$BaTiO_3$–$0.05CaZrO_3$–$0.04BaSnO_3$	R–O–T	560	3500	13.94	77	[88]
$BaTiO_3$–$0.16(0.4CaTiO_3$–$0.6BaSnO_3)$	R–O–T	570	5800	9.48	62	[95]
$(Ba_{0.85}Ca_{0.15})(Ti_{0.9}Zr_{0.1})O_3$	R–O–T	572	3000	11.40	82	[96]
$(Ba_{0.94}Ca_{0.06})(Ti_{0.9}Sn_{0.1})O_3$	R–O–T	600	6700	10.05	50	[89]
$Ba(Ti_{0.89}Sn_{0.11})O_3$–$0.18(Ba_{0.7}Ca_{0.3})TiO_3$	R–O–T	700	5600	11.24	60	[47]
$Ba(Ti_{0.89}Sn_{0.11})O_3$	R–O–T–C	1120	12000	6.0	40	[97]

in the piezoelectric improvement was achieved in Ca^{2+} and Zr^{4+} co-doped $(1-x)Ba(Zr_{0.2}Ti_{0.8})O_3$–$x(Ba_{0.7}Ca_{0.3})TiO_3$ (BTZ–xBCT, BCTZx) ceramics via forming the declared R–T phase boundary [94]. As shown in Figure 4.14a, the R and T two-phase coexistence (R–T phase boundary) with vanishing O phase is observed in BT-based ceramics by regulating the relative content of $Ba(Ti_{0.8}Zr_{0.2})O_3$ and $(Ba_{0.7}Ca_{0.3})TiO_3$ components in the incipient report, and the phase boundary starts from the triple point where R, T, and C phases coexist. However, many subsequent studies reveal the presence of an intermediate phase in the R–T phase boundary of the BTZ–xBCT system and this intermediate phase is proved to be the

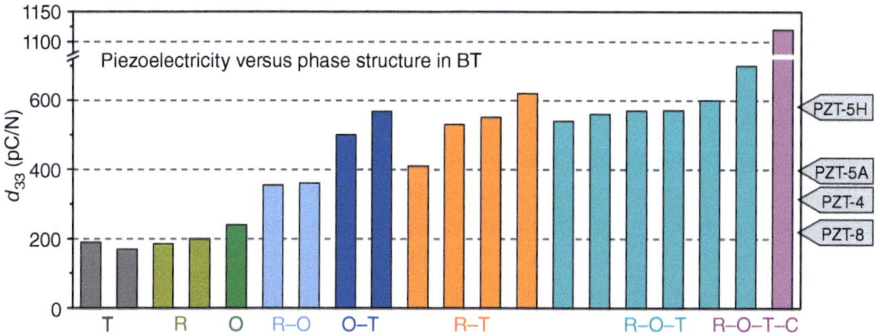

Figure 4.13 Comparison of d_{33} for BT-based ceramics with different phase structures.

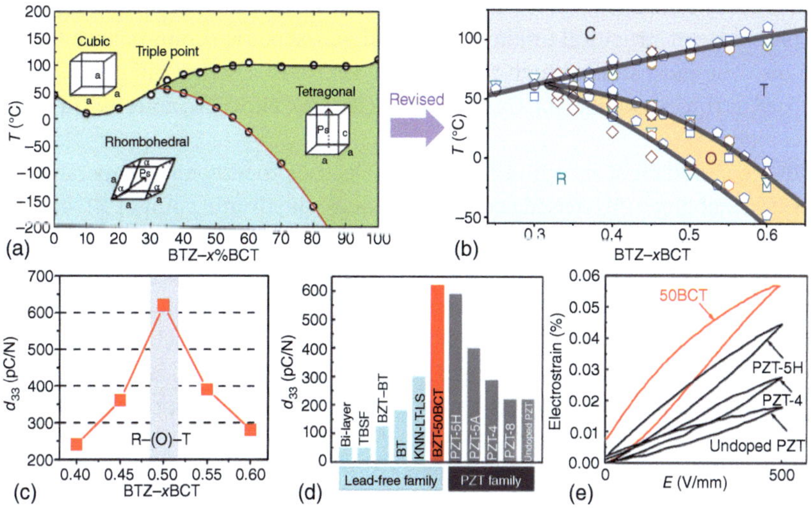

Figure 4.14 Phase boundary formation and piezoelectric properties of BTZ–xBCT ceramics: (a) Incipient phase diagram, (b) revised phase diagram, (c) composition dependence of d_{33}, comparison of (d) d_{33}, and (e) strain properties between BTZ–50BCT ceramic and some typical lead-based ceramics. Source: (a,c–e) Reproduced with permission from Ref. [94]; © 2009, American Physical Society.

O phase of BT by a series of methods, such as high-resolution XRD, Raman spectra, elastic compliance, and dielectric properties measurements, as well as theoretical calculations [100–105]. That is, the $T_{R\text{-}O}$ and $T_{O\text{-}T}$ of BT have not merged into one phase transition process, and the direct transition from R to T phase cannot be achieved in BT-based ceramics based on numerous studies. Meanwhile, the intermediate O phase is believed to play an important role in improving piezoelectric properties in this type of phase boundary in BT-based ceramics [106]. Therefore, the three-phase critical point declared in the BTZ–xBCT system is a four-phase convergent region from a quasi-quadruple point covering R, O, T, and C phases, as shown in Figure 4.14b [100–106]. With changing composition to $x = 0.5$, the R–(O)–T phase boundary is obtained in BTZ–xBCT ceramics near room temperature. Then the d_{33} of BTZ–xBCT ceramics rapidly increases to 620 pC/N, which is far higher than that of other lead-free ceramic systems at that time, and also higher than that of many typical lead-based ceramics (e.g. PZT-8, PZT-4, PZT-5A, and PZT-5H). At the same time, BTZ–0.5BCT ceramic also has high strain properties and a large converse piezoelectric coefficient d_{33}^* of ~1100 pm/V (5 kV/cm). This high strain value and converse piezoelectric coefficient are also higher than PZT-4, PZT-5H, and other lead-based piezoelectric ceramics, as shown in Figure 4.14c–e.

In the same way, the similar BT-based system, such as $Ba(Sn_{0.12}Ti_{0.88})O_3$–$x(Ba_{0.7}Ca_{0.3})TiO_3$ and $Ba(Hf_{0.2}Ti_{0.8})O_3$–$x(Ba_{0.7}Ca_{0.3})TiO_3$ with declared R–T phase boundary have been designed, and the high d_{33} of 530 and 550 pC/N are obtained, respectively [92, 93]. Besides, the obtained R–T phase boundary is also confirmed to be the R–(O)–T phase boundary with the presence of an intermediate O phase. After the abovementioned pioneering works, serious studies on chemical modification, phase boundary formation, piezoelectric improvement, and corresponding microscopic physical and structural origin of BT-based piezoelectric ceramics have been carried out. The most studied high-piezoelectric BT-based systems can be roughly generalized as three systems: (Ba, Ca)(Ti, Zr)O_3, (Ba, Ca)(Ti, Sn)O_3, and (Ba, Ca)(Ti, Hf)O_3 ceramics [79].

In 2018, to simultaneously optimize the piezoelectric property and temperature/composition stability in BT-based ceramics, the quasi-quadruple point of BT-based ceramics was employed to construct a broad phase-boundary zone with close and continuous R–O and O–T phase transitions, then to achieve the high piezoelectric properties and good temperature stability in a broad composition range. A NPB design strategy was employed in a designed $(1-x)Ba(Ti_{1-y}Sn_y)O_3$-$x(Ba_{1-z}Ca_z)TiO_3$ (when $y = 0.11$, $z = 0.30$, abbreviated as $BTS_{0.11}$–xBCT) system [47]. In this system, the left terminal $Ba(Ti_{1-y}Sn_y)O_3$ has a quasi-quadruple point comprising R, O, T, and C phases at $y = 0.11$ and ~40 °C, and the right terminal $(Ba_{1-z}Ca_z)TiO_3$ is added to facilitate the stability of ferroelectric phases of BT. By regulating these two compositions, the continuous phase transitions of adjacent R–O and O–T are achieved in a wide composition and temperature range. As shown in Figure 4.15a, the $T_{R\text{-}O}$ and $T_{O\text{-}T}$ of $BTS_{0.11}$ ceramic gradually separate from the quasi-quadruple point with increasing the adding content of $(Ba_{0.7}Ca_{0.3})TiO_3$. In this process, the R–O, O–T, and T–C phase transitions are slowly separated and extend to a broad temperature and composition range, then the convergent and successive R–O and O–T phase transition leads to multiphase coexistence (R–O–T phase boundary) in a wide composition and temperature range. In this broad region, excellent piezoelectric properties with large d_{33} of 700 ± 30 pC/N are achieved in $BTS_{0.11}$-xBCT at $x = 0.18$. More importantly, the high d_{33} of >600 pC/N can be attained over a wide composition scope of $0.05 < x \leq 0.22$ and a broad temperature range (10–40 °C, $x = 0.18$), as presented in Figure 4.15b. This highest piezoelectric property exceeds the performance of all the other reported lead-free piezoceramic systems and the majority of lead-based piezoceramic systems at that time. Besides, the high piezoelectric coefficient is also superior to previous BT-based ceramics and some typical soft lead-based ceramics in the temperature range of 10–40 °C, as displayed in Figure 4.15c,d. This work affords a new paradigm to optimize piezoelectricity and stability of BT-based and similar-type piezoceramics even with low Curie temperature.

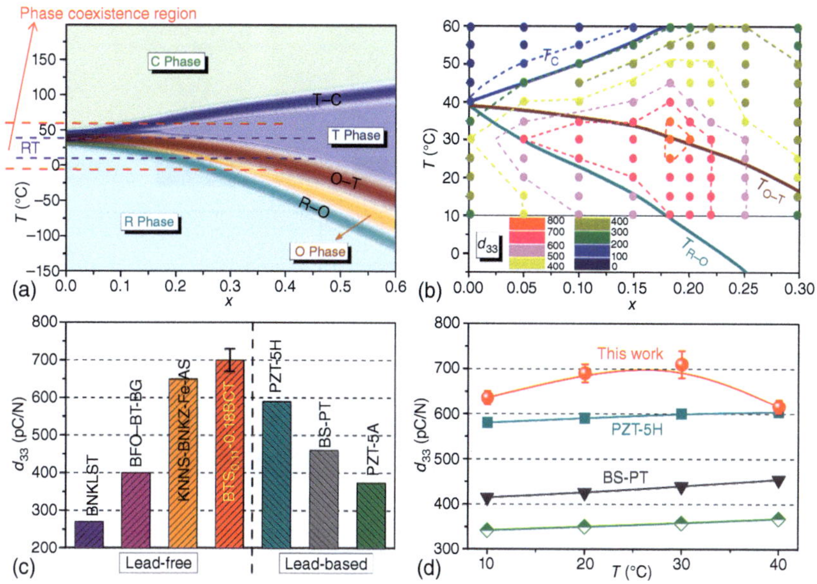

Figure 4.15 Phase boundary formation and piezoelectric properties of $BTS_{0.11}$−0.18BCT ceramics: (a) schematic showing phase coexistence region based on the phase diagram, (b) distribution of d_{33} as a function of temperature and composition, (c) comparison of the reported largest d_{33} among lead-free and typical lead-based ceramics, (d) comparison of temperature-dependent d_{33} for $BTS_{0.11}$−0.18BCT and lead-based ceramics. Source: (a–d) Reproduced with permission from Zhao et al. [47]; © 2018, American Chemical Society.

Until now, the physical and structure mechanisms associated with piezoelectricity improvement in phase boundary zone were extensively revealed in BT-based ceramics. For BTZ–xBCT-type systems, it has been proposed that the ferroelectric domain size at the phase boundary of BTZ–0.5BCT can be greatly miniaturized due to the coexistence of ferroelectric phases. As shown in Figure 4.16a–d, the ferroelectric domains in R- and T-phase compositions have big domain widths (~200 nm) and large domain lengths (up to >1 μm) for BTZ–xBCT ceramics [107, 108]. However, when the composition is close to the phase boundary, both the domain width and length can decrease to less than 100 nm. Besides, the domain architecture comprising hierarchical domain sizes from microscale to nanoscale is observed in the phase-boundary compositions, and the R-type domains comprise the T-type domains in the nanoscale. This domain configuration increases the density of domain walls, which can reduce the anisotropy energy of domain orientation and makes the ferroelectric domains of different phases easily switch. For piezoelectric ceramics, the size and activity of ferroelectric domains play an important role in piezoelectric properties. Therefore, the domain switching under the external electric field can be much easier in the BTZ–xBCT ceramics with phase boundary compared to single phase, resulting in high piezoelectric properties in phase-boundary compositions. The observed domain structure in BTZ–xBCT ceramics is quite similar to that of high-piezoelectric lead-based materials at the MPB.

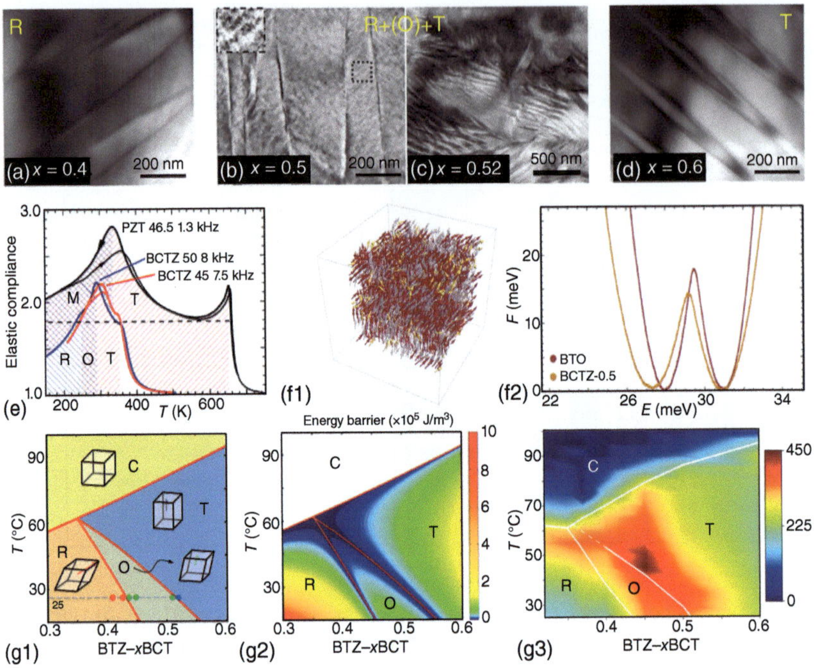

Figure 4.16 Mechanism interpretation of high piezoelectricity in BTZ–xBCT ceramics. (a)–(d) TEM images of domain structure in BTZ–xBCT ceramics with $x = 0.4$ (R phase), $x = 0.5$–0.52 (R–(O)–T phase boundary) and $x = 0.6$ (T phase), (e) comparison of elastic compliance between BCTZ (BTZ–xBCT) and PZT near phase boundary, (f1) three-dimensional R- and O-phase clusters in BCTZ0.5 (BTZ–0.5BCT) and (f2) comparison free energy curves between BT and BCTZ0.5, (g1) calculated phase diagram and (g2) corresponding distribution of energy barriers, (g3) d_{33} distribution in the composition-temperature phase diagram. Source: (a,b,d) Reproduced with permission from Gao et al. [107]; © 2011, American Institute of Physics; (c) Reproduced with permission from Zhakozheva et al. [108]; © 2015 American Physical Society; (e) Reproduced with permission from Cordero et al. [101]; © 2014 American Institute of Physics; (f1,f2) Reproduced with permission from Nahas et al. [106]; © 2017 Springer Nature; (g1,g2) [109] / Springer Nature / CC BY 4.0; (g3) Reproduced with permission from Acosta et al. [105]; © 2015 American Physical Society.

As a result, the miniaturized and hierarchical domain structure is identified as the microscopic structure origin of the high piezoelectric response in BTZ–xBCT-type ceramics.

BTZ–0.5BCT (BCTZ50 or BCTZ0.5) ceramic with high piezoelectricity has been proven to show R, O, and T three-phase coexistence; the high piezoelectric performance should be because in the multiphase coexistence region. The physical origin is assigned as the low polarization anisotropy, which is conducive to polarization rotation along the external electric field. Figure 4.16e shows the BCTZ45 composition with O–T phase boundary and BCTZ50 composition with R–O–T phase boundary [101]. Compared with the PZT with intermediate M phase, the temperature range of high elastic compliance in BCTZ45 and BCTZ50 is narrower than that of PZT with phase boundary, so the piezoelectric response of BCTZ50 will be higher than that of PZT. At the same time, it can be found that the elastic compliance peak of BCTZ45/50 appears in the transition temperature zone between the T phase and intermediate O phase, in which the ceramic has minimum anisotropic energy, indicating an extremely unstable phase transition state. Under this state, the polarization vector is easy to rotate. As shown in Figure 4.16f1,f2, the polarization vector of BCTZ0.5 gradually rotates, resulting in the unstable state of the polarization vector between the R, O, and T phases. This state is further promoted by intermediate O throughout the entire ceramic, finally resulting in a significant reduction of the polarization energy barrier compared to single-phase BT ceramic [106]. This conclusion has also been verified by Landau free energy simulation. Corresponding to the phase diagram shown in Figure 4.16g1,g2, the energy barrier at the R–O and O–T phase-transition zone is significantly lower than that in the single-phase region, and the energy barrier around the O–T phase transition is lower than at the R–O phase transition [109]. In the temperature-composition plane, the d_{33} of the BTZ–xBCT system reaches the maximum value near the R–O and O–T phase transition zones, and the largest piezoelectric response occurs in the region where the R, O, and T three-phase converge, which is accordance with the results of theoretical calculation of piezoelectricity [105], as displayed in Figure 4.16g3. As a result, previous studies attribute the physical mechanism of improving piezoelectric performance to the low anisotropy energy and small energy barrier in phase boundary for high-piezoelectric BT-based ceramics, which is conducive to the polarization rotation, finally resulting in the improvement of piezoelectric performance.

The atomic structure and physical origins of high piezoelectric properties with multiphase coexistence in BT-based ceramics are further revealed in the studies of $BTS_{0.11}$–0.18BCT ceramic. It is observed that the convergent multiphase coexistence will further miniaturize ferroelectric domains down to a smaller size and lead to a more complex hierarchical domain architecture, as shown in Figure 4.17a. Quite small domain size leads to high domain density and extremely low domain-wall energy, which can facilitate polarization orientation by easy domain switching and extension, and finally induce high piezoelectricity as an extrinsic contribution. The STEM was used for the atomic-scale polarization rotation state between different phases on the nanoscale. As shown in Figure 4.17b1–b3, it is observed that the R, O, and T three phases of high-performance $BTS_{0.11}$–0.18BCT ceramic coexist not only inside nanodomains but also at the atomic scale. A two-dimensional visual Ti–O displacement vectors map reveals that the distribution of spontaneous polarization vectors corresponding to R, O, and T phases is not homogeneous, and the rotation of polarization vectors among them is found to be a gradual process, which is similar to a static polling state. A similar result can be obtained from phase-field simulation. Figure 4.17c gives phase field simulation concerning the high-piezoelectric $BTS_{0.11}$–0.18BCT ceramic with convergent R–O–T phase boundary. It is observed that the spontaneous polarizations of R, O, and T phases coexist inside nanoregions and permeate into each other with gradual and ubiquitous polarization rotations, indicating the weak energy barrier and easy polarization rotation between different phases. As shown in Figure 4.17d, the DFT calculation shows that the energy

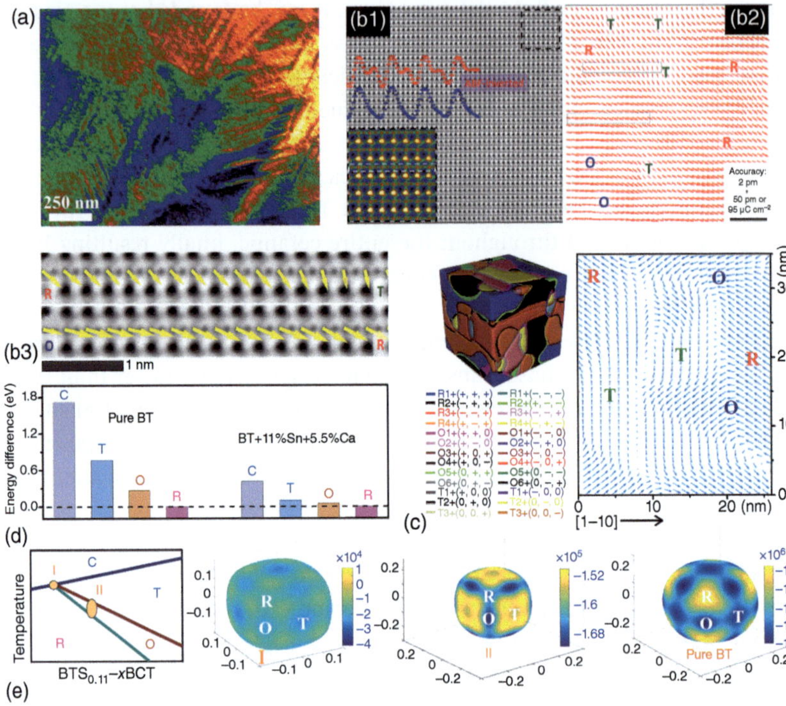

Figure 4.17 Mechanism interpretation of high piezoelectricity in $BTS_{0.11}-xBCT$ ceramics: (a) TEM images showing hierarchical domain structure, (b1) STEM ABF lattice image with an enlarged contrast-inverted image, (b2) the δ_{Ti-O} displacement vector maps based on (b1), (b3) enlarged image showing gradual polarization rotation from R to T phase and from O to R phase, (c) phase-field simulation showing R, O, and T three phases coexistence at $x = 0-0.2$, and polarization projection on the (110) plane, (d) DFT calculated total energy difference of C, T, O phases based on the most stable R phase for pure BT and (Ca^{2+}, Sn^{4+}) co-doped BT, (e) schematic phase diagram of constructing the convergent R–O–T phase boundary starting from a quasi-quadruple point, and Landau free energy modeling of BTS0.11−xBCT with quadruple critical point and convergent R–O–T phase boundary, and pure BT with T phase. Source: (a–e) Reproduced with permission from Zhao et al. [47]; © 2018, American Chemical Society.

difference among R, O, and T three phases are very small after the introduction of Ca^{2+} and Sn^{4+} to form the convergent R–O–T phase boundary, and it is possible to allow R, O, and T three phases to coexist in the nanodomains and transform them between any two phases in the nanoscale. Compared to pure BT, Landau free energy simulation also indicates that the R–O–T phase boundary extending from the quasi-quadruple point has extremely low polarization anisotropic energy due to the near isotropy free energy profile in the quasi-quadruple point, as shown in Figure 4.17e. As a result, the spontaneous polarization rotation at the nanoscale and polarization easy rotation promoted by an extremely low energy barrier, as the atomic structure origin and physical origin, contribute to high piezoelectric properties in high-performance BT-based piezoelectric ceramics through phase boundary formation via chemical modification [47].

4.3 Summary

The past 20 years have witnessed the rapid development of lead-free piezoelectric materials, and subsequently, we will summarize the development and the existing problems of KNN- and BT-based ceramics. For KNN-based piezoelectric materials, especially in phase boundary engineering, high piezoelectric properties, temperature stability, and physical mechanisms have made great progress. Even with these remarkable achievements, some challenges still exist. The first is the lack of deep understanding of how the chemical dopants (especially some key elements) alter the structure and performance. Currently, we only empirically choose the chemical dopants and change their content to modulate the phase structure and performance without mastering the inherent relationship between the chemical dopants and the structure. The second one is the development of overall performance for practical applications. Most studies focused on improving one property (e.g. piezoelectricity or temperature stability) of KNN-based materials. However, the practical applications need good overall performance, including the mechanical property, fatigue, aging, electromechanical quality factor (Q_m), loss factor (tan δ), and so on, which are not fully considered at present.

Therefore, future research of KNN-based materials can focus on the following investigations: (1) deeply decoding the relationship of "chemical dopants-structure-performance-physical mechanisms" and mastering how the chemical dopants alter the structure and performance, (2) normalizing the preparation of high-performance KNN-based materials and improving the reproducibility, (3) focusing on improving the overall performance of KNN-based materials, (4) tentatively preparing the incipient devices using KNN-based materials, (5) applying KNN-based materials to other fields, such as energy storage, piezoelectric nanogenerators, ferroelectric photovoltaic, piezo-catalytic, and biomedicine.

BT-based ceramics have been widely studied so far because of their simple structure and high piezoelectric/ferroelectric/dielectric properties. In recent years, great breakthroughs have been made in piezoelectric optimization via phase boundary formation and the exploration of physical mechanisms. However, some key issues still need to be solved before their actual application. Except for normal phase boundary design, elaborate control of phase content in phase boundary and regulation of domain structure and grain size may afford a new chance to further improve the piezoelectricity of BT-based ceramics. The temperature stability can be optimized by phase boundary design; however, the T_C of high-piezoelectric BT-based ceramics is still low (<100 °C), which greatly limits the temperature range of piezoelectric applications. It is still a great challenge to explore new modification methods or structural design strategies to improve the piezoelectric properties and maintain Curie temperature at a relatively high value. For example, introducing other perovskite additives with high T_C or defect regulation may bring new opportunities for the performance optimization of BT. Although the raw material of BT ceramic is cheap, the sintering temperature is too high, especially for high-performance BT-based ceramics whose sintering temperatures are often higher than 1450 °C, which will bring about a high production cost. At present, the general method is to introduce sintering aids to reduce the sintering temperature while it is still high (~1300–1400 °C) and may lead to the degeneration of piezoelectricity. Therefore, exploring new low-temperature sintering technology is necessary to prepare high-piezoelectric BT-based ceramics. The low T_C is not conducive to piezoelectric applications for BT-based piezoceramics. However, the low T_C, high ferroelectric polarization, and small coercive field make them very suitable in electrocaloric materials for electrocaloric refrigeration applications. In addition, BT-based ceramics usually have a high electrostrictive coefficient and low-hysteresis strain, and a good electrostrictive strain can be maintained over T_C. Therefore, they are also suitable for device applications based on electrostrictive strain. However, current studies on the abovementioned properties are not systematic and deep enough. For the applications of electrocaloric refrigeration or electrostrictive devices, multilayer ceramic structure may be a promising field for extending the application of BT-based piezoceramics.

References

1 Li, J.-F., Wang, K., Zhu, F.-Y. et al. (2013). (K, Na)NbO_3-based lead-free piezoceramics: fundamental aspects, processing technologies, and remaining challenges. *J. Am. Ceram. Soc.* 96 (12): 3677–3696.

2 Shirane, G., Newnham, R., and Pepinsky, R. (1954). Dielectric properties and phase transitions of $NaNbO_3$ and (Na,K)NbO_3. *Phys. Rev.* 96 (3): 581.

3 Egerton, L. and Dillon, D.M. (1959). Piezoelectric and dielectric properties of ceramics in the system potassium-sodium niobate. *J. Am. Ceram. Soc.* 42 (9): 438–442.

4 Saito, Y., Takao, H., Tani, T. et al. (2004). Lead-free piezoceramics. *Nature* 432 (7013): 84–87.

5 Rödel, J., Webber, K.G., Dittmer, R. et al. (2015). Transferring lead-free piezoelectric ceramics into application. *J. Eur. Ceram. Soc.* 35 (6): 1659–1681.

6 Wu, J., Xiao, D., and Zhu, J. (2015). Potassium-sodium niobate lead-free piezoelectric materials: past, present, and future of phase boundaries. *Chem. Rev.* 115 (7): 2559–2595.

7 Popovič, A., Bencze, L., Koruza, J., and Malič, B. (2015). Vapour pressure and mixing thermodynamic properties of the $KNbO_3$-$NaNbO_3$ system. *RSC Adv.* 5 (93): 76249–76256.

8 Chen, K., Zhou, J., Zhang, F. et al. (2015). Screening sintering aids for $(K_{0.5}Na_{0.5})NbO_3$ ceramics. *J. Am. Ceram. Soc.* 98 (6): 1698–1701.

9 Thong, H.-C., Zhao, C., Zhu, Z.-X. et al. (2019). The impact of chemical heterogeneity in lead-free (K, Na)NbO_3 piezoelectric perovskite: ferroelectric phase coexistence. *Acta Mater.* 166: 551–559.

10 Thong, H.-C., Payne, A., Li, J.-W. et al. (2021). The origin of chemical inhomogeneity in lead-free potassium sodium niobate ceramic: competitive chemical reaction during solid-state synthesis. *Acta Mater.* 211: 116833.

11 Jaffe, B., Cook, W.R., and Jaffe, H. (1971). *Piezoelectric Ceramics* (ed. B. Jaffe, W.R. Cook, and H. Jaffe). Academic Press.

12 Wu, J., Wang, X., Cheng, X. et al. (2014). New potassium-sodium niobate lead-free piezoceramic: giant-d_{33} vs. sintering temperature. *J. Appl. Phys.* 115 (11): 114104.

13 Zheng, T., Wu, J., Xiao, D., and Zhu, J. (2015). Giant d_{33} in nonstoichiometric (K,Na)NbO_3-based lead-free ceramics. *Scr. Mater.* 94: 25–27.

14 Zhen, Y. and Li, J.-F. (2006). Normal sintering of (K,Na)NbO_3-based ceramics: influence of sintering temperature on densification, microstructure, and electrical properties. *J. Am. Ceram. Soc.* 89 (12): 3669–3675.

15 Zheng, T. and Wu, J. (2015). Enhanced piezoelectricity over a wide sintering temperature (400–1050 °C) range in potassium sodium niobate-based ceramics by two step sintering. *J. Mater. Chem. A* 3 (13): 6772–6780.

16 Yao, W., Zhang, J., Wang, X. et al. (2019). High piezoelectric performance and domain configurations of $(K_{0.45}Na_{0.55})_{0.98}Li_{0.02}Nb_{0.76}Ta_{0.18}Sb_{0.06}O_3$ lead-free ceramics prepared by two-step sintering. *J. Eur. Ceram. Soc.* 39 (2): 287–294.

17 Du, H., Tang, F., Luo, F. et al. (2007). Effect of poling condition on piezoelectric properties of $(K_{0.5}Na_{0.5})NbO_3$-$LiNbO_3$ lead-free piezoelectric ceramics. *Mater. Sci. Eng. B* 137 (1): 175–179.

18 Tao, H. and Wu, J. (2017). New poling method for piezoelectric ceramics. *J. Mater. Chem. C* 5 (7): 1601–1606.

19 Sun, Y., Karaki, T., and Yamashita, Y. (2022). Recent progress on AC poling of relaxor-$PbTiO_3$ ferroelectric single crystals: a review. *Jpn. J. Appl. Phys.* 61 (SB): SB0802.

20 Ghosh, D., Sakata, A., Carter, J. et al. (2014). Domain wall displacement is the origin of superior permittivity and piezoelectricity in $BaTiO_3$ at intermediate grain sizes. *Adv. Funct. Mater.* 24 (7): 885–896.

21 Yang, W., Li, P., Wu, S. et al. (2019). A study on the relationship between grain size and electrical properties in (K,Na)NbO_3-based lead-free piezoelectric ceramics. *Adv. Electron. Mater.* 5 (12): 1900570.

22 Jaffe, B., Roth, R.S., and Marzullo, S. (1954). Piezoelectric properties of lead zirconate-lead titanate solid-solution ceramics. *J. Appl. Phys.* 25 (6): 809–810.

23 Lv, X., Zhu, J., Xiao, D. et al. (2020). Emerging new phase boundary in potassium sodium-niobate based ceramics. *Chem. Soc. Rev.* 49 (3): 671–707.

24 Lv, X., Zhang, X.-X., and Wu, J. (2020). Nandomains in lead-free piezoceramics: a review. *J. Mater. Chem. A* 8 (20): 10026–10073.

25 Lin, D., Li, Z., Xu, Z., and Yao, X. (2009). Characterization of KNN single crystals by slow-cooling technique. *Ferroelectrics* 381 (1): 1–8.

26 Lv, X., Zheng, T., Zhao, C. et al. (2022). Multiscale structure engineering for high-performance Pb-free piezoceramics. *Acc. Mater. Res.* 3 (4): 461–471.

27 Zuo, R., Fu, J., and Lv, D. (2009). Phase transformation and tunable piezoelectric properties of lead-free $(Na_{0.52}K_{0.48\text{-}x}Li_x)(Nb_{1\text{-}x\text{-}y}Sb_yTa_x)O_3$ system. *J. Am. Ceram. Soc.* 92 (1): 283–285.

28 Zuo, R., Ye, C., and Fang, X. (2007). Dielectric and piezoelectric properties of lead free $Na_{0.5}K_{0.5}NbO_3$-$BiScO_3$ ceramics. *Jpn. J. Appl. Phys.* 46 (10R): 6733.

29 Wang, R., Bando, H., Katsumata, T. et al. (2009). Tuning the orthorhombic–rhombohedral phase transition temperature in sodium potassium niobate by incorporating barium zirconate. *Phys. Status Solidi-Rapid Res. Lett.* 3 (5): 142–144.

30 Zuo, R. and Fu, J. (2011). Rhombohedral-tetragonal phase coexistence and piezoelectric properties of (NaK)(NbSb)O_3-$LiTaO_3$-$BaZrO_3$ lead-free ceramics. *J. Am. Ceram. Soc.* 94 (5): 1467–1470.

31 Liang, W., Wu, W., Xiao, D. et al. (2011). Construction of new morphotropic phase boundary in $0.94(K_{0.4-x}Na_{0.6}Ba_xNb_{1-x}Zr_x)O_3$-$0.06LiSbO_3$ lead-free piezoelectric ceramics. *J. Mater. Sci.* 46 (21): 6871–6876.

32 Wang, X., Wu, J., Xiao, D. et al. (2014). Giant piezoelectricity in potassium–sodium niobate lead-free ceramics. *J. Am. Chem. Soc.* 136 (7): 2905–2910.

33 Wu, B., Wu, H., Wu, J. et al. (2016). Giant piezoelectricity and high curie temperature in nanostructured alkali niobate lead-free piezoceramics through phase coexistence. *J. Am. Chem. Soc.* 138 (47): 15459–15464.

34 Xu, K., Li, J., Lv, X. et al. (2016). Superior piezoelectric properties in potassium–sodium niobate lead-free ceramics. *Adv. Mater.* 28 (38): 8519–8523.

35 Zheng, T., Wu, H., Yuan, Y. et al. (2017). The structural origin of enhanced piezoelectric performance and stability in lead free ceramics. *Energy Environ. Sci.* 10 (2): 528–537.

36 Tao, H., Wu, H., Liu, Y. et al. (2019). Ultrahigh performance in lead-free piezoceramics utilizing a relaxor slush polar state with multiphase coexistence. *J. Am. Chem. Soc.* 141 (35): 13987–13994.

37 Rödel, J. (2020). Crystallographic design for energy storage. *Nat. Mater.* 19 (9): 932–934.

38 Li, P., Zhai, J., Shen, B. et al. (2018). Ultrahigh piezoelectric properties in textured (K,Na)NbO_3-based lead-free ceramics. *Adv. Mater.* 30 (8): 1705171.

39 Liu, Y., Chang, Y., Sun, E. et al. (2018). Significantly enhanced energy-harvesting performance and superior fatigue-resistant behavior in $[001]_c$-textured $BaTiO_3$-based lead-free piezoceramics. *ACS Appl. Mater. Interfaces* 10 (37): 31488–31497.

40 Zhou, X., Xue, G., Luo, H. et al. (2021). Phase structure and properties of sodium bismuth titanate lead-free piezoelectric ceramics. *Prog. Mater. Sci.* 122: 100836.

41 Yan, Y., Geng, L.D., Liu, H. et al. (2022). Near-ideal electromechanical coupling in textured piezoelectric ceramics. *Nat. Commun.* 13 (1): 3565.

42 Yan, Y., Zhou, J.E., Maurya, D. et al. (2016). Giant piezoelectric voltage coefficient in grain-oriented modified $PbTiO_3$ material. *Nat. Commun.* 7 (1): 13089.

43 Hao, J., Ye, C., Shen, B., and Zhai, J. (2012). Enhanced piezoelectric properties of <001> textured lead-free $(K_xNa_{1-x})_{0.946}Li_{0.054}NbO_3$ ceramics with large strain. *Phys. Status Solidi A* 209 (7): 1343–1349.

44 Yang, W., Wang, Y., Li, P. et al. (2020). Improving electromechanical properties in KNANS-BNZ ceramics by the synergy between phase structure modification and grain orientation. *J. Mater. Chem. C* 8 (18): 6149–6158.

45 Li, F., Lin, D., Chen, Z. et al. (2018). Ultrahigh piezoelectricity in ferroelectric ceramics by design. *Nat. Mater.* 17 (4): 349–354.

46 Damjanovic, D. (2005). Contributions to the piezoelectric effect in ferroelectric single crystals and ceramics. *J. Am. Ceram. Soc.* 88 (10): 2663–2676.

47 Zhao, C., Wu, H., Li, F. et al. (2018). Practical high piezoelectricity in barium titanate ceramics utilizing multiphase convergence with broad structural flexibility. *J. Am. Chem. Soc.* 140 (45): 15252–15260.

48 Wu, J. (2018). *Advances in Lead-Free Piezoelectric Materials*. Springer.

49 Gruverman, A., Alexe, M., and Meier, D. (2019). Piezoresponse force microscopy and nanoferroic phenomena. *Nat. Commun.* 10 (1): 1661.

50 Ding, Y., Zheng, T., Zhao, C., and Wu, J. (2019). Structure and domain wall dynamics in lead-free KNN-based ceramics. *J. Appl. Phys.* 126 (12): 124101.

51 Fu, J., Zuo, R., and Xu, Z. (2011). High piezoelectric activity in (Na,K)NbO_3 based lead-free piezoelectric ceramics: contribution of nanodomains. *Appl. Phys. Lett.* 99 (6): 062901.

52 Liu, Q., Li, J.-F., Zhao, L. et al. (2018). Niobate-based lead-free piezoceramics: a diffused phase transition boundary leading to temperature-insensitive high piezoelectric voltage coefficients. *J. Mater. Chem. C* 6 (5): 1116–1125.

53 López-Juárez, R., Novelo-Peralta, O., González-García, F. et al. (2011). Ferroelectric domain structure of lead-free potassium-sodium niobate ceramics. *J. Eur. Ceram. Soc.* 31 (9): 1861–1864.

54 Schmitt, L.A., Schönau, K.A., Theissmann, R. et al. (2007). Composition dependence of the domain configuration and size in $Pb(Zr_{1-x}Ti_x)O_3$ ceramics. *J. Appl. Phys.* 101 (7): 074107.

55 Lv, X., Ma, Y., Zhang, J. et al. (2022). Coexisting multi-phase and relaxation behavior in high-performance lead-free piezoceramics. *Acta Mater.* 238: 118221.

56 Li, F., Zhang, S., Yang, T. et al. (2016). The origin of ultrahigh piezoelectricity in relaxor-ferroelectric solid solution crystals. *Nat. Commun.* 7 (1): 13807.

57 Gao, X., Cheng, Z., Chen, Z. et al. (2021). The mechanism for the enhanced piezoelectricity in multi-elements doped (K,Na)NbO_3 ceramics. *Nat. Commun.* 12 (1): 881.

58 Zhang, M.-H., Shen, C., Zhao, C. et al. (2022). Deciphering the phase transition-induced ultrahigh piezoresponse in (K,Na)NbO3-based piezoceramics. *Nat. Commun.* 13 (1): 3434.

59 Zhang, M.-H., Zhang, Q., Yu, T.-T. et al. (2021). Enhanced electric-field-induced strains in (K,Na)NbO3 piezoelectrics from heterogeneous structures. *Mater. Today* 46: 44–53.

60 Yao, F.-Z., Wang, K., Jo, W. et al. (2016). Diffused phase transition boosts thermal stability of high-performance lead-free piezoelectrics. *Adv. Funct. Mater.* 26 (8): 1217–1224.

61 Zheng, T., Yu, Y., Lei, H. et al. (2022). Compositionally graded KNN-based multilayer composite with excellent piezoelectric temperature stability. *Adv. Mater.* 34 (8): 2109175.

62 Zhang, Y., Yu, Y., Zhang, N. et al. (2023). Simultaneous realization of good piezoelectric and strain temperature stability via the synergic contribution from multilayer design and rare earth doping. *Adv. Funct. Mater.* 33 (11): 2211439.

63 Song, A., Liu, Y., Feng, T. et al. (2022). Simultaneous enhancement of piezoelectricity and temperature stability in KNN-based lead-free ceramics via layered distribution of dopants. *Adv. Funct. Mater.* 32 (34): 2204385.

64 Sun, X.-X., Li, R., Yang, Z. et al. (2022). Modulating polarization rotation to stimulate the high piezocatalytic activity of (K, Na)NbO_3 lead-free piezoelectric materials. *Appl. Catal. B Environ.* 313: 121471.

65 Wang, S., Chen, C., Wang, J. et al. (2022). Synergetic chemo-piezodynamic therapy of osteosarcoma enabled by defect-driven lead-free piezoelectrics. *Adv. Funct. Mater.* 32 (44): 2208128.

66 Wang, Y., Wu, H., Qin, X. et al. (2019). Outstanding piezoelectric performance in lead-free 0.95(K,Na)(Sb,Nb)O_3-0.05(Bi,Na,K)ZrO_3 thick films with oriented nanophase coexistence. *Adv. Electron. Mater.* 5 (4): 1800691.

67 Wang, Y., Yao, K., Qin, X. et al. (2017). High piezoelectric performance and phase transition in stressed lead-free (1−x)(K, Na)(Sb, Nb)O_3-x(Bi, Na, K)ZrO_3 thin films. *Adv. Electron. Mater.* 3 (7): 1700033.

68 Lin, D., Li, Z., Zhang, S. et al. (2010). Influence of MnO2 doping on the dielectric and piezoelectric properties and the domain structure in $(K_{0.5}Na_{0.5})NbO_3$ single crystals. *J. Am. Ceram. Soc.* 93 (4): 941–944.

69 Chen, K., Xu, G., Yang, D. et al. (2007). Dielectric and piezoelectric properties of lead-free 0.95$(K_{0.5}Na_{0.5})NbO_3$-0.05$LiNbO_3$ crystals grown by the Bridgman method. *J. Appl. Phys.* 101 (4): 044103.

70 Zhou, Z., Li, J., Tian, H. et al. (2009). Piezoelectric properties of the lead-free $K_{0.95}Li_{0.05}Ta_{0.61}Nb_{0.39}O_3$ single crystal. *J. Phys. D. Appl. Phys.* 42 (12): 125405.

71 Meng, X., Hu, C., Park, D.-S. et al. (2020). Ultra-high piezoresponse in tantalum doped potassium sodium niobate single crystal. *Appl. Phys. Lett.* 116 (11): 112902.

72 Liu, H., Veber, P., Rödel, J. et al. (2018). High-performance piezoelectric (K,Na,Li)(Nb,Ta,Sb)O_3 single crystals by oxygen annealing. *Acta Mater.* 148: 499–507.

73 Jiang, M., Zhang, J., Rao, G. et al. (2019). Ultrahigh piezoelectric coefficient of a lead-free K0.5Na0.5NbO3-based single crystal fabricated by a simple seed-free solid-state growth method. *J. Mater. Chem. C* 7 (47): 14845–14854.

74 Hu, C., Meng, X., Zhang, M.-H. et al. *Sci. Adv.* 6 (13): eaay5979.

75 von Hippel, A. (1950). Ferroelectricity, domain structure, and phase transitions of barium titanate. *Rev. Mod. Phys.* 22 (3): 221–237.

76 Bechmann, R. (1956). Elastic, piezoelectric, and dielectric constants of polarized barium titanate ceramics and some applications of the piezoelectric equations. *J. Acoust. Soc. Am.* 28 (3): 347–350.

77 Ravel, B., Stern, E.A., Vedrinskii, R.I., and Kraizman, V. (1998). Local structure and the phase transitions of $BaTiO_3$. *Ferroelectrics* 206 (1): 407–430.

78 Zhao, C., Huang, Y., and Wu, J. (2020). Multifunctional barium titanate ceramics via chemical modification tuning phase structure. *InfoMat.* 2 (6): 1163–1190.

79 Acosta, M., Novak, N., Rojas, V. et al. (2017). $BaTiO_3$-based piezoelectrics: fundamentals, current status, and perspectives. *Appl. Phys. Rev.* 4 (4): 041305.

80 Feng, Y., Wu, J., Chi, Q. et al. (2020). Defects and aliovalent doping engineering in electroceramics. *Chem. Rev.* 120 (3): 1710–1787.

81 Shu, C., Reed, D., and Button, T.W. (2018). A phase diagram of $Ba_{1-x}Ca_xTiO_3$ ($x = 0$–0.30) piezoceramics by Raman spectroscopy. *J. Am. Ceram. Soc.* 101 (6): 2589–2593.

82 Cao, G.P., Huang, H.B., Liang, D.S., and Ma, X.Q. (2017). A phenomenological potential and ferroelectric properties of $BaTiO_3$-$CaTiO_3$ solid solution. *Ceram. Int.* 43 (9): 6671–6676.

83 Kalyani, A.K., Brajesh, K., Senyshyn, A., and Ranjan, R. (2014). Orthorhombic-tetragonal phase coexistence and enhanced piezo-response at room temperature in Zr, Sn, and Hf modified $BaTiO_3$. *Appl. Phys. Lett.* 104 (25): 252906.

84 Yao, Y., Zhou, C., Lv, D. et al. (2012). Large piezoelectricity and dielectric permittivity in $BaTiO_3$-x$BaSnO_3$ system: the role of phase coexisting. *Europhys. Lett.* 98 (2): 27008.

85 Zhao, C., Wu, B., Thong, H.-C., and Wu, J. (2018). Improved temperature stability and high piezoelectricity in lead-free barium titanate-based ceramics. *J. Eur. Ceram. Soc.* 38 (16): 5411–5419.

86 Zhao, C., Wu, W., Wang, H., and Wu, J. (2016). Site engineering and polarization characteristics in $(Ba_{1-y}Ca_y)(Ti_{1-x}Hf_x)O_3$ lead-free ceramics. *J. Appl. Phys.* 119 (2): 024108.

87 Huang, Y., Zhao, C., Lv, X. et al. (2017). Multiphase coexistence and enhanced electrical properties in (1−x-y)$BaTiO_3$-x$CaTiO_3$-y$BaZrO_3$ lead-free ceramics. *Ceram. Int.* 43 (16): 13516–13523.

88 Zhao, C., Wu, B., and Wu, J. (2019). Composition-driven broad phase boundary for optimizing properties and stability in lead-free barium titanate ceramics. *J. Am. Ceram. Soc.* 102 (6): 3477–3487.

89 Zhao, C., Wang, H., Xiong, J., and Wu, J. (2016). Composition-driven phase boundary and electrical properties in $(Ba_{0.94}Ca_{0.06})(Ti_{1-x}M_x)O_3$ (M = Sn, Hf, Zr) lead-free ceramics. *Dalton Trans.* 45 (15): 6466–6480.

90 Zhu, L.-F., Zhang, B.-P., Zhao, X.-K. et al. (2013). Enhanced piezoelectric properties of $(Ba_{1-x}Ca_x)(Ti_{0.92}Sn_{0.08})O_3$ lead-free ceramics. *J. Am. Ceram. Soc.* 96 (1): 241–245.

91 Wang, D., Jiang, Z., Yang, B. et al. (2014). Phase transition behavior and high piezoelectric properties in lead-free $BaTiO_3$-$CaTiO_3$-$BaHfO_3$ ceramics. *J. Mater. Sci.* 49 (1): 62–69.

92 Xue, D., Zhou, Y., Bao, H. et al. (2011). Large piezoelectric effect in Pb-free $Ba(Ti,Sn)O_3$-x$(Ba,Ca)TiO_3$ ceramics. *Appl. Phys. Lett.* 99 (12): 122901.

93 Zhou, C., Liu, W., Xue, D. et al. (2012). Triple-point-type morphotropic phase boundary based large piezoelectric Pb-free material-$Ba(Ti_{0.8}Hf_{0.2})O_3$-$(Ba_{0.7}Ca_{0.3})TiO_3$. *Appl. Phys. Lett.* 100 (22): 222910.

94 Liu, W. and Ren, X. (2009). Large piezoelectric effect in Pb-free ceramics. *Phys. Rev. Lett.* 103 (25): 257602.

95 Zhu, L.-F., Zhang, B.-P., Zhao, L., and Li, J.-F. (2014). High piezoelectricity of $BaTiO_3$-$CaTiO_3$-$BaSnO_3$ lead-free ceramics. *J. Mater. Chem. C* 2 (24): 4764–4771.

96 Tian, Y., Chao, X., Jin, L. et al. (2014). Polymorphic structure evolution and large piezoelectric response of lead-free (Ba,Ca)(Zr,Ti)O_3 ceramics. *Appl. Phys. Lett.* 104 (11): 112901.

97 Wang, D., Fan, Z., Rao, G. et al. (2020). Ultrahigh piezoelectricity in lead-free piezoceramics by synergistic design. *Nano Energy* 76: 104944.

98 Chen, M., Xu, Z., Chu, R. et al. (2014). Enhanced piezoelectricity in broad composition range and the temperature dependence research of $(Ba_{1-x}Ca_x)(Ti_{0.95}Sn_{0.05})O_3$ piezoceramics. *Phys. B* 433: 43–47.

99 Zhao, C., Yin, J., Huang, Y., and Wu, J. (2019). Polymorphic characteristics challenging electrical properties in lead-free piezoceramics. *Dalton Trans.* 48 (30): 11250–11258.

100 Keeble, D.S., Benabdallah, F., Thomas, P.A. et al. (2013). Revised structural phase diagram of $(Ba_{0.7}Ca_{0.3}TiO_3)$-$(BaZr_{0.2}Ti_{0.8}O_3)$. *Appl. Phys. Lett.* 102 (9): 092903.

101 Cordero, F., Craciun, F., Dinescu, M. et al. (2014). Elastic response of $(1-x)Ba(Ti_{0.8}Zr_{0.2})O_3$-$x(Ba_{0.7}Ca_{0.3})TiO_3$ (x = 0.45-0.55) and the role of the intermediate orthorhombic phase in enhancing the piezoelectric coupling. *Appl. Phys. Lett.* 105 (23): 232904.

102 Xue, D., Gao, J., Zhou, Y. et al. (2015). Phase transitions and phase diagram of $Ba(Zr_{0.2}Ti_{0.8})O_3$-$x(Ba_{0.7}Ca_{0.3})TiO_3$ Pb-free system by anelastic measurement. *J. Appl. Phys.* 117 (12): 124107.

103 Xue, D., Zhou, Y., Bao, H. et al. (2011). Elastic, piezoelectric, and dielectric properties of $Ba(Zr_{0.2}Ti_{0.8})O_3$-$50(Ba_{0.7}Ca_{0.3})TiO_3$ Pb-free ceramic at the morphotropic phase boundary. *J. Appl. Phys.* 109 (5): 054110.

104 Zhang, L., Zhang, M., Wang, L. et al. (2014). Phase transitions and the piezoelectricity around morphotropic phase boundary in $Ba(Zr_{0.2}Ti_{0.8})O_3$-$x(Ba_{0.7}Ca_{0.3})TiO_3$ lead-free solid solution. *Appl. Phys. Lett.* 105 (16): 162908.

105 Acosta, M., Khakpash, N., Someya, T. et al. (2015). Origin of the large piezoelectric activity in $(1-x)Ba(Zr_{0.2}Ti_{0.8})O_3$-$x(Ba_{0.7}Ca_{0.3})TiO_3$ ceramics. *Phys. Rev. B* 91 (10): 104108.

106 Nahas, Y., Akbarzadeh, A., Prokhorenko, S. et al. (2017). Microscopic origins of the large piezoelectricity of leadfree (Ba,Ca)(Zr,Ti)O_3. *Nat. Commun.* 8 (1): 15944.

107 Gao, J., Xue, D., Wang, Y. et al. (2011). Microstructure basis for strong piezoelectricity in Pb-free $Ba(Zr_{0.2}Ti_{0.8})O_3$-$(Ba_{0.7}Ca_{0.3})TiO_3$ ceramics. *Appl. Phys. Lett.* 99 (9): 092901.

108 Zakhozheva, M., Schmitt, L.A., Acosta, M. et al. (2015). Wide compositional range in situ electric field investigations on lead-free $Ba(Zr_{0.2}Ti_{0.8})O_3$-$x(Ba_{0.7}Ca_{0.3})TiO_3$ piezoceramic. *Phys. Rev. Appl.* 3 (6): 064018.

109 Yang, T., Ke, X., and Wang, Y. (2016). Mechanisms responsible for the large piezoelectricity at the tetragonal-orthorhombic phase boundary of $(1-x)BaZr_{0.2}Ti_{0.8}O_3$-$xBa_{0.7}Ca_{0.3}TiO_3$ system. *Sci. Rep.* 6 (1): 33392.

5

Lead-Free Perovskite Piezoelectric Materials – Part Two

Jiagang Wu

Sichuan University, College of Materials Science and Engineering, No. 24 South Section 1, Yihuan Road 610065, Chengdu, PR China

5.1 Sodium Bismuth Titanate

5.1.1 Phase Boundaries Versus Piezoelectricity

Sodium bismuth titanate $\{(Bi_{0.5}Na_{0.5})TiO_3$, BNT$\}$ is one of the most promising lead-free ferroelectrics (FEs) with perovskite structure, which was first discovered by Smolenskii et al. in 1960s [1]. As for its ground-state structure, the decades-long issue reaches a general consensus until affirmation of a monoclinic (*Cc*) phase [2, 3], which arises from Na/Bi positional disorder. By suppressing this disorder via poling, the average *Cc* phase irreversibly transforms into the *R*3*c* phase [4]. Apart from the *Cc* phase, coexisting rhombohedral (*R*3*c*, $a^-a^-a^-$) and ordered tetragonal (*P*4*bm*, $a^0a^0c^+$) phases were also reported for the unpoled NBT. The pure BNT has a large remnant polarization ($P_r = 38\,\mu C/cm^2$) and high Curie temperature ($T_c = 320\,°C$). However, due to serious issues of the large coercive field ($E_c = 73\,kV/cm$) and high conductivity, it is difficult to fully pole the pure BNT materials, leading to undesirable piezoelectricity ($d_{33} = 73–95\,pC/N$) for this system [5–8]. To reduce the coercive field of BNT materials, strategies of using the composition design have been intensively investigated, such as chemical doping, binary system, and ternary system construction.

Through solid solution with other systems, for example, $BaTiO_3$ (BT) or $Bi_{0.5}K_{0.5}TiO_3$, the transition lines further extend in the phase diagram of (1–*x*)BNT–*x*BT (BNT–BT) or (1–*x*)BNT–*x*BKT (BNT–BKT), in which a rhombohedral–tetragonal (R–T) morphotropic phase boundary (MPB) can identified at $x \approx 0.06$ for BNT–BT or $x \approx 0.20$ for BNT–BKT systems [9–11]. BNT–BKT and BNT–BT systems have been studied extensively, and the piezoelectric charge coefficient of pure BNT was increased from 73–95 to 207 pC/N and 125 pC/N for BNT–BKT and BNT–BT, respectively (Figure 5.1a) [10, 16]. The detailed relationship between the phase boundary and the piezoelectricity of BNT-based materials is systematically reviewed in our previous review work of reference [17].

Piezoelectric Materials: From Fundamentals to Emerging Applications, First Edition. Edited by Jiagang Wu.

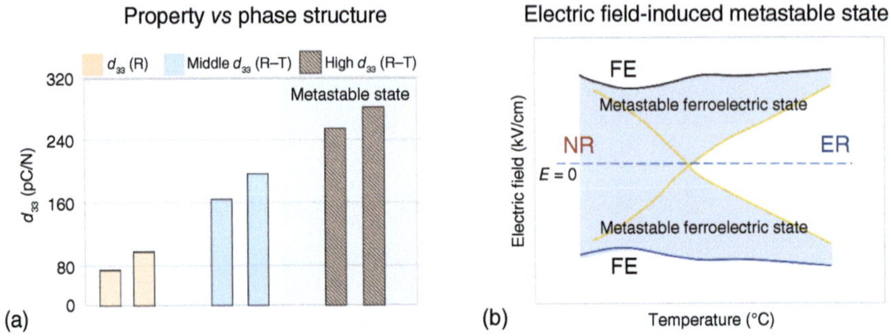

Figure 5.1 The piezoelectric property of BNT-based materials with different phase structures: (a) increased piezoelectric charge coefficient of BNT-based materials by establishing R–T phase boundary. Source: Adapted from [6, 12–15], (b) electric field-induced metastable ferroelectric state can further enhance piezoelectricity.

In addition to the phase boundary strategy, to further enhance the piezoelectric properties of BNT-based materials, it is also effective to drive BNT-based material systems to the critical state with the coexistent nonergodic relaxor (NR) and ergodic relaxor (ER) phases (Figure 5.1b). BNT-based samples exhibit the relaxor phase structure at the NR state when the *E*-field is not applied. After applying and then removing the *E*-field, normal FE behaviors can be observed for these NR–BNT-based samples. In the ER state, the samples undergo the *E*-field-induced relaxor–ferroelectric (RE–FE) phase transition, while the FE behaviors cannot be well kept after the removal of the *E*-field [18, 19]. It is worth noting that the samples can undergo the NR–ER transition at a temperature region (metastable FE region), where the balanced R/T ratios can be helpful for the enhanced dielectric and piezoelectric responses. In this metastable state, the RE–FE phase transition can be beneficial for obtaining the higher piezoelectricity in BNT-based materials. For example, by synergistically using the phase boundary and metastable state strategy, a large piezoelectric strain constant of up to 271 pC/N has been achieved in Li, Sr-doped BNT–BKT system [15]. Notably, Piezo Technology (Physik Instrumente company), a commercial supplier specialized in piezoelectric ceramics, has recently released a line of lead-free BNT-based ceramic products, named PIC700, for use in ultrasonic transducers operating in the MHz range, as well as sonar and hydrophone applications [20]. However, despite its low piezoelectricity ($d_{33} < 300$ pC/N), PIC700 is not yet considered a viable replacement for PZT-based piezoelectric ceramics.

5.1.2 Depolarization Phenomenon

Due to the complex phase structure of BNT-based materials, there are many interesting phenomena that cannot be well explained until now. Since BNT-based materials lose their piezoelectricity around the depolarization temperature (T_d), a deep understanding of the depolarization phenomenon in BNT-based materials is essential (Figure 5.2). The first phase transition of BNT occurs around 200 °C, as evidenced by the appearance of pinched *P*–*E* loops and a sharp reduction in P_r.

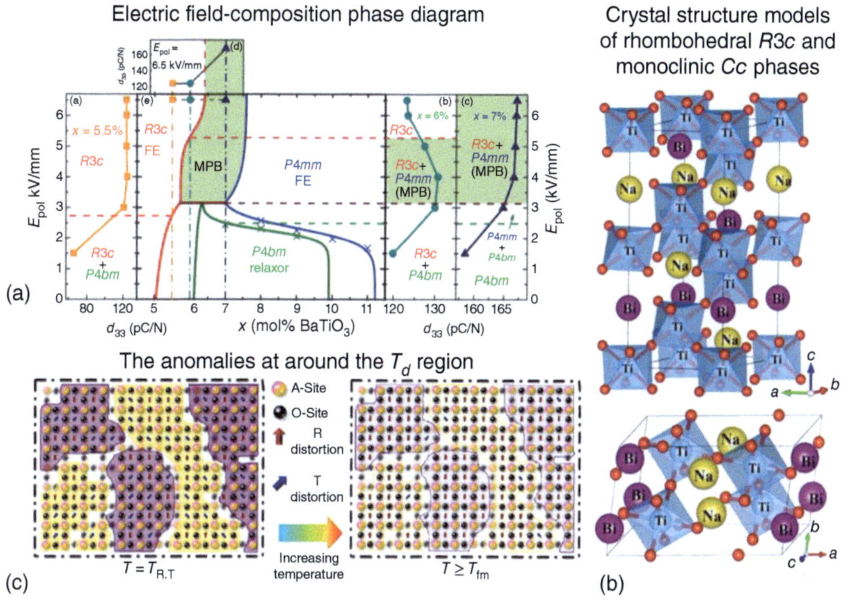

Figure 5.2 Composition-, *E* field-, and temperature-dependent structural evolution of BNT-based materials: (a) the electric field-composition phase diagram for (1−*x*)BNT−*x*BT ceramics, (b) crystal structure models of rhombohedral *R*3*c* (hexagonal setting) and monoclinic *Cc* phases, (c) conceptual domain wall (DW) model utilized to illustrate the anomalies around the T_d region for BNT−BKT ceramics. Source: (a) Reproduced with permission from Ma et al. [17]; © 2012, American Physical Society. (b) Reproduced with permission from Rao et al. [21]; © 2013, American Physical Society; (c) Reproduced with permission from Yin et al. [22]; © 2019, Wiley.

Some researchers consider that the depolarization of BNT at T_d can be ascribed to incompatible in-phase tilted octahedral regions in the field-stabilized long-range rhombohedral distortion [21]. This in-phase tilt disorder can be revealed by the superlattice reflections of $1/2\{ooe\}_{pc}$. The T_d of BNT system is regarded as the characteristic temperature of the transition from FE state to relaxor state, which can be clearly observed in the dielectric permittivity variation with temperature. However, studies have revealed that the depolarization and FE–RE transition are two separate and discrete processes [23]. The increase in temperature first leads to the loss of FE texture, that is, depolarization. Subsequently, with the further increase in temperature, detextured FE domains dissociate into nanoscale entities, resulting in the FE–RE transition. Reducing the anisotropy field and/or increasing the random field will both cause the depolarization phenomenon in the BNT-based relaxor system [24].

There are three effective methods that can be employed to regulate T_d: the ions doping/oxides addition, the introduction of second phases, and improvements in processing (Figure 5.3). Han et al. significantly increased the T_d of Pb-doped BNT ceramics to 390 °C by introducing a small amount of Pb^{2+} to the BNT–BKT matrix ceramics, which attributed to the hybridization of Pb^{2+} and O^{2-} [25]. The introduction of Zn^{2+} into BNT–BT/ZnO composites can also stabilize the FE state of

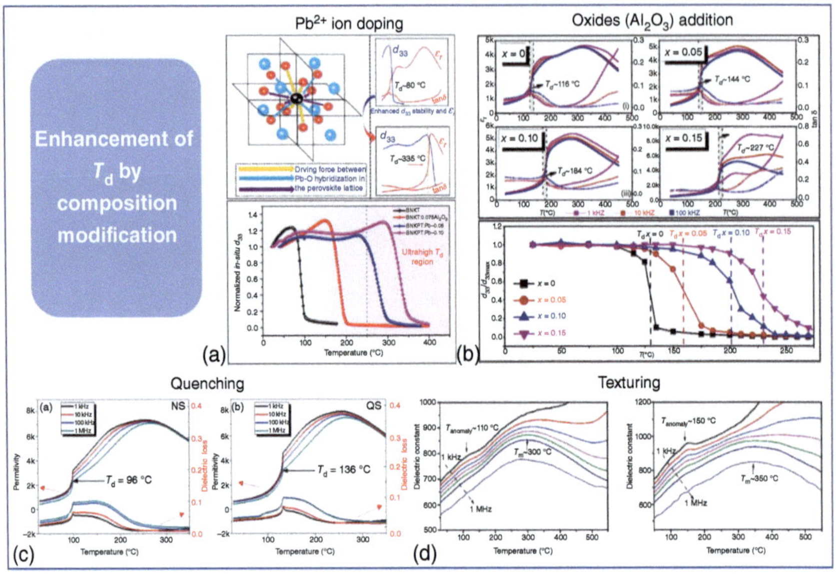

Figure 5.3 Depolarization phenomenon in BNT-based materials: (a) temperature-dependent d_{33} for poled Pb^{2+} doped BNT–BKT ceramics, (b) temperature-dependent dielectric constant (ε_r) and loss tangent (tan δ) of poled BNKT:$x$$Al_2O_3$ (x = 0–0.15) composites (1, 10, and 100 kHz), and the relatively retained d_{33} as a function of annealing temperatures, (c) temperature-dependent dielectric permittivity and dielectric loss of normal sintered and quenched BNT–BT samples, (d) temperature-dependent permittivity spectra of nontextured (left) and textured (right) BNT–BT thin film. Source: (a) Reproduced with permission from Han et al. [25]; © 2020, Elsevier. (b) Reproduced with permission from Yin et al. [26]; © 2018, Elsevier. (c) Reproduced with permission from Li et al. [27]; © 2018, Wiley. (d) Reproduced with permission from Zhou et al. [28]; © 2018, Wiley.

BNT-based materials to a higher temperature [29]. Three typical models have been used to explain the improved thermal stability of BNT-based materials: (i) Stabilization force from the partial compensation of depolarization fields, (ii) stabilization force from the thermal deviatoric stresses between BNT matrix and oxides, (iii) stabilization force from the ions to diffusion-induced stabilized FE long-range order, yielding to a more stabilized FE phase after the poling process. Apart from ZnO, the thermal depolarization process can be deferred to a high temperature (227 °C) by varying the amount of Al_2O_3 [26]. Considering the effect of the second phases, introducing the non-FE phases effectively destabilizes the polarization, which destroys the long-range FE order and thus decreases the T_d. Introducing the FE phases changes local electric and strain fields, while the introduced FE symmetry also provides the enhanced polarization field. If the introduced polarization field is stronger than the introduced local random electric and strain fields, it will help form the long-range FE order, and thus the deferred T_d. For example, for BNT–BT or BNT–BKT system, T_d reaches the lowest value around the MPB; however, when the content of BT or BKT is increased, the depolarization temperature can be higher than 200 °C [30].

In addition to the chemical design, researchers reported that the T_d of BNT-based materials can be increased or decreased by modifying heat treatment processes, such as quenching and annealing [27]. For example, the T_d of pure BNT–0.06BT is approximately 96 °C; however, the T_d of a quenched BNT sample at 1100 °C is increased to 136 °C. Moreover, the <001>-textured BNT–0.0BT thin films is successfully fabricated by the sol–gel method, the T_d of which can be increased from 110 to 150 °C [28].

5.1.3 Depolarization Temperature T_d Versus Piezoelectricity d_{33}

Around the T_d of $Bi_{0.5}Na_{0.5}TiO_3$(BNT)-based solid solutions ($T_d \sim 160$ °C for pure BNT), the macroscopic piezoelectricity ($d_{33} \sim 60$ pC/N for pure BNT) will disappear, which is an obstacle for their practical applications. Suppressing or even eliminating the T_d of high-performance BNT is a creative endeavor for researchers; however, the critical issue is that there is an inverse relationship between T_d and d_{33}, regardless of the fabrication method (using bulk ceramics, textured ceramics, and single crystals). That is, a relatively higher T_d is often accompanied by an inferior piezoelectricity, while lower T_d can obtain a higher piezoelectricity (Figure 5.4). When T_d further decreases to lower temperatures, the NR to ER state transition will lead to the disappearance of the macroscopic piezoelectricity. However, reducing the T_d to around room temperature will be beneficial for enhancing the electro-strain property of BNT-based materials, and an even higher $d_{33}{}^*$ can be achieved.

5.1.4 Large and Reversible Electro-strain

The large and reversible electro-strain ($S > 0.3\%$) of BNT-based materials is regarded as one of the most attractive phenomena in lead-free piezoelectric materials. However, it is still controversial to clearly explain the origin of this large

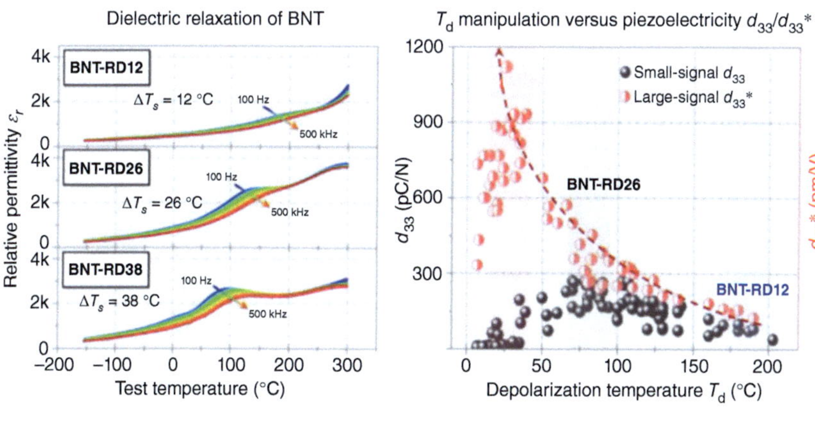

Figure 5.4 Dielectric relaxor behaviors of different relaxor degree (RD) BNT-based materials. Compilation of the electromechanical properties (d_{33} and d_{33}^*) against depolarization temperature (T_d) for the BNT-based system. Source: Reproduced with permission from Yin et al. [24]; © 2022, Springer Nature.

electro-strain. To further develop the electromechanical property of lead-free FEs, it is required to have a clear understanding of underlying mechanisms behind this large electro-strain. Therefore, in this part, we will review the most recent investigations on the underlying mechanisms and new designing strategies of this large electro-strain.

5.1.4.1 Underlying Mechanism Behind the Large Electro-strain

When BNT-based materials go through the relaxor–FE crossover region (around T_d), the electro-strain of BNT-based materials reaches the maximum, which was generally attributed to the state transition between ER and FE states [5]. Manipulating the dynamics of the ER–FE transition can be effective for regulating the electro-strain of BNT-based materials, which demand for the judicious tuning of the critical equilibrium of local anisotropy field and random field.

Currently, chemical substitution is a widely applied method to regulate the electro-strain of BNT-based materials, and it effectively modifies the random field by means of additives with different structures. For example, Liu et al. introduced the isovalent Ta^{5+} into the BNKT matrix, and the introduced random field effect shifted the relaxor/FE crossover to room temperature, which induced a large d^*_{33} of 1240 pm/V [31]. Hussain et al. utilized the ABO_3-type perovskite ($SrZrO_3$) to modify the strain performance of BNKT-based materials [32]. The introduced random elastic field effect also tailors the relaxation characteristics. Around critical composition with 3 mol% $SrZrO_3$, a large unipolar strain of 0.37% can be obtained. Herein, typical chemical modifying components and strain regulation results are summarized in Table 5.1. The more detailed strain regulation mechanism and components can be viewed in our latest research [43].

Recently, based on representative compositions of BNT-based ceramics, our group constructed a "composition–structure–performance" relationship, providing a better understanding (anisotropy field and local random field) from atomic

Table 5.1 Strain performance of BNT-based materials at room temperature.

Composition (References)	Strain	E_{max} (kV/cm)	d^*_{33} (pm/V)
BNT–0.25 mol% Mn [33]	0.24	80	300
0.93BNT–0.07BT [34]	0.24	50	475
0.92BNT–0.06BT–0.02KNN [35]	0.45	80	563
0.79BNT–0.20BKT–0.01KNN [36]	0.48	80	600
NBKT–5FN/NKBT [37]	0.36	50	720
0.92(0.935BNT–0.065$BaTiO_3$)–0.02$SrZrO_3$ [38]	0.39	55	722
0.94(0.75BNT–0.25BKT)–0.06$BiAlO_3$ + 20%BNT [39]	0.37	40	930
NBKT–7BA/NBT [40]	0.37	40	930
NBT–7BT (textured) [41]	0.7	70	1000
NBT–KBT–BT (textured) [42]	0.41	40	1030

scale about the large electro-strain in BNT-based materials [24]. The deviated polarizations with low-angle walls play a vital role in regulating the anisotropy field. By reducing the antiphase oxygen octahedron tilt, the magnitude of polarization is reduced, and more polarizations with low-angle walls can deviate from their surroundings with the decreased wall angle, leading to an effectively reduced anisotropy field and increased dielectric permittivity (Figure 5.5a). On the other hand, as for the deviated polarizations with high-angle walls, the introduction of doping heterovalent defects effectively regulates the random field, providing the restoring force from *E*-field-induced large-size FE domains to small-size multidomains (Figure 5.5b). By applying an *E*-field, the weakened anisotropy field facilitates the polarization rotation from *P*4*bm* to *R*3*c* (accompanied by the polarization extension) (Figure 5.5c). In this case, the optimized electro-strain can only be achieved when the anisotropy field and random field reach a critical equilibrium, where the competitive and synergetic effects between the anisotropy and random fields should be considered for designing the electro-strain of BNT-based materials (Figure 5.5d).

In addition to chemical modifications, it is worth noting that modifying the material synthesis conditions (such as heat treatment and sintering processing) is also the common approach to manipulate the local structural features, providing more freedom to optimize the electro-strain response in BNT-based materials. As mentioned above, the giant strain originates from the relaxor–FE transition, and in essential, the reversible transition corresponds to the percolation of polar entities and the relief of mechanical mismatch stress. Tailoring internal stress can effectively modify strain parameters (such as remanent strain and poling strain), and most important, grain size related to the sintering process/heat treatment is a critical factor for internal stress regulation. Our recent work has reported a large electro-strain level ($S > 0.7\%$) in the BNKLT system via regulating sintering conditions, including sintering time and sintering temperature [44]. Moreover, Dung et al. promoted d_{33}^* up to 800 pm/V on the BNKT matrix via optimizing sintering temperature and composition [45]. Tong et al. successfully prepared BNT–ST actuators, and through reconstructed inactive regions and sintering parameters, the active strain can arrive at 0.168% under the electric field of 2 kV/mm [46].

5.1.4.2 New Designing Strategy for Enhancing the Electro-strain

Except for conventional designing strategies (e.g. metal ion doping and perovskite additives), some new strain designing strategies were also proposed in recent years, with the help of the developing simulation technology. MPB was generally realized to effectively facilitate the polarization rotation due to its low energy barriers, inducing a large strain [47]. Moreover, disrupting the FE order to relaxor state via chemical modifications was believed to decrease the strain hysteresis performance [48]. However, owing to the complexity of the composition design for a trade-off between FE and relaxor states, it was very difficult to simultaneously achieve both large strain and low hysteresis. Recently, Li et al. utilized the phase simulation to design the BNT-based materials with excellent strain performance (i.e. giant electro-strain and low hysteresis), as exhibited in Figure 5.6a [49]. At certain doping contents, the

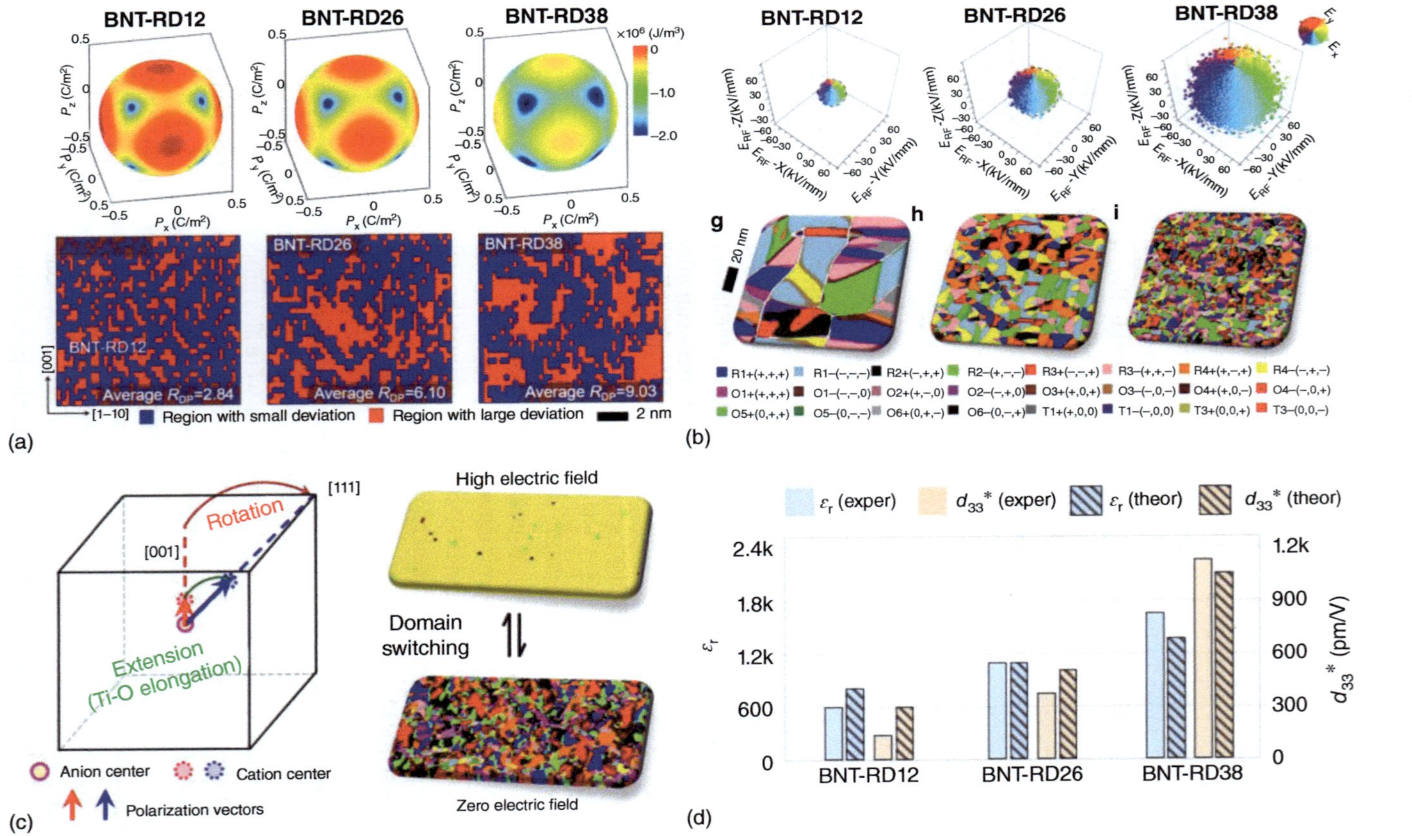

Figure 5.5 "Composition–structure–performance" relationship in BNT-based bulk ceramics: (a) the reduced anisotropy-induced increasing density of low-angle walls in BNT-based bulk materials, (b) the increased random fields induced increasing density of high angle walls in BNT-based bulk ceramics, (c) the proposed mechanism concerning the competitive and synergetic effect of polarization rotation, polarization extension, and domain switching process in BNT-based bulk ceramics, (d) the realized enhanced electro-strain in BNT-based bulk ceramics, which can be evidenced from both experimental and theoretical perspectives. Source: Reproduced with permission from Yin et al. [24]. © 2022, Springer Nature.

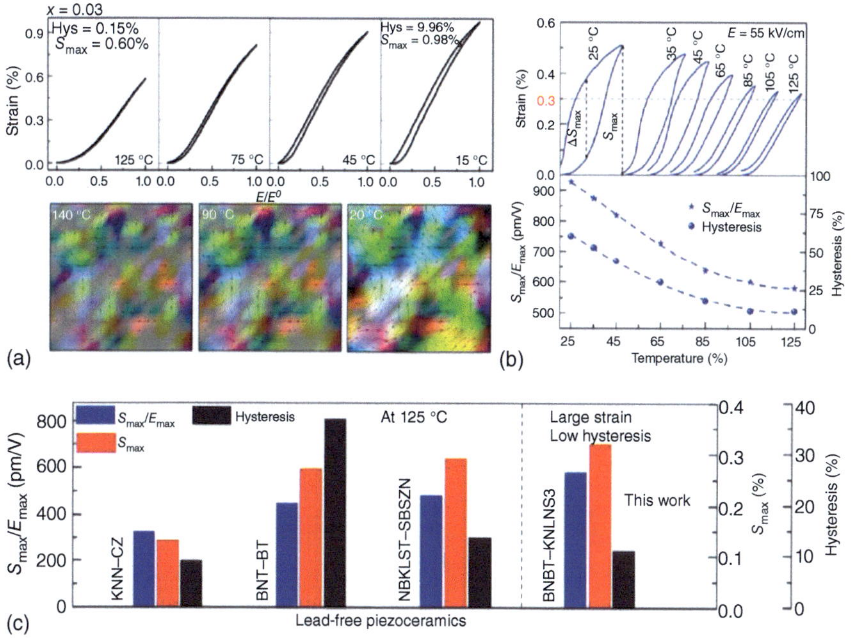

Figure 5.6 Theoretical design (phase field simulation): (a) microstructural evolution of the BNT-based system ($x = 0.03$) with correspondingly calculated P–E and S–E loops at various temperature (25, 90, and 140 °C), (b) unipolar S–E curves with S_{max}/E_{max} ratio and hysteresis of the BNBT–KNLKS3 ceramic for temperature range of 25–125 °C, (c) comparison of BNT-based materials with other perovskite systems. Source: (a–c) Reproduced with permission from Li et al. [49]. © 2022, Wiley.

relaxor status can exist over a broad range of temperatures and be coupled with the MPB composition, contributing to both large strain and low hysteresis. Based on simulation results, Li et al. selected FE state KNLNS ($K_{0.47}Na_{0.47}Li_{0.06}Nb_{0.99}Sb_{0.01}O_{2.99}$) to replace the BNBT matrix, and significantly, BNBT–KNLNS3 presented a large strain (>0.32%) and low strain hysteresis (down to 11.1%) over a broad temperature range (25–125 °C) (Figure 5.6b,c).

Except for conventional chemical design, Feng et al. also reported a new strategy to enhance the strain performance of BNT-based ceramics via designing "heterostrain" [50]. They combined the field-induced phased transition and non-180° domain switching and utilized the atomic-scale defect engineering and mesoscale domain engineering to further modify the strain performance (i.e. improving the contribution from domain switching while maintaining low crystalline symmetry (Figure 5.7a,c)). The prepared BNT-based ceramics can achieve an ultra-high strain of 2.3% at high temperatures (220 °C), which is the highest value reported over other piezoelectric ceramics and piezoelectric single crystals (Figure 5.7b). These new designing strategies provide researchers with more freedom to develop the electro-strain of BNT-based materials further, and more interesting phenomena could be expected in this material system.

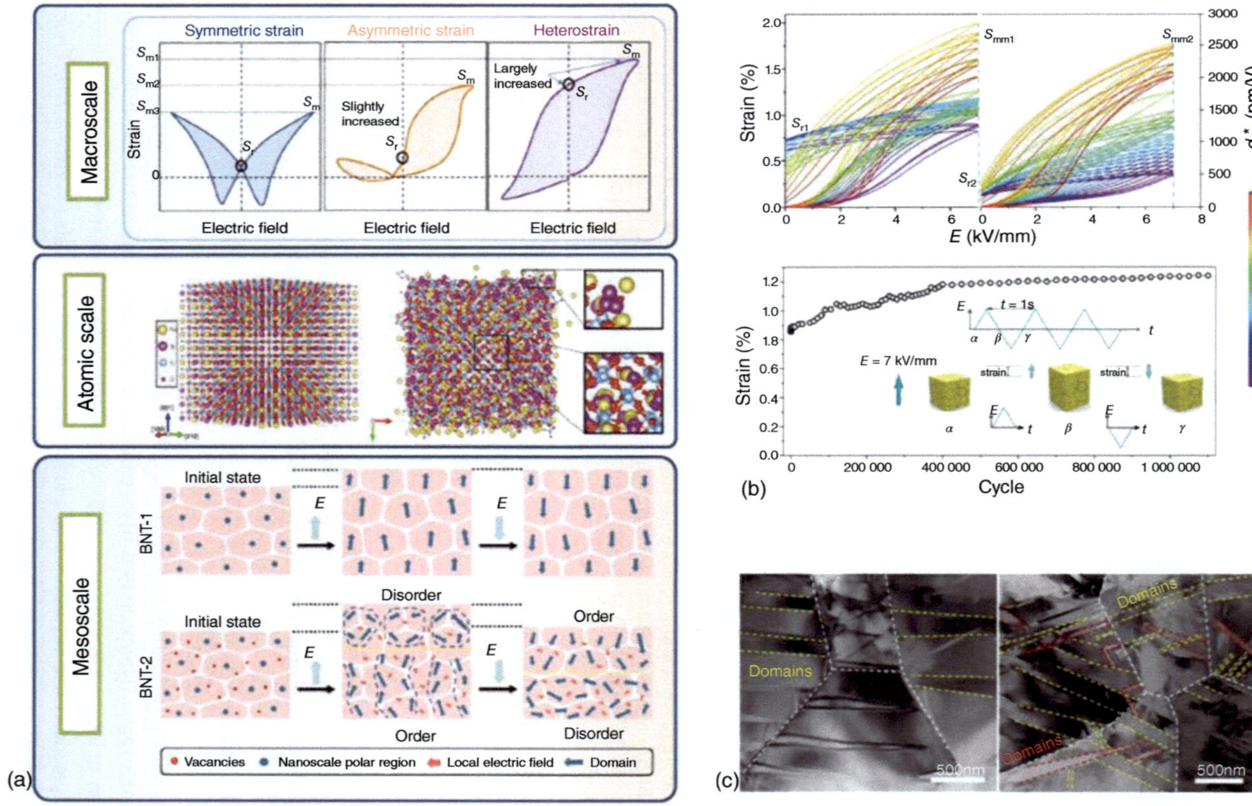

Figure 5.7 Theoretical design (construction of heterostrain): (a) design of piezoelectric with heterostrain from different dimensions, (b) piezoelectric performances of BNT polycrystalline, including temperature-dependent strain curves and cyclical stability, (c) comparison of BNT-based materials with other perovskite systems. Source: (a–c) Reproduced with permission from Feng et al. [50]. © 2022, Springer Nature.

5.1.5 Emerging Novel Research Topics in BNT-Based Materials

In addition to the abovementioned widely investigated topics, there are also several newly emerging areas of research for BNT-based materials. For example, our group has recently reported controllable domain morphologies in BNT-based ceramics, ranging from labyrinthine and refined stripe to bubble domains [51], as shown in Figure 5.8a, by chemically tuning the local structures. Using atomic-scale STEM, intriguing polar skyrmion-like structures with either flux-closure circinate or hedgehog-like geometries were observed in bulk BNT-based ceramics, similar to topological configurations reported in $(PbTiO_3)_m/(SrTiO_3)_n$ superlattices [54]. As mentioned in Section 5.1.1, pure BNT ceramics have a square-shaped *P–E* hysteresis loop with a maximum polarization of around 40 μC/cm^2, and the polarization of BNT-based films can be doubled [55]. For a representative example, by means of chemical modification, such as dilute Mn doping, Yang et al. reported that the maximum polarization and dielectric breakdown strength can be increased to 113 μC/cm^2 and 2.3 MV/cm in Mn: BNT–BT–$BiFeO_3$ thin films [52], resulting in a large recoverable energy density (W_{rec} = 81.9 J/cm^3) with η of 64.4%.

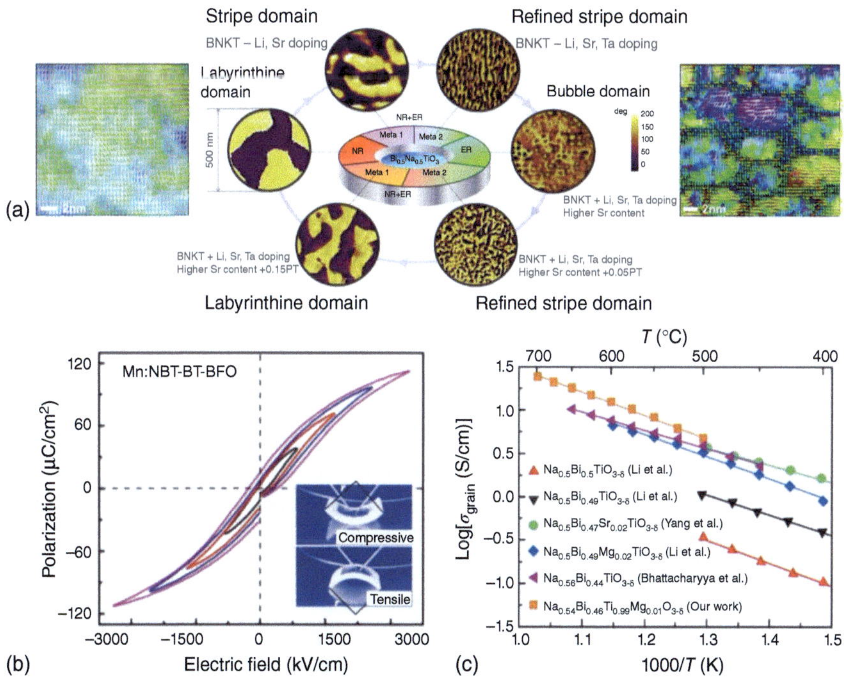

Figure 5.8 Emerging novel research topics in BNT-based materials: (a) PFM and STEM phase images of chemically (K, Li, Sr, Ta doping) driven domain evolution in BNT-based bulk ceramics, (b) *P–E* loops of Au/Mn : NBT–BT–BFO/Pt capacitors on mica substrates, (c) comparison of ionic conductivity of $Na_{0.5}Bi_{0.5}TiO_{2.965}$ with different kinds of chemical doping in the range of 400–700 °C. Source: (a) Reproduced from Yin et al. [51]; 2021/Springer Nature/CC BY 4.0; (b) Reproduced with permission from Yang et al. [52], © 2019/John Wiley & Son; (c) Reproduced with permission from Bhattacharyya et al. [53]; © 2018, Elsevier.

Although smaller than the record-high W_{rec} of 154 J/cm^3 and η of 97% [56], the Mn- and BFO-modified NBT–BT system exhibits superior frequency stability, a broad operating window, and anti-fatigue properties (1×10^9 cycles), even under bending states, as shown in Figure 5.8b. Some experiences from BNT-based thin films are also valid in guiding the performance design in BNT-based bulk capacitors [57]. The excellent electrical performances correlate intimately with the conductivity of BNT, as indicated by transport property studies. Sensitivity to Bi–O vacancies enables vacancy-doped BNT to deliver an ultrahigh ionic conductivity of ~14.3 S/cm at 600 °C in grains (Figure 5.8c), which is three times higher than that of 8 mol% Y_2O_3 stabilized ZrO_2 (8YSZ) and similar to that of $1CeO_2$–$9Sc_2O_3$–$90ZrO_2$ (1Ce9ScSZ) [53]. Moreover, the polar topological structures offer a pathway for enhancing its electronic conductivity, as evidenced in FEs such as $BiFeO_3$ [58]. Therefore, the synergistic ionic and electronic conductivity yields close-to-ideal interfacial polarization screening [59], which implies that improved electrical performances (e.g. electric conductivity and energy storage) could be assisted by electric-field driven electrochemical activities, similar to Au/$SrTiO_3$/LSMO capacitors. These novel observations should open new directions in designing BNT-based materials toward many multifunctional applications, such as high-density FE memories, high-power energy-storage capacitors, and intermediate-temperature SOFCs.

5.2 Bismuth Ferrite

5.2.1 Crystal Structure

Bismuth ferrite ($BiFeO_3$, BFO) has been widely studied due to high theoretical polarization intensity and high Curie temperature [60]. BFO has multiferroic properties because of its unique crystal structure. Michel and Kubel confirmed that BFO has a rhombohedral distorted perovskite structure and belongs to *R*3*c* space group by XRD test and neutron diffraction analysis [61, 62]. To quantitatively study the crystal structure, Rietveld refinement was performed by XND software, as shown in Figure 5.9a [63]. The cell structure was visualized according to the refinement results. The rhombohedral distorted structure of BFO is determined by the antiphase rotation of the adjacent FeO_6 oxygen octahedra around the $[111]_c$ direction relative to the $Pm\bar{3}m$ cubic cell, and the displacement of Fe^{3+} and Bi^{3+} cations along the central symmetric position in the same $[111]_c$ direction. The atomic structure and electric-field dipole of BFO were analyzed by the experimental device of 4D scanning electron diffraction using an AC–STEM, as shown in Figure 5.9b [64]. DFT calculation shows that the electrical polarization in BFO is due to the deformation of Bi 6*s* lone pair of electrons, and it can drive the rotation of the FeO_6 octahedron along the (111) axis, and then the Fe atoms are moved away from the center of the surrounding oxygen octahedron cage [65]. The HAADF image of BFO in the direction of (100) shows that Bi atoms are shifted from the geometric center closest to the FeO column to the upper right, and the dipole moment in BFO is related to the displacement of

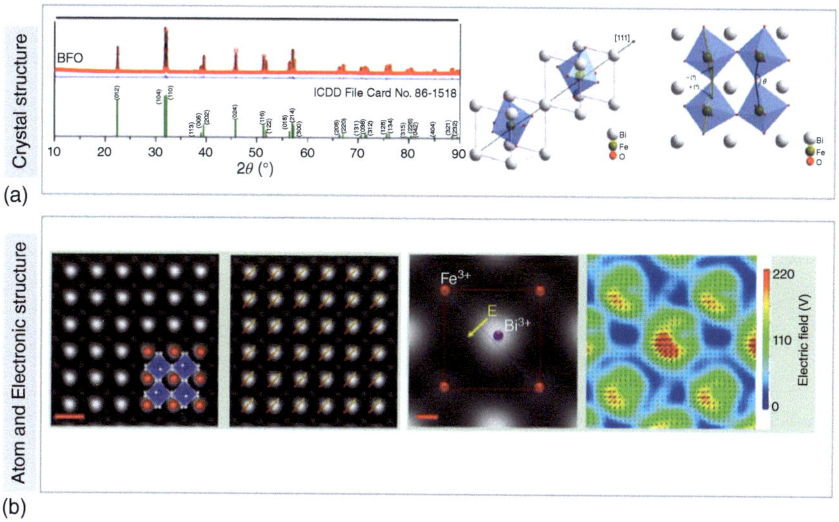

Figure 5.9 The multilevel structure of the BFO: (a) the crystal structure, (b) atom and electronic structure. Source: (a) Reproduced with permission from Wei et al. [63]; © 2017, American Chemical Society; (b) Reproduced with permission from Gao et al. [64], © 2019/Springer Nature.

Bi atoms [66]. The O column in BFO is elliptical due to the large rotation of the O octahedron causing the O column to split in the 2D projection. The charge profile at the Fe site is triangular due to the deviation of the O atom from the main axis and the formation of the anisotropy of the Fe—O bond in the d orbital of the Fe atom. The positive charge density pocket of Bi is partially connected to the nearest O-column, and there are intense red pockets between cations and O, indicating that the Bi—O and Fe—O bonds in BFO have stronger covalent bond properties.

Ion substitution of Bi and Fe in BFO results in different phase transitions. For example, when Sm^{3+} is used to replace Bi^{3+} at the A site [67], three phases with different crystal structures can be formed. When the content of Sm^{3+} is small, BSFO is a polar rhombohedrally distorted perovskite with the space group *R*3*c*. When Sm^{3+} content increases, BSFO evolves into a coexistence phase of *Pbam* and *R*3*c*. With further increase in the Sm^{3+} content, BSFO forms a nonpolar, orthorhombic structure with space groups *Pnma*, or *Pbnm*. While Co^{3+} is used to replace Fe^{3+} of B site [68], BFO undergoes transition from the rhombohedral (R) phase to the monoclinic (M) phase, and finally to the tetragonal (T) phase. The lack of a symmetry axis in the M phase structure allows the rotation of the FE polarization vector, which is considered as the origin of the piezoelectric enhancement [69, 70]. XANES test indicates a small structural difference in the local structure around the Co/Fe ions of the T and M phases, and the instability of Fe ions plays an important role in the sequential structural distortion of the BCFO system.

Considering the relationship between magnetism and crystal symmetry, the low-symmetry structure of BFO allows for the appearance of weak ferromagnetism. BFO has space-modulated magnetic structure of cycloidal type, where

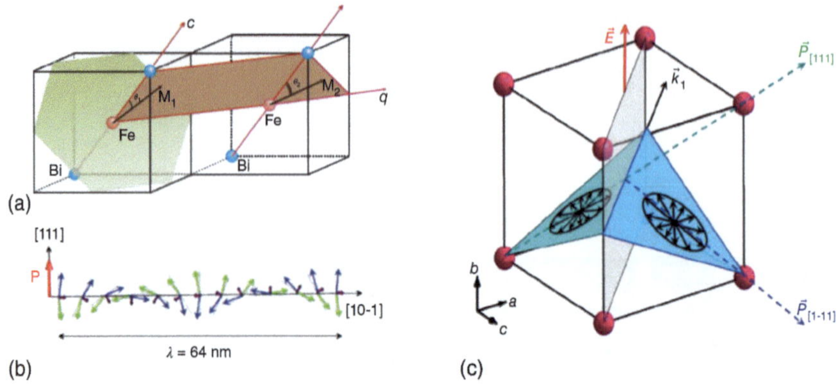

Figure 5.10 Schematic diagram of magnetic structure of the BFO: (a) diagram of the magnetization rotation plane, (b) schematic representation of the spin cycloid, (c) relationship between the magnetic easy plane containing the spins, the vector of ferroelectric polarization, and the vector of cycloid propagation. Source: (a) Reproduced with permission from Gippius et al. [71]; © 2010, Wiley; (b,c) Reproduced with permission from Lebuegle et al. [72]; © 2008, American Physical Society.

Fe^{3+} are turned in the plane perpendicular to the hexagonal basal plane along the propagation direction of the modulated wave, as shown in Figure 5.10a [71]. Each Fe^{3+} spin is surrounded by six antiparallel spins on the nearest Fe neighbors; however, the spins are not perfectly antiparallel, as there is a weak canting moment caused by local magnetoelectric coupling to the polarization. The magnetic easy plane (the plane within which the spins rotate) is defined by the propagation vector and the polarization vector (Figure 5.10b) [72]. The canted antiferromagnetic spins (blue and green arrows) give rise to a net magnetic moment (purple arrows) that is spacially averaged out to zero due to the cycloidal rotation. Figure 5.10c shows the magnetic moments rotate within the plane defined by the polarization (P // $[111]_{PC}$) and the cycloid propagation vector (k // $[1,0,-1]_{PC}$) [72], and if the direction of polarization changes, the magnetic susceptibility plane will also change.

5.2.2 Domain Configuration

DW is a special type of interface naturally formed in ferroic materials, which separate distinct domains with a different orientation in the order parameter (spontaneous polarization, strain, or magnetization). High electric conductivity of the FE and ferroelastic DWs opens up a whole new and emerging field of domain-wall nanoelectronics. The high electric conductivity of the domain walls in polycrystalline $BiFeO_3$ results in an unexpectedly large effect on the macroscopic piezoelectric response. Even if the local conductivity within the DWs does average macroscopically to a small value, the dynamics of DWs may still be strongly affected by their conductivity. Such an effect would be important because DW motion has a dominant influence on the dielectric and piezoelectric properties of FE materials, contributing up to 70% of the total macroscopic response. Rojac et al. provided the local current–voltage (I–V) response measurements by positioning the AFM tip either on a domain or a

DW for the existence of spontaneously formed conductive DWs in polycrystalline $BiFeO_3$ (Figure 5.11a) [73]. The current is measured by exposing the domain or the DW to a constant DC bias of +33 V for ≈3 minutes, which confirms the true steady conduction behavior of the walls (Figure 5.11c). The observed enhanced current signal at the DWs was found not to correlate with the sample topography and show that these conducting interfaces persist in poled material and move significantly under sub-switching driving fields (Figure 5.11b), contributing strongly and in an unusual way to the macroscopic piezoelectric response (Figure 5.11e,f). Several emerging features as a result of the dynamics of conducting DWs, including the low-frequency nonlinear piezoelectric Maxwell–Wagner (M–W) mechanism and the associated large enhancement of the piezoelectricity, which have not been demonstrated so far in FEs.

The precise mechanism behind the conductive behavior of DW in FEs has been controversial since the discovery of DW conductivity in $BiFeO_3$ thin films by Seidel et al. [75]. Resolving the mechanism of DW conductivity is crucial to exploit the functional potential of DW. If accumulated defects are the main cause of DW conductivity, a relationship between the type of carriers (oxygen vacancies, cation vacancies, electrons, or holes) and DW motion under applied electric fields can be foreseen. Rojac et al. provide atomic-scale chemical and structural analyses showing the accumulation of charged defects at domain walls in $BiFeO_3$ [74]. The accumulation of charged defects at DWs in $BiFeO_3$ was identified as Fe^{4+} cations and bismuth vacancies (Figure 5.11d), revealing *p*-type hopping conduction at domain walls caused by the presence of electron holes associated with Fe^{4+}.

Domain configuration and regulation of domain walls electric conductivity are prominent for the piezoelectric and FE property modulation of bismuth ferrate-based ceramics.

5.2.2.1 Domain Structure Versus Electrical Properties

Zheng et al. intend to investigate the piezoelectric and FE properties of BFO ceramic samples prepared by the rapid hot-press sintering method [76]. They found that the enhanced piezoelectric property is ascribed to the two-level hierarchical stripe-like domain structure (Figure 5.12a), which may respond more easily to electrical and strain stimuli than those irregular dendrite-like domains. The enhanced remnant polarization can be due to the improved sample quality and large grains in the properly hot-pressed sintered samples. Zheng et al. proposed a new phase diagram established by temperature dependence of dielectric properties, temperature dependence of piezoelectric coefficient, and the evolution of their properties to explain the contribution of piezoelectric and strain response by comparing FE and relaxor ferroelectric (RFE) compositions (Figure 5.12b) [78]. They found that (i) macrodomains and multipolar phase coexistence contribute to the piezoelectricity improvement. (ii) Nanodomains and an unstable depolarization temperature (T_d) region engineer the strain enhancement, which can promote electric-field-induced domain switching, lattice strain, and irreversible phase transition. In another work, Zheng et al. realized the dielectric relaxation behavior of bismuth ferrite-based FEs by designing polar nanostructures [78]. Their research showed that accompanied

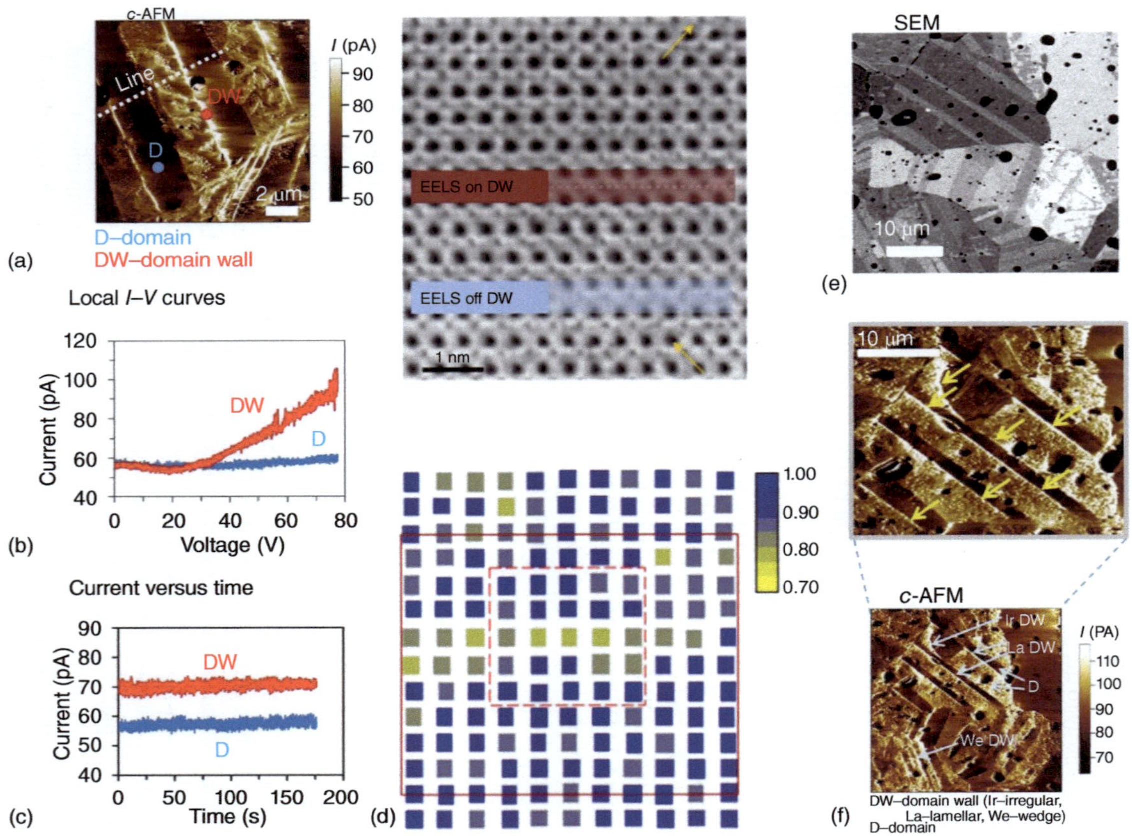

Figure 5.11 (a) c-AFM image of a region with lamellar-like domains [73], (b) local $I-V$ curves obtained by positioning the tip in the domain area (point D, panel A) and at the DW (point DW, panel A), (c) time dependence of current, measured by positioning the tip at the domain or DW [73], (d) accumulation of charged defects (Fe^{4+} and bismuth vacancies) at 109° DW in polycrystalline $BiFeO_3$ [74], (e) BE SEM micrograph, (f) c-AFM image, the inset of panel F shows an enlarged view of the conductive DWs (see arrows) [73]. Source: (a–c,e,f) Reproduced with permission from Rojac et al. [73]; © 2015/John Wiley & Sons; (d) Reproduced with permission from Rojac et al. [74]; © 2017/Springer Nature.

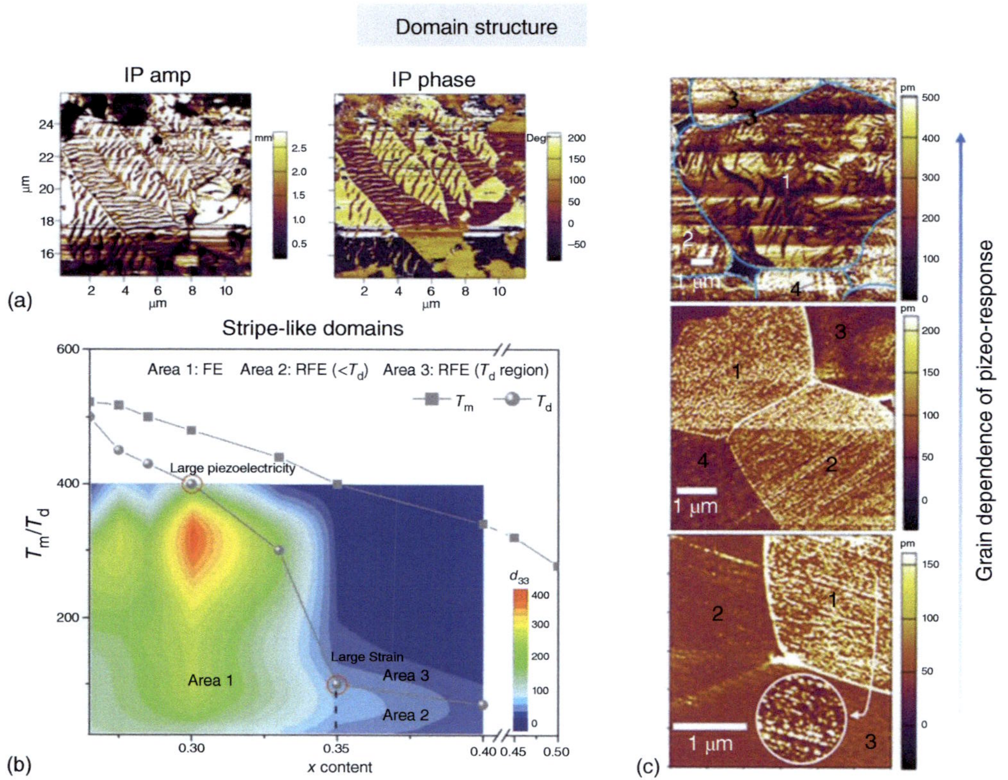

Figure 5.12 (a) The high-resolution PFM images for the stripe-like domain structure inside a coarse grain, [76] (b) a new phase diagram established by temperature dependence of dielectric properties, temperature dependence of piezoelectric coefficient, and the evolution of their properties, [77] (c) sketch map to reveal the composition-induced grain dependence of piezo-response. [78] Source: (a) Reproduced with permission from Zheng et al. [76]; © 2018/AIP Publishing LLC; (b) Reproduced with permission from Zheng and Wu [77]; © 2020, Wiley; (c) Reproduced with permission from Zheng and Wu [78]; © 2020/Royal Society of Chemistry.

by the strengthened random field and the degenerated octahedral distortion, the grain inhomogeneity featuring the coexistence of strong piezoresponse polar cluster grains and weak piezo-response polar entity grains (Figure 5.12c) contribute to the relaxation behavior.

5.2.2.2 Domain Walls Conductivity Versus Electrical Properties

Bencan et al. confirm on the atomic scale that the quenching leads to the depletion of the defects from the DWs and, inversely, that these defects tend to reaccumulate at the DWs by slow cooling (aging). Note that the quenched sample still contains a measurable fraction of conductive DWs, which could be attributed to the imperfect disordering induced by the quenching and/or partial electron-hole re-accumulation at the DWs occurring to a smaller degree at room temperature after the sample was quenched. From a relative perspective, however, the decreased fraction in the quenched sample, relative to the pristine state, and its increase upon aging satisfactorily agrees with the expected electron-hole ordering degree. They found that the presence of defects at the wall has a measurable effect on the lattice-strain distribution inside and outside the DW region. Compared to the DW depleted of defects (quenched), an enhanced shear strain is clearly observed in the pristine sample. In other words, accumulated defects release the shear strain on average inside the DW. Rojac et al. show that the local DW conductivity can be tailored by controlling the atmosphere during high-temperature annealing. The reduction of the DW conductivity by annealing $BiFeO_3$ in N_2 and the recovery of the elevated conductivity by annealing in O_2 confirm the *p*-type conduction behavior of the DWs. Their results confirm the key role of the defect chemistry and processing conditions in the conduction behavior at DWs. In the presence of electrostatic and elastic driving forces at DWs, the conductivity can be mediated by charge carriers closely related to stable ionic oxidation states in the perovskite. Besides, Zhang et al. have demonstrated extremely robust nanoscale FE domains that remain almost unchanged over one year in defect-engineered T′-BFO films [79]. Extrapolation of the domain sizes yields an expected lifetime of well over 10 years. The origin of such long polarization retention is proposed to be the result of designer-defect nanoregions, which are present throughout the volume of the film. These nanoregions, formed during the film fabrication, preserve strain coherence throughout the film. They found that (i) DW movement is inhibited, as shown by significantly increased activation fields for DW movement. (ii) The polarization retention behavior is independent of the initial domain size, a feature that is highly desirable toward the goal of increasing storage density while maintaining good retention.

5.2.3 Electrical Properties

5.2.3.1 Piezoelectricity

The difficulty of achieving breakthroughs in the piezoelectric properties of BF-based ceramics has been a long-standing problem, and a large number of studies have been carried out from different aspects, hoping to improve their piezoelectric performance. The volatile nature of bismuth and multiple valences of iron make sintering bulk BF-based ceramics difficult, and the high leakage currents in ions-modified

BF ceramics are the main obstacles to obtaining good piezoelectric properties [80–82]. The means that have been proven effective can be divided into the following categories:

BF Ions Yang et al. briefly review the structural and electromechanical properties of rare earth ($RE^{3+} = La^{3+}$, Sm^{3+}, and Dy^{3+}) substituted BFO by means of a thin-film composition spread technique [83]. By using first-principles calculations, they propose that the origin of the double hysteresis loop and the accompanying piezoelectric coefficient enhancement are the transition from the paraelectric orthorhombic phase to the polar rhombohedral phase induced by the electric field. Figure 5.13a,b is a generic phase diagram and the evolution of room-temperature P–E hysteresis loops across the structural boundary for the $RE = Sm^{3+}$, and La^{3+} cases and the piezoelectric coefficient d_{33} values as a function of the dopant composition for the $RE = Dy^{3+}$, Sm^{3+}, and La^{3+}, respectively based on the abovementioned findings.

The piezoelectric properties were evaluated in various BF-based systems. Fujii et al. reported the piezoelectricity of a BNT–BF system and the large signal piezoelectric response (d_{33}^*) exhibited from 32 to 182 pm/V as a function of BF concentration. However, an asymmetric piezoelectric response was exhibited in a wide range of compositions due to a strong internal bias by defect dipoles, which might be a hindrance to applying for applications [88]. In 2015, Lee et al. reported the high piezoelectric performance (402 pC/N) of a $BiFeO_3$–$BaTiO_3$ (BF–BT)-based system [89], which led to various studies and developments on lead-free BF–BT-based piezoelectric systems.

Several techniques have been developed to fabricate BFO ceramic bulks to overcome issues such as the formation of secondary phases as well as the occurrence of high leakage current. In 2004, a rapid liquid-phase sintering method developed by Wang et al. was used to prepare BFO ceramics with single-phase structure and high resistivity [90]. Recently, Heo et al. revealed the overall impact of light on largely enhanced piezoelectricity in BFO single crystals. The spectroscopic measurements of current–voltage (I–V) curves and PFM amplitude hysteresis loops under illumination show a close link between light effects and enhanced piezoelectricity, primarily due to the bulk photovoltaic (BPV) effect and the Schottky barrier effect [91].

BF–BT Solid Solution Zheng et al. systematically studied the effects of different equivalent ion doping on the piezoelectric properties of BF–BT ceramics, including Sc, Ga, Al, In, Co, Ni plasma substituting Fe, and found that Sc, Ga, or Al-doped BF–BT ceramics had the best piezoelectric performance, and their piezoelectric constants were all above 160 pC/N, which the authors believed was related to the phase structure of ceramics in which three sides and pseudo-cubic coexisted [92]. Besides, BF–BT-based ceramics are significantly enhanced by the quenching or rapid cooling process compared to the non-quenching process (commonly referred to as furnace cooling, slow cooling, or as-sintered). Therefore, a quenching process is found to be one of the successful methods to improve piezoelectricity in BF-based lead-free systems, and this has led to further research on the effect of quenching processes, such as quenching temperature or cooling rate [85, 93].

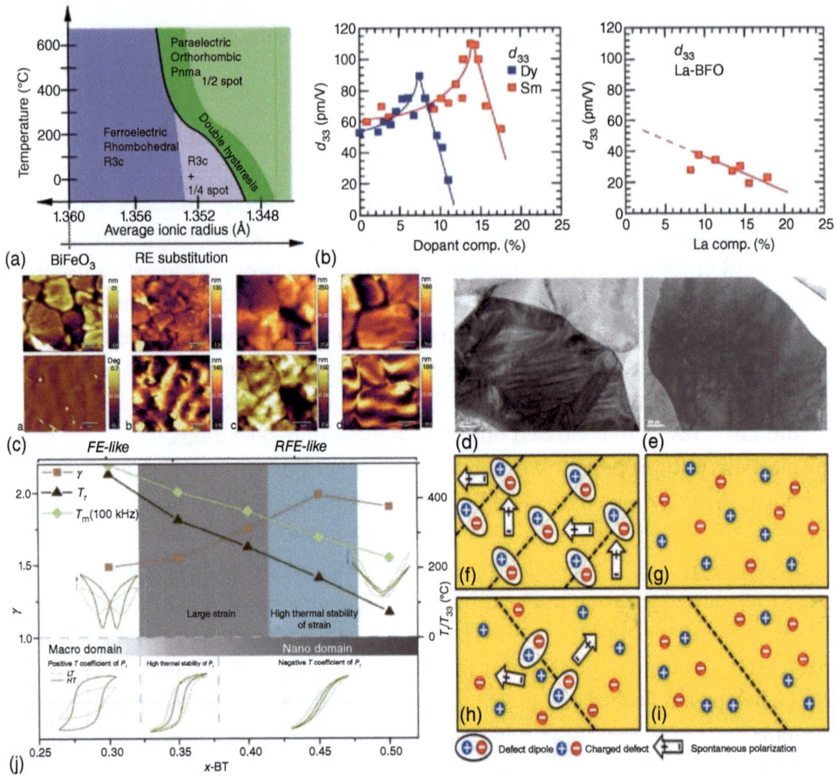

Figure 5.13 (a) Universal phase diagram of RE-substituted $BiFeO_3$ films. The black line represents the structural phase boundary between the rhombohedral (in blue) and the orthorhombic (in green) structural phases. At the lower temperature side, the 1/4{011} spot due to the minority phase with a $PbZrO_3$-type structure is observed in the region in light blue. The double hysteresis loop behavior emerges in the region in dark green, (b) piezoelectric coefficient d_{33} at 400 kV/cm as a function of RE substitution for RE = Dy^{3+}, Sm^{3+}, and La^{3+} cases, (c) evolution of magnetic domain structures in $Bi_{1-x}Nd_xFeO_3$ ceramics with increasing x observed by MFM; Bright-field transmission electron microscope of 0.75BF–0.25BT ceramic: (d) air quenching, (e) slow cooling at 2 °C/min; schematic of defect structure in BF–BT ceramics: (f) as-sintered, (g) at high temperature, (h) furnace cooling after annealing, and (i) quenching after annealing; (j) relation diagram together with P–E loops with three modes of temperature coefficient, temperature-dependent S–E loops below/above freezing temperature, freezing/maximum (100 kHz) temperature, and γ curve with increasing BT content. Source: (a,b) Reproduced with permission from Yang et al. [83]; © 2012, Royal Society of Chemistry; (c) Reproduced with permission from Chen et al. [84]; © 2018/John Wiley & Sons; (d,e) Reproduced with permission from Kim et al. [85]; © 2016/AIP Publishing LLC; (f–i) Reproduced with permission from Kim et al. [86]; © 2017, AIP Publishing; (j) Reproduced with permission from Li et al. [87]; © 2020, Elsevier.

5.2.3.2 Ferroelectricity

In 2018, Chen et al. reported improved magnetoelectric coupling by destroying the cycloidal state and switching to the weak ferromagnetic state through symmetry modulation. The structure is tuned from polar *R3c* to polar $Pna2_1$ and finally to nonpolar *Pbnm* by forming $Bi_{1-x}Nd_xFeO_3$ solid solutions, where two MPBs are detected. Greatly enhanced FE polarization is obtained together with the desired weak ferromagnetic characteristics in $Bi_{1-x}Nd_xFeO_3$ ceramics at the compositions near MPBs [84], the evolution of magnetic domain in $Bi_{1-x}Nd_xFeO_3$ ceramics is evaluated by magnetic force microscope (MFM) as shown in Figure 5.13c. The quenching method has a positive effect on improving the piezoelectric and FE properties of BFO and BFO-based ceramics. Figure 5.13d,e explain the effect of cooling rates on the FE properties in a microscopic view. The domain pattern of the quenched sample is well-aligned, while the slow-cooled sample is irregular. The freezing of aimlessly oriented defect dipoles and the depression to phase decomposition during cooling in the thermodynamically unstable temperature region can explain the improvement of the FE and piezoelectric effects [85]. In addition, Kim et al. proposed that distribution of charged defects can be controlled by a quenching process [86]. As shown in Figure 5.13f–i, the defect dipoles cannot be formed and moved in the DW by the quenching process; the leakage current is reduced, and the electrical properties are thus enhanced. Li et al. analyzed the temperature-dependent electrical characteristics of BF–BT ceramics from static structure and domain dynamics. A new relation diagram is constructed, as shown in Figure 5.13j. Temperature coefficient of remanent polarization (P_r) and negative strain (S_{neg}) changes from positive to negative as BT component increases, leading to a nearly zero temperature coefficient of P_r at critical component (BT = 0.35) [87].

5.2.3.3 Electro-strain

With the increase of $BaTiO_3$ content, the negative strain of butterfly-shaped *S*–*E* curves of (1−*x*)BF–*x*BT decrease. This is because the addition of BT is good for domain switching. The reduction of negative strain is related to the domain back-switching during bipolar cycling [94]. After doping Nd and BZN into BF–BT, Wang et al. obtained $S_{pos} = 0.463\%$ and $d_{33}^* = 424$ pm/V with *SH* (Strain hysteresis) of 37% for 0.5%BZN doped BF–BT [95]. The origin of the large strain was ascribed to a field-induced transition from short- to long-range dipolar order at the crossover from normal to a RFE behavior. Zheng et al. reported that the pseudo-cubic phase is a key factor to get a high strain response in nanodomain-engineered relaxor BTO–BFO-based FE ceramics [77]. Thermal depolarization could enhance strain in BFO-based perovskite RFEs, while the coexistence of the long-range ordered microdomains and multipolar phase is beneficial for increasing piezoelectric properties in BFO-based perovskite FEs [77]. A large strain of ~0.6% was found in $0.01Nd(Li_{0.5}Nb_{0.5})O_3$-doped BF–0.30BT ceramics at 150 kV/cm in *in situ* poling synchrotron XRD [96]. The pseudo-cubic structure was maintained without changing the full width at half maximum (FWHM) of the XRD peak before and after the application of the electric field. However, the strain anisotropy in the XRD peak (200 > 220 > 111), as shown in Figure 5.14a,b, indicates the possibility of the presence of multiple symmetrically polar nano regions that allow for a large average distortion in the applied electric field direction [97].

5.2.3.4 Energy Storage Property

Due to its high polarization of 100 μC/cm^2, $BiFeO_3$ (BF) is considered to have a large P_{max}, offering an enormous potential energy-storage density [99]. Recently, Yan et al. reported the ultrahigh energy density of 9.3 J/cm^3 and excellent efficiency with 90%,

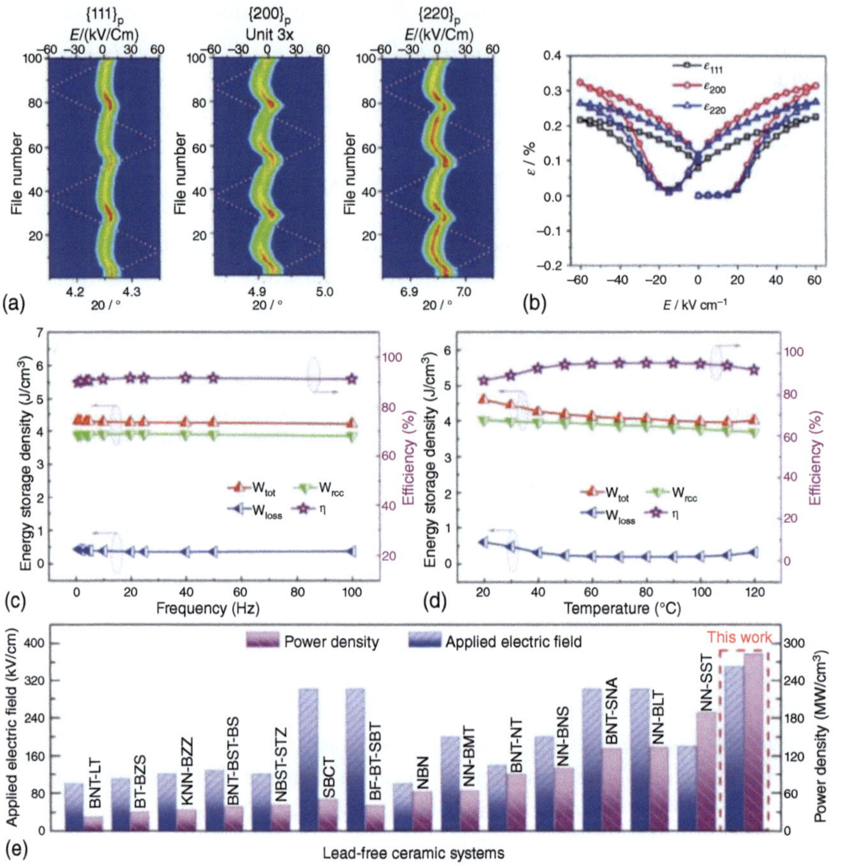

Figure 5.14 (a) 2D contour plot of (111), (200), and (220) peak profiles in $0.01Nd(Li_{0.5}Nb_{0.5})O_3$-doped BF−0.30BT under the electric field, (b) lattice strain anisotropic from (111), (200), and (220) peaks, respectively, (c) frequency, (d) temperature for 0.35BF−0.65ST ceramics, (e) the power density and applied electric field between other lead-free system and $0.35BiFeO_3$−$0.65SrTiO_3$ ceramics. Source: (a, b) Reproduced with permission from Wang et al. [97]; © 2019, Royal Society of Chemistry; (c–e) Reproduced with permission from Yan et al. [98]; © 2022, Wiley.

as shown in Figure 5.14c,d, which is an excellent energy density property compared to the that of other lead-free systems, as shown in Figure 5.14e [98].

5.3 Summary

Here, we briefly summarize the development and the perspectives of BNT- and BFO-based materials. BNT is a promising lead-free FE system with a high piezoelectric constant of over 200 pC/N. However, the loss of piezoelectricity at the depolarization temperature (T_d) due to evolving short-range-order local structures hinders its further development in the piezoelectric direction. Nevertheless, electro-strain is significantly enhanced at T_d, due to the dynamic balance within these evolving local structures. Beyond conventionally investigated electromechanical properties and FE structures, BNT-based materials have also been studied for their various electrical properties and novel local structures, including energy storage, ionic conductivity, and topological polar morphologies. Moreover, controlling their underlying structures has been demonstrated to be effective in improving the electrical performance of BNT-based RFEs. This review provides insights into the representative research directions for BNT-based materials, and also guides the way to design these kind of materials for further enhance their electrical performance.

BFO-based materials have acquired significant attention from researchers due to its high theoretical polarization intensity and high Curie temperature. For single-component bismuth ferrite, unique crystal structure leads to multiferroic properties and ionic substitution-induced phase transition can change the symmetry of the crystal, which is an effective approach to modulate its performance (FE, piezoelectric, and ferromagnetism). As the study progressed, the electrical properties of bismuth ferrate have been gradually regulated from various perspectives, including DW motion, domain wall conductivity, valence defects, domain structure, preparation process, and ion substitution. The gradual improvement of its comprehensive performance makes it more prominent in emerging nano-electronic, bulk applications, and other functional devices.

References

1 Smolenskii, G.A., Isupov, V.A., Agranovskaya, A.I., and Krainik, N.N. (1961). New ferroelectrics of complex composition 4. *Sov. Phys. Solid State* 2 (11): 2651–2654.

2 Gorfman, S. and Thomas, P.A. (2010). Evidence for a non-rhombohedral average structure in the lead-free piezoelectric material $Na_{0.5}Bi_{0.5}TiO_3$. *J. Appl. Crystallogr.* 43 (6): 1409–1414.

3 Rao, B.N., Fitch, A.N., and Ranjan, R. (2013). Ferroelectric-ferroelectric phase coexistence in $Na_{1/2}Bi_{1/2}TiO_3$. *Phys. Rev. B* 87 (6): 060102.

4 Rao, B.N., Olivi, L., Sathe, V., and Ranjan, R. (2016). Electric field and temperature dependence of the local structural disorder in the lead-free ferroelectric $Na_{0.5}Bi_{0.5}TiO_3$: an EXAFS study. *Phys. Rev. B* 93 (2): 024106.

5 Jo, W., Dittmer, R., Acosta, M. et al. (2012). Giant electric-field-induced strains in lead-free ceramics for actuator applications-status and perspective. *J. Electroceram.* 29 (1): 71–93.

6 Hiruma, Y., Nagata, H., and Takenaka, T. (2009). Thermal depoling process and piezoelectric properties of bismuth sodium titanate ceramics. *J. Appl. Phys.* 105 (8): 084112.

7 Naderer, M., Kainz, T., Schütz, D., and Reichmann, K. (2014). The influence of Ti-nonstoichiometry in $Bi_{0.5}Na_{0.5}TiO_3$. *J. Eur. Ceram. Soc.* 34 (3): 663–667.

8 Pal, V. and Dwivedi, R.K. (2012). Effect of rare earth gadolinium substitution on the structural, microstructure and dielectric properties of lead free BNT ceramics. *Adv. Mater. Res.* 585: 200–204.

9 Das Adhikary, G., Mahale, B., Rao, B.N. et al. (2021). Depoling phenomena in $Na_{0.5}Bi_{0.5}TiO_3$-$BaTiO_3$: a structural perspective. *Phys. Rev. B* 103 (18): 184106.

10 Takenaka, T., Maruyama, K.-I., and Sakata, K. (1991). $(Bi_{1/2}Na_{1/2})TiO_3$-$BaTiO_3$ system for lead-free piezoelectric ceramics. *Jpn. J. Appl. Phys.* 30 (9S): 2236.

11 Cordero, F., Craciun, F., Trequattrini, F. et al. (2010). Phase transitions and phase diagram of the ferroelectric perovskite $(Na_{0.5}Bi_{0.5})_{1-x}Ba_xTiO_3$ by anelastic and dielectric measurements. *Phys. Rev. B* 81 (14): 144124.

12 Herabut, A. and Safari, A. (1997). Processing and electromechanical properties of $(Bi_{0.5}Na_{0.5})_{(1-1.5x)}La_xTiO_3$ ceramics. *J. Am. Ceram. Soc.* 80 (11): 2954–2958.

13 Li, A., Wu, J., Qiao, S. et al. (2012). Microstructure and electrical properties of $(Bi_{0.50+x}Na_{0.50})_{0.94}Ba_{0.06}TiO_3$ ceramics with bismuth nonstoichiometry. *Phys. Status Solidi A* 209 (7): 1213–1218.

14 Han, J., Yin, J., and Wu, J. (2020). BNT-based ferroelectric ceramics: electrical properties modification by Ta_2O_5 oxide addition. *J. Am. Ceram. Soc.* 103 (1): 412–422.

15 Yin, J., Zhao, C., Zhang, Y., and Wu, J. (2017). Composition-induced phase transitions and enhanced electrical properties in bismuth sodium titanate ceramics. *J. Am. Ceram. Soc.* 100 (12): 5601–5609.

16 Zhang, Y.-R., Li, J.-F., Zhang, B.-P., and Peng, C.-E. (2008). Piezoelectric and ferroelectric properties of Bi-compensated $(Bi_{1/2}Na_{1/2})TiO_3$-$(Bi_{1/2}K_{1/2})TiO_3$ lead-free piezoelectric ceramics. *J. Appl. Phys.* 103 (7): 074109.

17 Ma, C., Guo, H., Beckman, S.P., and Tan, X. (2012). Creation and destruction of morphotropic phase boundaries through electrical poling: a case study of lead-free $(Bi_{1/2}Na_{1/2})TiO_3$-$BaTiO_3$ piezoelectrics. *Phys. Rev. Lett.* 109 (10): 107602.

18 Ehara, Y., Novak, N., Yasui, S. et al. (2015). Electric-field-temperature phase diagram of Mn-doped $Bi_{0.5}(Na_{0.9}K_{0.1})_{0.5}TiO_3$ ceramics. *Appl. Phys. Lett.* 107 (26): 262903.

19 Ehara, Y., Novak, N., Ayrikyan, A. et al. (2016). Phase transformation induced by electric field and mechanical stress in Mn-doped $(Bi_{1/2}Na_{1/2})TiO_3$-$(Bi_{1/2}K_{1/2})TiO_3$ ceramics. *J. Appl. Phys.* 120 (17): 174103.

20 PI Company. https://www.piceramic.com/en/#ce145667__scrollanchor.

21 Rao, B.N., Datta, R., Chandrashekaran, S.S. et al. (2013). Local structural disorder and its influence on the average global structure and polar properties in $Na_{0.5}Bi_{0.5}TiO_3$. *Phys. Rev. B* 88 (22): 224103.

22 Yin, J., Tao, H., Liu, G., and Wu, J. (2020). Domain-scale imaging to dispel the clouds over the thermal depolarization of Bi0.5Na0.5TiO3-based relaxor ferroelectrics. *J. Am. Ceram. Soc.* 103 (3): 1881–1890.

23 Zhou, X., Xue, G., Luo, H. et al. (2021). Phase structure and properties of sodium bismuth titanate lead-free piezoelectric ceramics. *Prog. Mater. Sci.* 122: 100836.

24 Yin, J., Shi, X., Tao, H. et al. (2022). Deciphering the atomic-scale structural origin for large dynamic electromechanical response in lead-free $Bi_{0.5}Na_{0.5}TiO_3$-based relaxor ferroelectrics. *Nat. Commun.* 13 (1): 6333.

25 Han, J., Yin, J., and Wu, J. (2020). $(Bi_{0.5}Na_{0.5})TiO_3$ ferroelectric ceramics: achieving high depolarization temperature and improved piezoelectric properties. *J. Eur. Ceram. Soc.* 40 (15): 5392–5401.

26 Yin, J., Wang, Y., Zhang, Y. et al. (2018). Thermal depolarization regulation by oxides selection in lead-free BNT/oxides piezoelectric composites. *Acta Mater.* 158: 269–277.

27 Li, Z., Liu, H., Thong, H. et al. (2019). Enhanced temperature stability and defect mechanism of BNT-based lead-free piezoceramics investigated by a quenching process. *Adv. Electron. Mater.* 5: 1800756.

28 Zhou, Z., Zhang, S.-W., Zheng, Y. et al. (2018). Textured $Bi_{1/2}Na_{1/2}TiO_3$-$BaTiO_3$ lead-free films with enhanced piezoelectric property and depolarization temperature. *Adv. Electron. Mater.* 4: 1800351.

29 Riemer, L.M., Lalitha, K.V., Jiang, X. et al. (2017). Stress-induced phase transition in lead-free relaxor ferroelectric composites. *Acta Mater.* 136: 271–280.

30 Yang, D., Wu, X., Wang, X. et al. (2022). New strategy for simultaneously achieving enhanced piezoelectricity and deferred T_d in $Bi_{0.5}Na_{0.5}TiO_3$-based relaxor ferroelectrics. *J. Appl. Phys.* 132 (11): 115103.

31 Liu, X. and Tan, X. (2016). Giant strain with low cycling degradation in Ta-doped $[Bi_{1/2}(Na_{0.8}K_{0.2})_{1/2}]TiO_3$ lead-free ceramics. *J. Appl. Phys.* 120 (3): 034102.

32 Hussain, A., Rahman, J.U., Zaman, A. et al. (2014). Field-induced strain and polarization response in lead-free $Bi_{1/2}(Na_{0.80}K_{0.20})_{1/2}TiO_3$-$SrZrO_3$ ceramics. *Mater. Chem. Phys.* 143 (3): 1282–1288.

33 Guo, Y., Fan, H., Long, C. et al. (2014). Electromechanical and electrical properties of $Bi_{0.5}Na_{0.5}Ti_{1-x}Mn_xO_{3-\delta}$ ceramics with high remnant polarization. *J. Alloys Compd.* 610: 189–195.

34 Chen, P.-Y., Chen, C.-S., Tu, C.-S. et al. (2014). Relaxor effect on electric field induced large strain in $(1-x)(Bi_{0.5}Na_{0.5})TiO_3-xBaTiO_3$ lead-free piezoceramics. *Ceram. Int.* 40 (4): 6137–6142.

35 Zhang, S.-T., Kounga, A.B., Aulbach, E. et al. (2007). Giant strain in lead-free piezoceramics $Bi_{0.5}Na_{0.5}TiO_3$-$BaTiO_3$-$K_{0.5}Na_{0.5}NbO_3$ system. *Appl. Phys. Lett.* 91 (11): 112906.

36 Seifert, K.T.P., Jo, W., and Rödel, J. (2010). Temperature-insensitive large strain of $(Bi_{1/2}Na_{1/2})TiO_3$-$(Bi_{1/2}K_{1/2})TiO_3$-$(K_{0.5}Na_{0.5})NbO_3$ lead-free piezoceramics. *J. Am. Ceram. Soc.* 93 (5): 1392–1396.

37 Zhang, Y., Fan, P., Fan, H. et al. (2020). Tailoring the strain performance of lead-free relaxor/ferroelectric-layered composites. *J. Electroceram.* 44 (1): 32–40.

38 Maqbool, A., Hussain, A., Ur Rahman, J. et al. (2014). Enhanced electric field-induced strain and ferroelectric behavior of $(Bi_{0.5}Na_{0.5})TiO_3$-$BaTiO_3$-$SrZrO_3$ lead-free ceramics. *Ceram. Int.* 40 (8, Part A): 11905–11914.

39 Lee, D.S., Jeong, S.J., Kim, M.S., and Kim, K.H. (2013). Effect of sintering time on strain in ceramic composite consisting of $0.94Bi_{0.5}(Na_{0.75}K_{0.25})_{0.5}TiO_3$-$0.06BiAlO_3$ with $(Bi_{0.5}Na_{0.5})TiO_3$. *Jpn. J. Appl. Phys.* 52 (2R): 021801.

40 Su Lee, D., Jong Jeong, S., Soo Kim, M., and Hyuk Koh, J. (2012). Electric field induced polarization and strain of Bi-based ceramic composites. *J. Appl. Phys.* 112 (12): 124109.

41 Jiang, C., Zhou, X., Zhou, K. et al. (2016). Grain oriented $Na_{0.5}Bi_{0.5}TiO_3$-$BaTiO_3$ ceramics with giant strain response derived from single-crystalline $Na_{0.5}Bi_{0.5}TiO_3$-$BaTiO_3$ templates. *J. Eur. Ceram. Soc.* 36 (6): 1377–1383.

42 Zhao, Z.-H., Ye, M.-Y., Ji, H.-M. et al. (2018). Enhanced piezoelectric properties and strain response in <001> textured BNT-BKT-BT ceramics. *Mater. Des.* 137: 184–191.

43 Wu, X., Wen, L., Wu, C. et al. (2023). Insight into the large electro-strain in bismuth sodium titanate-based relaxor ferroelectrics: from fundamentals to regulation methods. *J. Mater. Sci. Technol.* 150: 27–48.

44 Yin, J., Liu, G., Lv, X. et al. (2019). Superior and anti-fatigue electro-strain in $Bi_{0.5}Na_{0.5}TiO_3$-based polycrystalline relaxor ferroelectrics. *J. Mater. Chem. A* 7 (10): 5391–5401.

45 Dung, D.D., Van Quyet, N., and Bac, L.H. (2015). Role of sintering temperature on giant field-induced strain in lead-free $Bi_{0.5}(Na,K)_{0.5}TiO_3$-based ceramics. *Ferroelectrics* 474 (1): 113–119.

46 Tong, X.-Y., Du, Z.-Z., Yang, Y.-T. et al. (2022). Effect of stress on the phase transition and optimizing electric-induced strain behavior of $Bi_{0.5}Na_{0.5}TiO_3$-$SrTiO_3$ lead-free co-fired multilayer piezoactuators. *J. Eur. Ceram. Soc.* 42 (5): 2157–2169.

47 Waqar, M., Wu, H., Chen, J. et al. (2022). Evolution from lead-based to lead-free piezoelectrics: engineering of lattices, domains, boundaries, and defects leading to giant response. *Adv. Mater.* 34 (25): 2106845.

48 Wu, H., Zhang, Y., Wu, J. et al. (2019). Microstructural origins of high piezoelectric performance: a pathway to practical lead-free materials. *Adv. Funct. Mater.* 29 (33): 1902911.

49 Li, T., Liu, C., Shi, P. et al. (2022). High-performance strain of lead-free relaxor-ferroelectric piezoceramics by the morphotropic phase boundary modification. *Adv. Funct. Mater.* 32 (32): 2202307.

50 Feng, W., Luo, B., Bian, S. et al. (2022). Heterostrain-enabled ultrahigh electrostrain in lead-free piezoelectric. *Nat. Commun.* 13 (1): 5086.

51 Yin, J., Zong, H., Tao, H. et al. (2021). Nanoscale bubble domains with polar topologies in bulk ferroelectrics. *Nat. Commun.* 12 (1): 3632.

52 Yang, C., Lv, P., Qian, J. et al. (2019). Fatigue-free and bending-endurable flexible Mn-doped $Na_{0.5}Bi_{0.5}TiO_3$-$BaTiO_3$-$BiFeO_3$ film capacitor with an ultrahigh energy storage performance. *Adv. Energy Mater.* 9 (18): 1803949.

53 Bhattacharyya, R., Das, S., and Omar, S. (2018). High ionic conductivity of Mg^{2+}-doped non-stoichiometric sodium bismuth titanate. *Acta Mater.* 159: 8–15.

54 Yadav, A.K., Nelson, C.T., Hsu, S.L. et al. (2016). Observation of polar vortices in oxide superlattices. *Nature* 530 (7589): 198–201.

55 Song, J., Gao, J., Zhang, S. et al. (2019). Structure and electrical properties of $Na_{0.5}Bi_{0.5}TiO_3$ epitaxial films with (110) orientation. *Crystals* 9 (11): 558.

56 Peng, B., Zhang, Q., Li, X. et al. (2015). Giant electric energy density in epitaxial lead-free thin films with coexistence of ferroelectrics and antiferroelectrics. *Adv. Electron. Mater.* 1 (5): 1500052.

57 Li, J., Shen, Z., Chen, X. et al. (2020). Grain-orientation-engineered multilayer ceramic capacitors for energy storage applications. *Nat. Mater.* 19 (9): 999–1005.

58 Yang, W., Tian, G., Zhang, Y. et al. (2021). Quasi-one-dimensional metallic conduction channels in exotic ferroelectric topological defects. *Nat. Commun.* 12 (1): 1306.

59 Stengel, M., Vanderbilt, D., and Spaldin, N.A. (2009). Enhancement of ferroelectricity at metal–oxide interfaces. *Nat. Mater.* 8 (5): 392–397.

60 Wu, J., Fan, Z., Xiao, D. et al. (2016). Multiferroic bismuth ferrite-based materials for multifunctional applications: ceramic bulks, thin films and nanostructures. *Prog. Mater. Sci.* 84: 335–402.

61 Michel, C., Moreau, J.-M., Achenbach, G.D. et al. (1969). The atomic structure of $BiFeO_3$. *Solid State Commun.* 7 (9): 701–704.

62 Kubel, F. and Schmid, H. (1990). Structure of a ferroelectric and ferroelastic monodomain crystal of the perovskite $BiFeO_3$. *Acta Crystallogr., Sect. B* 46 (6): 698–702.

63 Wei, J., Wu, C., Liu, Y. et al. (2017). Structural distortion, spin-phonon coupling, interband electronic transition, and enhanced magnetization in rare-earth-substituted bismuth ferrite. *Inorg. Chem.* 56 (15): 8964–8974.

64 Gao, W., Addiego, C., Wang, H. et al. (2019). Real-space charge-density imaging with sub-ångström resolution by four-dimensional electron microscopy. *Nature* 575 (7783): 480–484.

65 Seshadri, R. and Hill, N.A. (2001). Visualizing the role of Bi 6*s* "Lone Pairs" in the off-center distortion in ferromagnetic $BiMnO_3$. *Chem. Mater.* 13 (9): 2892–2899.

66 Kim, Y.-M., Morozovska, A., Eliseev, E. et al. (2014). Direct observation of ferroelectric field effect and vacancy-controlled screening at the $BiFeO_3$/$La_xSr_{1-x}MnO_3$ interface. *Nat. Mater.* 13 (11): 1019–1025.

67 Walker, J., Simons, H., Alikin, D.O. et al. (2016). Dual strain mechanisms in a lead-free morphotropic phase boundary ferroelectric. *Sci. Rep.* 6 (1): 19630.

68 Ishimatsu, N., Watanabe, T., Oka, K. et al. (2015). Differences in local structure around Co and Fe of the $BiCo_{1-x}Fe_xO_3$ system determined by X-ray absorption fine structure. *Phys. Rev. B* 92 (5): 054108.

69 Bellaiche, L., Garcia, A., and Vanderbilt, D. (2000). Finite-temperature properties of $Pb(Zr_{1-x}Ti_x)O_3$ alloys from first principles. *Phys. Rev. Lett.* 84 (23): 5427–5430.

70 Vanderbilt, D. and Cohen, M.H. (2001). Monoclinic and triclinic phases in higher-order Devonshire theory. *Phys. Rev. B* 63 (9): 094108.

71 Gippius, A.A., Khozeev, D.F., Morozova, E.N., and Zalessky, A.V. (2003). Observation of spin modulated magnetic structure at Bi- and Fe-sites in $BiFeO_3$ by nuclear magnetic resonance. *Phys. Status Solidi A* 196 (1): 221–224.

72 Lebeugle, D., Colson, D., Forget, A. et al. (2008). Electric-field-induced spin flop in $BiFeO_3$ single crystals at room temperature. *Phys. Rev. Lett.* 100 (22): 227602.

73 Rojac, T., Ursic, H., Bencan, A. et al. (2015). Mobile domain walls as a bridge between nanoscale conductivity and macroscopic electromechanical response. *Adv. Funct. Mater.* 25 (14): 2099–2108.

74 Rojac, T., Bencan, A., Drazic, G. et al. (2017). Domain-wall conduction in ferroelectric $BiFeO_3$ controlled by accumulation of charged defects. *Nat. Mater.* 16 (3): 322–327.

75 Seidel, J., Martin, L.W., He, Q. et al. (2009). Conduction at domain walls in oxide multiferroics. *Nat. Mater.* 8 (3): 229–234.

76 Zheng, S.H., Li, Z.W., Zhang, C.X. et al. (2018). Two-level hierarchical stripe domains and enhanced piezoelectricity of rapid hot-press sintered $BiFeO_3$ ceramics. *J. Appl. Phys.* 124 (19): 194104.

77 Zheng, T. and Wu, J. (2020). Perovskite $BiFeO_3$-$BaTiO_3$ ferroelectrics: engineering properties by domain evolution and thermal depolarization modification. *Adv. Electron. Mater.* 6 (5): 2000079.

78 Zheng, T. and Wu, J. (2020). Mesoscale origin of dielectric relaxation with superior electrostrictive strain in bismuth ferrite-based ceramics. *Mater. Horiz.* 7 (11): 3011–3020.

79 Zhang, D., Sando, D., Sharma, P. et al. (2020). Superior polarization retention through engineered domain wall pinning. *Nat. Commun.* 11 (1): 349.

80 Kumar, M.M., Palkar, V.R., Srinivas, K., and Suryanarayana, S.V. (2000). Ferroelectricity in a pure $BiFeO_3$ ceramic. *Appl. Phys. Lett.* 76 (19): 2764–2766.

81 Palai, R., Katiyar, R.S., Schmid, H. et al. (2008). β Phase and γ-β metal-insulator transition in multiferroic $BiFeO_3$. *Phys. Rev. B* 77 (1): 014110.

82 Rojac, T., Kosec, M., Budic, B. et al. (2010). Strong ferroelectric domain-wall pinning in $BiFeO_3$ ceramics. *J. Appl. Phys.* 108 (7): 074107.

83 Yang, C.-H., Kan, D., Takeuchi, I. et al. (2012). Doping $BiFeO_3$: approaches and enhanced functionality. *Phys. Chem. Chem. Phys.* 14 (46): 15953–15962.

84 Chen, J., Xu, B., Liu, X.Q. et al. (2019). Symmetry modulation and enhanced multiferroic characteristics in $Bi_{1-x}Nd_xFeO_3$ ceramics. *Adv. Funct. Mater.* 29 (3): 1806399.

85 Kim, D.S., Cheon, C.I., Lee, S.S., and Kim, J.S. (2016). Effect of cooling rate on phase transitions and ferroelectric properties in $0.75BiFeO_3$-$0.25BaTiO_3$ ceramics. *Appl. Phys. Lett.* 109 (20): 202902.

86 Kim, S., Khanal, G.P., Ueno, S. et al. (2017). Revealing the role of heat treatment in enhancement of electrical properties of lead-free piezoelectric ceramics. *J. Appl. Phys.* 122 (1): 014103.

87 Li, C.Y., Zheng, T., and Wu, J.G. (2021). Competitive mechanism of temperature-dependent electrical properties in $BiFeO_3$-$BaTiO_3$ ferroelectrics controlled by domain evolution. *Acta Mater.* 206: 116601.

88 Fujii, I., Ito, Y., Suzuki, T., and Wada, T. (2016). Ferroelectric and piezoelectric properties of $(Bi_{1/2}Na_{1/2})TiO_3$-$BiFeO_3$ ceramics. *J. Mater. Res.* 31 (1): 28–35.

89 Lee, M.H., Kim, D.J., Park, J.S. et al. (2015). High-performance lead-free piezoceramics with high curie temperatures. *Adv. Mater.* 27 (43): 6976–6982.

90 Wang, Y.P., Zhou, L., Zhang, M.F. et al. (2004). Room-temperature saturated ferroelectric polarization in $BiFeO_3$ ceramics synthesized by rapid liquid phase sintering. *Appl. Phys. Lett.* 84 (10): 1731–1733.

91 Heo, Y. and Alexe, M. (2022). Boosting piezoelectricity under illumination via the bulk photovoltaic effect and the Schottky barrier effect in $BiFeO_3$. *Adv. Mater.* 34 (5): 2105845.

92 Zheng, T., Jiang, Z., and Wu, J. (2016). Enhanced piezoelectricity in $(1-x)Bi_{1.05}Fe_{1-y}A_yO_3-xBaTiO_3$ lead-free ceramics: site engineering and wide phase boundary region. *Dalton Trans.* 45 (28): 11277–11285.

93 Nam, H., Kim, S., Aizawa, T. et al. (2018). Influence of quenching temperature on piezoelectric and ferroelectrics properties in $BaTiO_3$-$Bi(Mg_{1/2}Ti_{1/2})O_3$-$BiFeO_3$ ceramics. *Ceram. Int.* 44: S199–S202.

94 Kim, S., Khanal, G.P., Nam, H.-W. et al. (2017). Structural and electrical characteristics of potential candidate lead-free $BiFeO_3$-$BaTiO_3$ piezoelectric ceramics. *J. Appl. Phys.* 122 (16): 164105.

95 Wang, D., Fan, Z., Li, W. et al. (2018). High energy storage density and large strain in $Bi(Zn_{2/3}Nb_{1/3})O_3$-doped $BiFeO_3$-$BaTiO_3$ ceramics. *ACS Appl. Energy Mater.* 1 (8): 4403–4412.

96 Wang, D., Fan, Z., Zhou, D. et al. (2018). Bismuth ferrite-based lead-free ceramics and multilayers with high recoverable energy density. *J. Mater. Chem. A* 6 (9): 4133–4144.

97 Wang, G., Fan, Z., Murakami, S. et al. (2019). Origin of the large electrostrain in $BiFeO_3$-$BaTiO_3$ based lead-free ceramics. *J. Mater. Chem. A* 7 (37): 21254–21263.

98 Yan, F., Bai, H., Ge, G. et al. (2022). Composition and structure optimized $BiFeO_3$-$SrTiO_3$ lead-free ceramics with ultrahigh energy storage performance. *Small* 18 (10): 2106515.

99 Zheng, D.G. and Zuo, R.Z. (2017). Enhanced energy storage properties in $La(Mg_{1/2}Ti_{1/2})O_3$-modified $BiFeO_3$-$BaTiO_3$ lead-free relaxor ferroelectric ceramics within a wide temperature range. *J. Eur. Ceram. Soc.* 37 (1): 413–418.

6

Piezoelectricity in Molecular Ferroelectrics

Yu-Meng You

Southeast University, Jiangsu Key Laboratory for Science and Applications of Molecular Ferroelectrics, Nanjing, No. 2 Southeast University Road, 211189, China

6.1 Introduction

Of all the piezoelectrics, ferroelectrics stand out, as their polarization can be reversed by an external electric field. The history of ferroelectrics began in 1920, when J. Valasek discovered Rochelle Salt, the first ferroelectric material [1]. Industry-leading inorganic ferroelectrics, such as $BaTiO_3$ (BTO), $Pb(Zr,Ti)O_3$ (PZT), LNO, and $(K,Na)NbO_3$ (KNN), have been in wide application for their robust properties and great utilization potential [2–4]. Nowadays, molecular ferroelectrics with excellent piezoelectricity have emerged and been a key topic among researchers as they can be processed under a low temperature in an environmentally friendly, featuring lightweight, nontoxicity, biocompatibility, and mechanical flexibility [5–8].

The past decade has seen much improvements in chemical synthesis and state-of-the-art characterization methods. This has prompted more systematic and rational research on molecular ferroelectrics [9]. Various high-performance molecular ferroelectrics have been newly designed and synthesized that approach or even exceed some inorganic ferroelectrics [10]. For example, single-component molecular ferroelectric croconic acid demonstrated a high polarization of up to 23 $\mu C/cm^2$, comparable to that of BTO [11]. Diisopropylammonium bromide was found to have a Curie temperature (T_c) of 426 K, higher than that of BTO [12]. Besides, the organic–inorganic hybrid perovskite molecular ferroelectric $[(CH_3)_3NCH_2Cl]CdCl_3$ shows an excellent piezoelectric response d_{33} value of 383 pC/N [13, 14].

In addition, attributed to the structural tunability and chemical variability of molecular ferroelectrics, several practice-oriented theories of chemical design strategies have been proposed [15], including lowering molecular symmetry [16], introducing homochirality [17], and applying H/F substitution [18, 19]. New ferroelectric physics was discovered, such as supramolecular charge-transfer complexes [20], and light-induced *cis–trans* isomerism [21, 22]. Accordingly, molecular ferroelectrics are expected to be promising alternatives or supplements to conventional inorganic ferroelectric ceramics and piezoelectric devices.

Piezoelectric Materials: From Fundamentals to Emerging Applications, First Edition. Edited by Jiagang Wu.

In this chapter, we have reviewed the recent strides in molecular ferroelectrics with excellent piezoelectricity summarizing several piezoelectric forms of molecular ferroelectrics, focusing on practical application with better performance. We then introduce the roles of piezoresponse force microscopy (PFM) in assisting the characterization and design of piezoelectrics. These achievements have boosted the rapid development of the piezoelectric performance in molecular ferroelectrics. In the final section, we will summarize the potential application and challenges that molecular ferroelectrics face.

6.2 Piezoelectricity in Molecular Ferroelectrics

6.2.1 Piezoelectricity in Crystals

Over the past decade, two representative molecular ferroelectric crystals, croconic acid [11] and diisopropylammonium bromide [12], made significant strides in ferroelectric performance (T_c and P_s), catching up to traditional inorganic ferroelectrics. However, their piezoelectric performance d_{33} is even an order of magnitude lower than that of the single-crystal perovskite ferroelectric BTO, not to mention those of perovskite solid solutions such as PZT and $Pb(Mg_{1/3}Nb_{2/3})O_3$–$PbTiO_3$ (PMN–PT) [23, 24]. Considering that these outstanding inorganic piezoelectrics described previously all have a similar perovskite crystal structure [25–28], efforts have been made to improve the piezoelectric response of molecular ferroelectrics with organic–inorganic hybrid perovskite crystal structure, which seems to be a starting point to probe into deeper research.

In a hybrid perovskite crystal, the inorganic framework features high mechanical strength that is good for strain propagation through the entire material; the various organic cations render structural compatibility and compositional variability [8, 29, 30].

According to the defining equation: $d = (\partial D/\partial T)_E$ (T for stress, and D for dielectric displacement), the piezoelectric constant can be evaluated as a partial derivative at a constant external electric field, E. External stresses may rotate the polar cations and change the polarization direction, inducing a significant change of the polarization intensity along the direction of applied stresses and resulting in a large d_{33} [31].

In this way, researchers have reported several cases of molecular ferroelectric crystals with distinct piezoelectricity by introducing globular or quasi-spherical polar cations for easier polarization rotation under external strains.

The hybrid perovskite $[(CH_3)_4N]MnCl_3$, similar to BNO, has been intensively researched during the pursuit of hybrid perovskite molecular ferroelectrics [32]. Even though undergoing a non-ferroelectric phase transition for the motion of $[(CH_3)_4N]^+$ cation with the T_d molecular group, $[(CH_3)_4N]MnCl_3$, the prototypical system, shows a regulatable prototype for chemical design. Later, the prototypical system was successfully endowed with ferroelectricity and piezoelectricity by You et al., who successfully utilized the quasi-spherical theory in ferroelectrochemistry. After an H atom of $[(CH_3)_4N]^+$ cation was substituted by a Cl atom, a hybrid perovskite molecular ferroelectric $[(CH_3)_3NCH_2Cl]MnCl_3$ was obtained. It undergoes a $6/mmmFm$ phase transition induced by the order–disorder movement of the

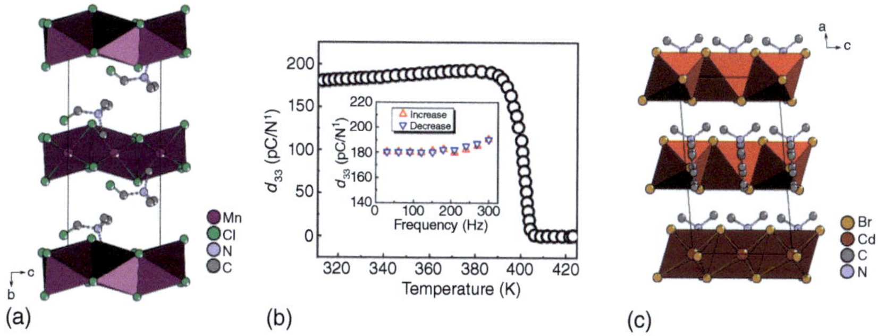

Figure 6.1 (a) Crystal structure of $[(CH_3)_3NCH_2Cl]MnCl_3$ in its ferroelectric phase. (b) d_{33} of $[(CH_3)_3NCH_2Cl]MnCl_3$ as a function of temperature and (inset) frequency. (c) Crystal structure confinement effect for ferroelectric/ferroelastic switching of $[C_6H_5N(CH_3)_3]CdBr_3$. Source: (a,c) Yu-Meng You, (b) You et al. [13]. Reproduced with permission of the American Association for the Advancement of Science.

quasi-spherical $[(CH_3)_3NCH_2Cl]^+$ cation (Figure 6.1a). In a paraelectric phase such as $[(CH_3)_4N]^+$, the $[(CH_3)_3NCH_2Cl]^+$ cation is isotropic and spherical in shape, while in a ferroelectric phase, its disorder is frozen, and its molecular symmetry is reduced from T_d to C_{3v}, leading to ferroelectric polarization [13].

Traditionally, most inorganic ferroelectric crystals are normally grown by the Czochralski method in paraelectric phase at a high temperature. The crystals exhibit multidomain structures and must be processed by high-voltage poling before their piezoelectric and ferroelectric applications. On the contrary, crystals of molecular ferroelectrics, including $[(CH_3)_3NCH_2Cl]MnCl_3$, grew in the ferroelectric phase at room temperature, exhibiting a monodomain structure without poling. Subsequently, a maximum d_{33} of 185 pC/N was measured along the [102] direction of the $[(CH_3)_3NCH_2Cl]MnCl_3$ crystal without poling (Figure 6.1b) [13].

Researchers have analyzed the large piezoelectricity in the $[(CH_3)_3NCH_2Cl]MnCl_3$ crystal for further mechanistic interpretation, proving that it is a fully ferroelectric and partially ferroelastic crystal. When an external stress switches the strain state, the associated polarization state on the same polar axis is likewise changed accordingly. Such polarization rotation actuated an anomalously large piezoelectric response, as described by the above-defining equation [13]. Remarkably, its particular organic–inorganic structure contributes a lot to the piezoelectricity; the electronegative Cl atom and moderate Cl···Cl halogen bonding interactions between inorganic frameworks and organic cations further activate the generation of the piezoelectric response [34]. Besides, first-principles simulations were made to predict that the anomalously large piezoelectric response in $[(CH_3)_3NCH_2Cl]MnCl_3$ and $[(CH_3)_3NCH_2Cl]CdCl_3$ is extrinsic and originated from the stress-driven phase switching between the ground phase and metastable phase, remarkably changing the spontaneous polarization [35].

Currently, the focus of investigations has been turned to explore more excellent functions of piezoelectric crystals, with their ferroelectricity retained. This can be achieved by further engineering through halogen/metal substitution and molecular chemical design. For example, researchers have designed and synthesized three isostructural hybrid perovskite molecular ferroelectric crystals with outstanding piezoelectric responses, $[(CH_3)_3NCH_2Cl]CdCl_3$ [13], $[(CH_3)_3NCH_2Cl]CdBr_3$ [36], and $[(CH_3)_3NCH_2Br]MnBr_3$ [37]. Their d_{33} values are 220, 139, and 112 pC/N, respectively, which approach or even exceed those of $[(CH_3)_3NCH_2Cl]MnCl_3$ and BTO.

Hu et al. demonstrated the $[C_6H_5N(CH_3)_3]CdCl_3$ crystal as a hybrid perovskite molecular ferroelectric with a ferroelastic shear strain of 21.5% [33]. Additionally, they observed a considerable shear piezoelectric coefficient d_{35} of up to 4830 pm/V in the $[C_6H_5N(CH_3)_3]CdBr_{3x}Cl_{3(1-x)}$ solid solution ($x = 0.9$). In the crystal, the confinement effect of the large $C_6H_5N(CH_3)_3$ cations hinders structural switching of 180° polarization (Figure 6.1c). The breaking and reformation of Cd–X bonds facilitate macroscopic ferroelastic switching. Furthermore, Br substitution of Cl effectively softens the Cd–X bonds and lowers the energy barrier of ferroelastic switching. These factors enhance not only intrinsic piezoelectricity but also extrinsic ferroelasticity. This thereby reveals the great potential of molecular ferroelectrics in electromechanical responses [33].

Later, as a continuation of this work, Hu et al. introduced the larger-size iodine to this series in order to weaken the Cd–X bonds effectively and reduce the polarization switching barrier [38]. They obtained the solid solution $[C_6H_5N(CH_3)_3]CdBr_2Cl_{0.75}I_{0.25}$ with excellent piezoelectric constants of $d_{33} = 367$ pm/V and $g_{33} = 3595 \times 10^{-3}$ V m/N. Their work provides a much more significant step forward than the previous ones in terms of practical value. Researchers employed a special confinement method to solve the problem of the time-consuming growth of large single crystals. In this method, they grew a single crystal wafer with a large cross-sectional area of up to 1 cm^2 perpendicular to the d_{33} direction. A giant voltage output of 210 V and a maximum power density of 134 W/m^2 were obtained on an energy harvester composed of a single crystal sandwiched by silver. Crystals with such excellent properties are proven to be promising sources of future soft and wearable piezoelectric devices [38].

To sum up, it remains a challenge for us to rationally design molecular ferroelectric crystals with high-performance intrinsic piezoelectricity. This is largely due to a lack of knowledge of the fundamentals governing the phase transition and piezoelectric response. Many reports have been made concerning two prototypes of hybrid perovskite molecular ferroelectrics that can be classified into quasi-planar five-membered ring and quasi-spherical organic cations [34]. Taking the abovementioned two prototypes as benchmarks, Yang et al. drew some design strategies from their first-principles computation for deeper understanding [34]. First, they summarized that the change from quasi-planar organic cations to quasi-spherical ones can give rise to better piezoelectricity, in accordance with the quasi-spherical theory. Besides, in order to obtain excellent piezoelectricity, it is expected that quasi-spherical organic cations be chemically modified as slightly as possible, but the lattice parameter β of the crystal should be as large as possible during the phase transition. Another two influential factors are how to balance the sizes of tripod and terminal group of quasi-spherical organic cations and how to balance subtle interactions between quasi-spherical organic cations and inorganic frameworks [34].

6.2.2 Piezoelectricity in Crystalline Thin Films

Under the trend toward flexibility and miniaturization during the last decade, booming needs have been emerged for thin-film, small-scale, and on-chip integration of ferroelectrics and piezoelectrics in microelectronics and optoelectronics industries [39]. Thus, the focus has shifted from bulk crystals and ceramics to thin films, leading to the development of multiple inorganic ferroelectric thin films. However, the worrying problem is that the lattice mismatch between substrates and crystalline thin films may result in strain and deteriorate the performance of films. Therefore, the substrate, a determining factor of the film quality, shall be properly selected [40]. In addition to the substrate restriction, fabricating inorganic ferroelectric thin films tends to be quite energy-intensive and time-consuming, with methods such as pulsed laser deposition, chemical vapor deposition, magnetron sputtering, and molecular beam epitaxy [41]. Fortunately, molecular ferroelectric thin films are emerging as viable alternatives or supplements to inorganic ferroelectric thin films for fulfilling practical demands due to their substrate independence as well as solution and low-temperature processing.

Zhang et al. utilized a convenient spin-coating approach to fabricate thin films of the molecular ferroelectric imidazolium perchlorate ($ImClO_4$) (Figure 6.2a) [44].

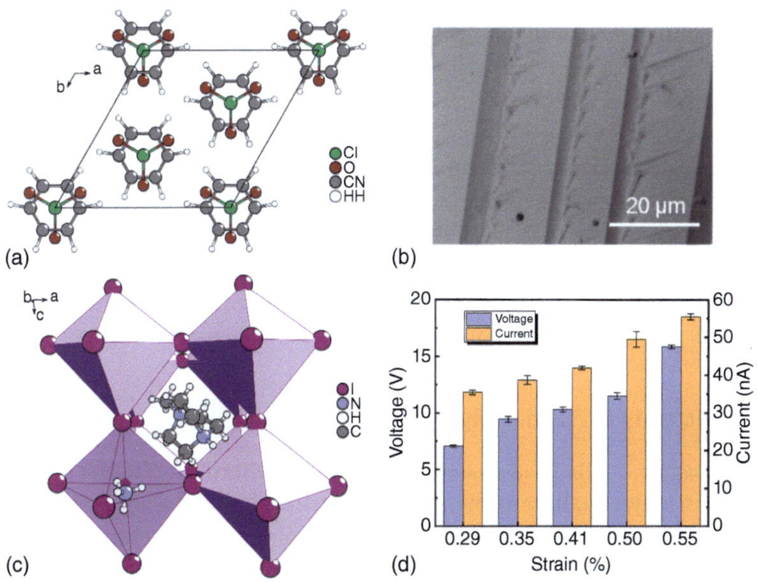

Figure 6.2 (a) Crystal structure of $ImClO_4$ in ferroelectric *R*3*m* phase, in which the imidazolium cation is dynamically disordered with three carbon and two nitrogen atoms indistinguishably distributed over the six vertices of the hexagon. (b) Optical image of the continuous and smooth surface morphology of the $ImClO_4$ thin film produced by in-plane and liquid-phase growth. (c) Crystal structure of $(MDABCO)NH_4I_3$ in ferroelectric *R*3 phase. (d) Output voltage and current performances of the piezoelectric nanogenerator of $(MDABCO)NH_4I_3$ under different applied strains. Source: (a,c) Yu-Meng You, (b) From Zhang et al. [42]/ Reproduced with permission from American Association for the Advancement of Science - AAAS, (d) Wu et al. [43]. Reproduced with permission of Wiley.

Millimeter-scale $ImClO_4$ dendritic crystals can grow on various substrates in a saturated solution, with stems constantly elongating in a variety of favored orientations, much like a vine clinging to a wall. As structured dendritic crystals were produced, and there was insufficient room left to lengthen stems and branches, they turned to continuously extend and amalgamate into a smooth, transparent, and crystalline film [44]. This film had a piezoelectric performance higher than PZT and poly(vinylidene fluoride-co-trifluoroethylene) (P(VDF-TrFE)). Zhang et al. also fabricated large-area $ImClO_4$ thin films (Figure 6.2b) using a novel water-based air-processable technique [42]. Under the conditions of external water partial pressure and temperature gradients, they managed to achieve supersaturation at the liquid–air interface. The translucent and bamboo-like films exhibit adjustable electro-optic activity as well as a reversible electromechanical response [42].

Metal-free organic perovskite molecular ferroelectrics have a variety of appealing properties, including enormous ferroelectric polarization, electro-optic modulation, and nonlinear optics [45–48]. Particularly, a representative compound $(MDABCO)NH_4I_3$ (Figure 6.2c) has a large P_s of 22 μC/cm, higher than that of BTO [49], and a large d_{x5} of 178 pC/N according to first-principles calculations [50]. This lightweight and lead-free piezoelectric material greatly suits the needs of flexible and wearable devices as well as biocompatible and implantable systems. Wu et al. prepared films of $(MDABCO)NH_4I_3$ by using a hot-casting method during the process of preheating/spin-coating/annealing [43]. They further integrated the film into a piezoelectric nanogenerator (PENG) and obtained an optimal output voltage of 15.9 V and a current of 54.5 nA (Figure 6.2d). Moreover, they demonstrated that this PENG can power an light-emitting diode (LED), charge a capacitor, and even function in human–machine interface applications [43].

Researchers applied these universal and practical strategies described above to accurately control the roughness, homogeneity, thickness, crystallinity, and crystal orientation of the crystalline thin film by adjusting the solution concentration and growth environment. This enables researchers to make ferroelectric sensors and energy harvesters with controllable polarization and adaptable piezoelectric.

6.2.3 Piezoelectricity in Plastic Polycrystals

Over the last decade, researchers have attempted to make multiaxial molecular ferroelectrics applied in polycrystalline forms, motivated by the plastic crystal and plastic phase transition [51]. In the plastic phase, globular molecules rotate rapidly and isotropically, and their orientations are disordered, similar to a liquid. However, each molecule retains its three-dimensional long-range positional order, constructing a crystal lattice similar to an ordinary solid crystal. Plastic crystals are those that possess plastic phases within a certain temperature range, although they are not in such phases under ambient conditions [52].

Plastic crystals provide many outstanding properties [53–56]. The most characteristic finding is that the orientationally disordered molecules in the plastic phase occupy more spherical and larger volumes than orientationally ordered ones in ordinary solid crystals. Therefore, the plastic phase usually has highly symmetric structures belonging to the cubic crystal system, together with more isotropic and weaker intermolecular interactions. Contrary to the typical ordinary solid crystals with brittleness and rigidity, plastic crystals exhibit mechanical deformability or so-called plasticity [57]. Besides, plastic crystals commonly undergo one or more solid–solid phase transitions upon cooling. If a crystal loses its orientational disorder during the transition and crystallizes in a polar space group, it may behave as ferroelectric with multiaxial characteristic [58, 59].

This combination of plasticity and multiaxial ferroelectricity contributes a lot to ferroelectric and piezoelectric applications with a low coercive field in bulk polycrystalline forms [58]. When the uniaxial pressure is applied to polycrystalline powders, the high-temperature sintering required by inorganic ferroelectric ceramics is avoided. Each microcrystal can be permanently deformed without fracture, uniformly extruded together, and finally cast into any desired polycrystalline shape of free-standing thin films, molecular ceramics, or something alike [58, 60].

In this section, we summarize the recent development of plastic molecular ferroelectrics, especially focusing on their piezoelectric properties in bulk polycrystalline forms.

$[C(NH_2)_3][ClO_4]$ (guanidinium perchlorate) was discovered as a plastic molecular ferroelectric with multiaxial $m\bar{3}m$F3m phase transition [61, 62]. Its high-temperature paraelectric phase crystallizes in the centrosymmetric cubic $Pm\bar{3}m$, indicating that both guanidinium and perchlorate spherical molecules are isotropically rotating in a plastic phase. Pan et al. fabricated the $[C(NH_2)_3][ClO_4]$ molecular ceramic from microcrystalline powders at room temperature by cold pressing method, superior to the sintering of inorganic ferroelectric ceramics that requires high energy intensity and high temperature. The obtained molecular ceramic shows a significant increase in the piezoelectric coefficient from 0 to 10 pC/N after poling, comparable to that of the LNO single crystal (8 pC/N). Due to the multiaxial polarization, each grain is paralleled to the direction applied in the poling field as closely as possible and displays a larger polarization and higher piezoelectric response [62].

Another family of $[(CH_3)_4N][MCl_4]$ (M = Fe [60] and Ga [63]) was plastic molecular ferroelectrics. Reportedly, they can achieve ferroelectric polarization switching and outstanding piezoelectric response in their free-standing polycrystalline films. Among the family mentioned above, both $[(CH_3)_4N][FeCl_4]$ (Figure 6.3a) and $[(CH_3)_4N][FeBrCl_3]$ synthesized by Harada et al. show several solid–solid phase transitions between their high-temperature plastic phase and low-temperature ferroelectric and piezoelectric phase [60]. Due to their plastic malleability and intrinsic multiaxial ferroelectricity, the translucent films are easily fabricated by pressing their powders, exhibiting d_{33} up to 80 and 110 pC/N, respectively [60]. In addition, Li et al. reported that $[(CH_3)_4N][GaCl_4]$ undergoes a plastic and multiaxial phase transition from cubic $Pm\bar{3}m$ to $Amm2$, and displays piezoelectricity d_{33} measured as 80 pC/N in the powder form after poling [63]. According to the research by Wang et al., an H of $[(CH_3)_4N][GaCl_4]$ was substituted by Cl, and another novel plastic molecular ferroelectric $[(CH_3)_3NCH_2Cl][GaCl_4]$ has been achieved [64]. Later, a multiaxial plastic ferroelectric phase transition occurred,

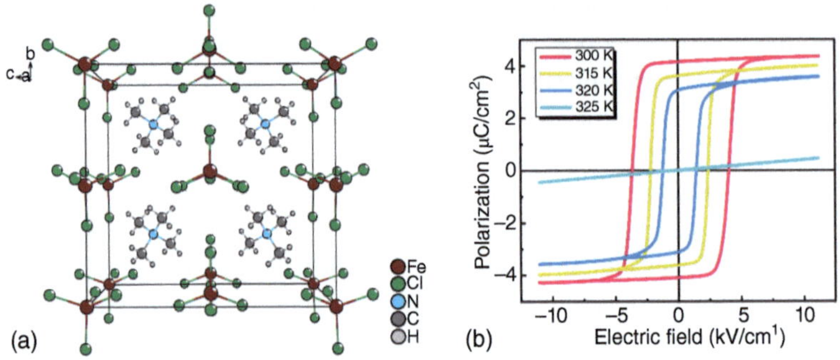

Figure 6.3 (a) Crystal structure of $[(CH_3)_4N][FeCl_4]$ in the paraelectric *Pbcm* phase. (b) *P*–*E* hysteresis loops of polycrystalline films of the plastic molecular ferroelectric $[AH][ReO_4]$. Source: (a) Yu-Meng You, (b) Adapted from Harada et al. [57]

with an Aizu notation of $m\bar{3}mFm$. Researchers prepared a polycrystalline film with a thickness of 250 μm through pressure casting and determined its room temperature g_{33} to be 1318×10^{-3} V m/N and d_{33} to be 226 pC/N, much higher than those of $[(CH_3)_4N][GaCl_4]$. The H/Cl substitution results in a larger dipole moment of $[(CH_3)_3NCH_2Cl]^+$ than that of the $[(CH_3)_4N]^+$ cation, and a weak Cl—Cl halogen-bonding favorable to the order–disorder motion of the $[(CH_3)_3NCH_2Cl]^+$ cation [64].

Typical representatives of plastic molecular ferroelectrics also include a series of quinuclidinium-related compounds, $[Q][XO_4]$ [55, 65, 66] and $[AH][XO_4]$ [57] (Q = quinuclidinium, AH = 1-azabicyclo[2.2.1]heptanium, X = Re, or I). These compounds present outstanding ferroelectric and piezoelectric performance similar to the aforementioned compounds in polycrystalline forms (Figure 6.3b).

6.2.4 Piezoelectricity in Composites

Recently, flexible piezoelectric energy harvesters (PEHs) and flexible pressure sensors have received intensive attention for their wide applications, say, smart and wearable sensors, low-power portable electronics, stretchable and bio-implanted electronic skins, human–machine interfaces, and healthcare monitoring [67–70]. Flexible molecular ferroelectrics with intrinsic piezoelectricity can greatly enhance the malleability and adaptability of these flexible devices due to the simple solution fabrication, low density, and lightweight nature.

However, the fracture tendency of molecular ferroelectric crystals under mechanical stress hinders their applications in bulk crystalline forms. Consequently, researchers attempt to embed molecular ferroelectrics into polymer matrix composites to obtain better mechanical properties, such as toughness, stiffness, and tensile strength.

Molecular ferroelectrics embedded into a polymer matrix are more flexible and stable. The polymer matrix can cushion against the excessive external stress applied on materials to avoid fracture. Besides, the matrix can also protect materials against ambient water and oxygen and offer strong chemical and thermal resistance.

In addition, polymer packages can build up practical devices without large single crystals for molecular ferroelectrics manufactured in tiny crystals, particles, or plastic polycrystalline forms aforementioned.

In this section, we review both the architecture fabrication and performance characterization of recent molecular ferroelectrics@polymer matrix composites for PEHs and sensors. Discussions have also been made on the design of future molecular ferroelectric systems that matter in performance enhancements.

6.2.4.1 PDMS-Based Composites

Polydimethylsiloxane (PDMS) is the most widely used silicon-based organic polymer in prototyped polymeric devices [71]. PDMS is optically transparent, generally inert, nontoxic, and nonflammable. Due to its versatility, its applications are widespread, ranging from medical devices to biomedical microelectromechanical systems. Therefore, researchers focus on PDMS for packaging molecular ferroelectric devices. With PDMS, devices can be fabricated conveniently. During the device fabrication, low-molecular-weight PDMS (liquid) mixed with a cross-linking agent can be heated and cured in a mold to achieve a desirable elastomer in any shape (cross-linked) [72].

Two key factors for the output performance of PEHs are the piezoelectric charge coefficient d_{33} and the piezoelectric voltage coefficient g_{33} [73]. The plastic molecular ferroelectric $[(CH_3)_3NCH_2Cl]GaCl_4$ with excellent d_{33} and g_{33} properties, as mentioned above, was further evaluated for its PEH performance in composite devices. Wang et al. fabricated a series of $[(CH_3)_3NCH_2Cl]GaCl_4$@PDMS composite devices with a thickness of 200 mm by dispersing $[(CH_3)_3NCH_2Cl]GaCl_4$ into PDMS and mechanically agitating the mixture [64]. As heterogeneous materials exhibit Maxwell–Wagner–Sillars-type polarization, the V_{oc} of the 10 wt% composite device peaks at 38.1 V at 1 N and 3 Hz. A 10 wt% sample lit up 18 LEDs, demonstrating outstanding PEH performance and mechanical durability [64].

Huang et al. proposed a molecular ferroelectric composite of $[(CH_3)_3NCH_2Cl]_2SnCl_6$@PDMS PEH [74]. $[(CH_3)_3NCH_2Cl]_2SnCl_6$ with a perovskite structure (Figure 6.4a) possesses a P_s of 8.7 μC/cm^2, a large d_{33} of 137 pC/N, and a g_{33} of 980×10^{-3} V m/N. By a combination of density functional theory (DFT) calculations and experimental investigations, researchers find that the ultrahigh piezoelectric coefficients of $[(CH_3)_3NCH_2Cl]_2SnCl_6$ are synergistically derived from the displacement of inorganic $SnCl_6$ octahedra and rotation of organic molecules. They prepared the piezoelectric composite devices through stirring and dispersing $[(CH_3)_3NCH_2Cl]_2SnCl_6$ nanoparticles evenly into PDMS. The output voltage V_{oc} and current I_{sc} of the PEH device are 81 V and 2 μA, respectively, at an applied mechanical excitation of 4.9 N and 40 Hz. Based on this work, novel molecular ferroelectrics with ultrahigh d_{33} and g_{33} are more likely to be applied in self-powered electronics [74].

Deswal et al. synthesized a 1D ABX_3 molecular ferroelectric $[(CH_3)_3NCH_2CH_2OH]CdCl_3$ with a perovskite-like structure. This molecular ferroelectric is composed of edge-sharing $CdCl_5O$ octahedra connected to the $(CH_3)_3NCH_2CH_2OH$ cation through its Cl···H—O hydrogen bond and Cd—O coordinate covalent bond of the

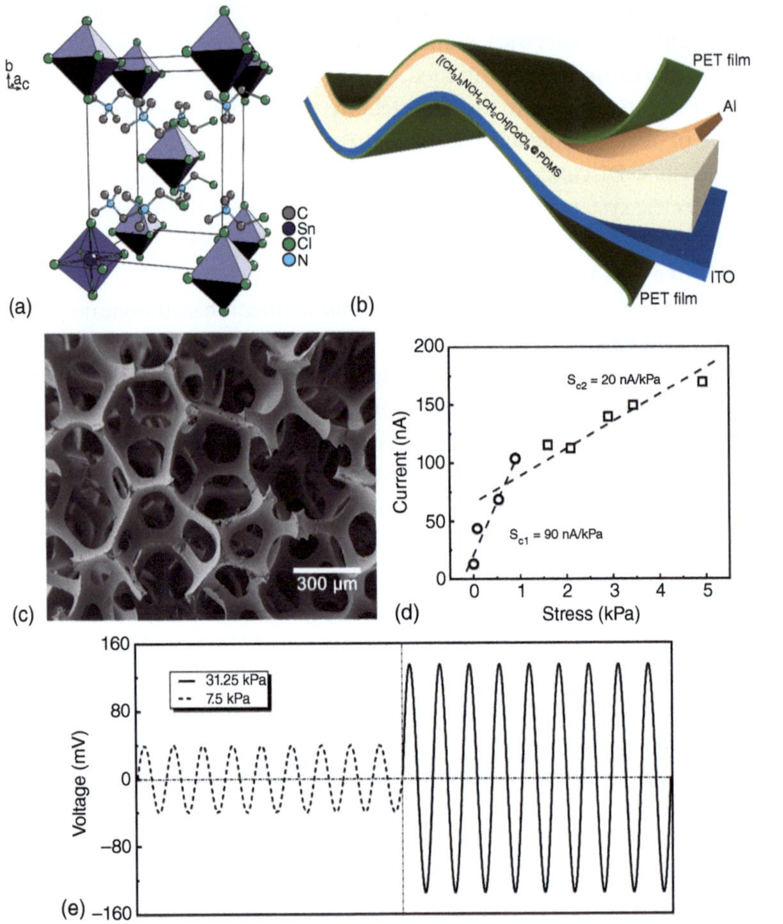

Figure 6.4 (a) Crystal structure of $[(CH_3)_3NCH_2Cl]_2SnCl_6$ in the ferroelectric *Imm*2 phase. (b) Schematic diagram of the as-fabricated flexible $[(CH_3)_3NCH_2CH_2OH]CdCl_3$@PDMS composite device. (c) SEM image of the $[C(NH_2)_3]ClO_4$@PU composite. (d) The sensitivity of the $[C(NH_2)_3]ClO_4$@PU composite characterized by the current under various stresses. (e) The piezoelectric voltage waveforms of the $ImClO_4$@BC hydrogel composite device under various out-of-plane compression pressures. Source: (a,b) Yu-Meng You, (c,d) Li et al. [75]. Reproduced with permission of Wiley, (e) Lu et al. [76]. Reproduced with permission of American Chemical Society.

hydroxyl group [77]. This arrangement facilitates the long-range ordering of the cations and anions along the polar axis and generates a high P_r of 17.1 μC/cm^2. Above 363 K, this compound undergoes a structural phase transition to its coordination isomer accompanied by corresponding bond breaking among the same polar *Pna*2_1 space group. During the phase transition, the framework consists of face-sharing $CdCl_6$ octahedra, and the cations become discrete and uncoordinated. Researchers fabricated and tested its $[(CH_3)_3NCH_2CH_2OH]CdCl_3$@PDMS composite devices (Figure 6.4b) with various weight percentages for PEH applications. The best-performing 15 wt% device obtained the highest V_{oc} of 55.2 V and a maximum power density of 70.9 μW/cm^2 at an optimal load of 4 MΩ. The generated energy can be a source for charging the 10 μF capacitor and lightening a set of 35 LEDs under the conditions of $F = 40$ N and $f = 25$ Hz. These findings encourage the design and development of new types of molecular ferroelectrics@PDMS composites and their utilization in self-powered electronics [77].

6.2.4.2 PU-Based Composites

Polyurethane (PU) is a class of thermoplastic polymers composed of organic units joined by carbamate (urethane) links. It possesses a wide variety of applications because of its high biocompatibility and strong moldability [78, 79]. Researchers take advantage of the high performance of PU to manufacture excellent molecular ferroelectric composites.

Vijayakanth et al. synthesized a ferroelectric organic–inorganic hybrid salt $[Ph_3MeP]_4[Ni(NCS)_6]$ with $P_r = 18.71$ μC/cm^2 [80]. They prepared and embedded this compound of various weight percentages into PU using the solution casting method. The $[Ph_3MeP]_4[Ni(NCS)_6]$@PU composite devices were investigated for PEH applications. A maximum output voltage of 19.29 V and a current of 14 μA was observed for the best performing 15 wt% composite device at an optimized mechanical force of 17 N and 8 Hz. Researchers further utilized this device for charging capacitors and obtained excellent energy storage with the highest stored energy of 198.8 μJ (200 μF) and a charge of 600 μC (1000 μF). Another evidence that shows the flexibility of the composites is that this device can be stretched up to 300% strain without any damage. Meanwhile, the effective stress transfer is compromised under the agglomeration of particles at a higher concentration. This results in the randomization of dipoles and worse performance in a 20 wt% device than a 15 wt% one. Recently, Ramamoorthy Boomishankar and coworkers recently have reported a series of molecular ferroelectrics@PU composites [67, 77, 81]. These discoveries may result in vigorous development of flexible electronics.

The plastic molecular ferroelectric $[C(NH_2)_3]ClO_4$ aforementioned was deposited onto a porous PU sponge by Li et al. to fabricate self-powered flexible mechanical sensors [75]. The $[C(NH_2)_3]ClO_4$ crystal was physically grafted onto the sponge frameworks by the solvent evaporation method at room temperature (Figure 6.4c). As a result, $[C(NH_2)_3]ClO_4$@PU composite, a 3D interconnected molecular ferroelectric has been fabricated, which retains the crystal structure of $[C(NH_2)_3]ClO_4$ and pore structure of the PU sponge. By contrast to the most advanced piezoelectric, triboelectric, and piezoresistive sensors, this composite device can detect strains of 1% and pressures of 3 Pa (Figure 6.4d) [75].

During a proof-of-concept study, researchers constructed a 4 × 4 sensor array with 20 × 20 mm^2 pixels by using $[C(NH_2)_3]ClO_4$@PU composites. They hope to discriminate the mechanical stimuli from various orientations during the research process. The array features great homogeneity, high sensitivity, and super flexibility. It can transform mechanical loads via twisting, bending, pressing, stretching, and shearing into a 2D voltage distribution map without fracture. Reportedly, the array has been able to identify the movements of a robot wrist, as well as monitor and recognize corresponding loading conditions. This research offers a new methodology for developing and utilizing molecule ferroelectrics in applications such as electric skins, conformable transducers, and energy-absorbing smart structures [75].

6.2.4.3 Hydrogel-Based Composites

In addition to artificially synthesized polymers such as PDMS and PU utilized above, researchers also pay attention to natural polysaccharide polymers such as cellulose to construct molecular ferroelectric composites.

Lu et al. reported an environmentally friendly and flexible piezoelectric sensor. It is composed of a bacterial cellulose (BC) hydrogel as the matrix and molecular ferroelectric $ImClO_4$ ($d_{33} = 46$ pC/N) as the functional element [76]. First, researchers incubated and treated *Gluconacetobacter xylinus* culture for purified BC hydrogels with a 3D interconnected network. Later, they embedded $ImClO_4$ homogeneously into the micropore network of BC hydrogels by evaporating the $ImClO_4$ solution. The fabricated $ImClO_4$@BC hydrogel composite formed a self-powered piezoelectric sensor that possesses a high sensitivity of 4 mV/kPa and a wide operational condition from 0.2 to 31.25 kPa (Figure 6.4e). The composites can also be assembled into practical electronic devices. For example, researchers utilized the composite array to fabricate a self-powered paper keyboard. When connected to a data acquisition card, the system allows input and display of letters and symbols on the screen [76].

More interestingly, $ImClO_4$@BC hydrogel composites are biodegradable and recyclable. The BC hydrogel can be thoroughly decomposed by cellulases into glucose and oligosaccharides, which are two small sugar molecules easily absorbed by microorganisms; $ImClO_4$ can be redissolved by water and recycled for environmentally friendly fresh devices. This work proves well that, by combining the simple manufacturing process of natural biomaterials with the advantages of molecular ferroelectrics, a series of recyclable and biodegradable piezoelectric devices can be designed and applied in practice [76].

Nevertheless, research on molecular PEHs and sensors is still in its early stages, with much space for more improvement. For instance, interfacial polarization, namely the interactions between the polymers and piezoelectric particles when utilized in composite forms, is a crucial element affecting the output performance of molecular PEHs and sensors. One way to enhance the output performance is to investigate the systems of such a composite after gaining a greater understanding of its interfacial chemistry.

6.2.5 Piezoelectricity in Molecular Solid Solutions

In 1954, Jaffe et al. developed one effective strategy for optimizing electromechanical response after a maximal piezoelectricity was realized in the PZT ceramic solid solution at the so-called morphotropic phase boundary (MPB) [82]. Recently, this strategy has been first recognized in molecular piezoelectric solid solutions that require a deep understanding of the complex distortions and spatial orientations between molecules. As a result, molecular ferroelectrics would be more promising in applications.

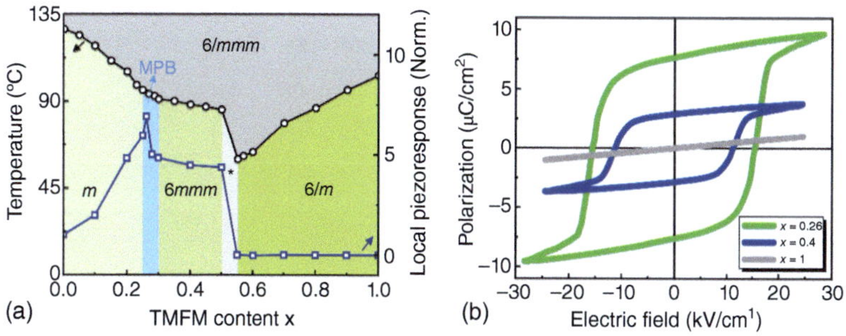

Figure 6.5 (a) Binary phase diagram (left) and local piezoelectric response (right) plotted as a function of the relative concentration x of TMFM cations for the molecular solid solution $(TMFM)_x(TMCM)_{1-x}CdCl_3$ $(0 \leq x \leq 1)$. Point groups for various phases marked on the diagram show the presence of an MPB region in the composition range of $0.25 \leq x \leq 0.3$. (*) Severe compositional segregation without compositionally homogeneous crystals occurs in the region $0.5 < x < 0.55$. (b) $P-E$ hysteresis loops of selected compositions of $(TMFM)_x(TMCM)_{1-x}CdCl_3$ $(0 \leq x \leq 1)$. Source: (a,b) Adapted from Liao et al. [83].

The molecular perovskite solid solution $(TMFM)_x(TMCM)_{1-x}CdCl_3$ $(0 \leq x \leq 1)$ was synthesized and characterized by Liao et al. [83]. They determined the phase diagram related to the x composition and observed an MPB region where both the ferroelectric hexagonal $6mm$ phase and ferroelectric monoclinic m phase coexist within $0.25 \leq x \leq 0.3$ (Figure 6.5a and b). Attributed to the fantastic MPB, a tremendous piezoelectric coefficient of 1540 pC/N was demonstrated for the $x = 0.26$ compound, seven times larger than that of the $(TMCM)CdCl_3$ counterpart (Figure 6.5a). It can be easily applied in flexible and wearable devices due to the attainable room-temperature solution process. Several other molecular piezoelectric solid solution systems have also been constructed, such as $(TMDFM)_x(TMCM)_{1-x}CdCl_3$ and $(TMFM)_x(TMCM)_{1-x}MnCl_3$. Both systems have similar phase diagrams and exhibit dramatic increases in piezoelectricity near an MPB region, demonstrating the universality of the MPB strategy [83].

6.2.6 Piezoelectricity in Nanoparticles

The first piezoelectric nanogenerator reported on ZnO nanowires was developed in 2006 [84]. Since then, developing novel piezoelectric nanoparticles has been promising for energy harvesters, self-powered actuators, and sensors due to their inherent electromechanical coupling and higher piezoelectric coefficients [85, 86]. Meanwhile, molecular ferroelectrics, as promising alternatives to piezoelectrics, have been explored for nano-piezoelectric applications.

A novel molecular ferroelectric $(ATHP)_2PbBr_2Cl_2$ (ATHP = 4-aminotetrahydropyran) was reported to exhibit superior transverse piezoelectricity by Khan et al. [87, 88]. This crystal has a 2D layered perovskite structure, where amine groups of ATHP cations are bonded to the PbX_6 (X = Cl or Br) octahedra by X···H—N hydrogen bonds (Figure 6.6a). Researchers adopted a modified ligand-assisted reprecipitation technique and successfully synthesized the nano-piezoelectric

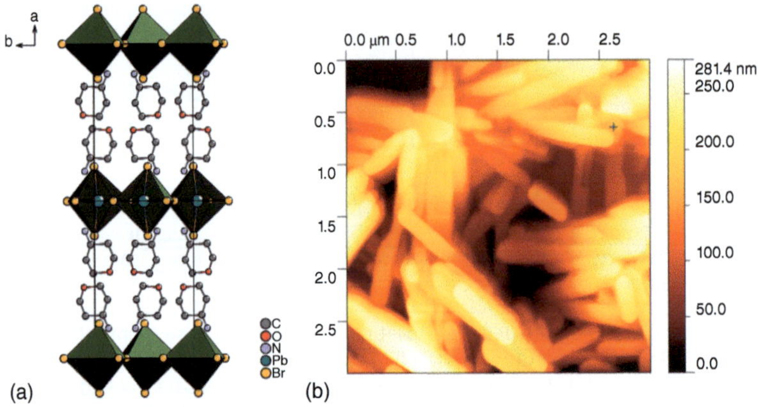

Figure 6.6 (a) Crystal structure of $(ATHP)_2PbBr_2$ in the ferroelectric $Cmc2_1$ phase. (b) AFM image of $(ATHP)_2PbBr_2$ nanorods. Source: (a) Yu-Meng You. (b) From Khan et al. [87]. Reproduced with permission from Elsevier.

nanorods of $(ATHP)_2PbBr_2Cl_2$ at room temperature. More specifically, due to a lower concentration of octylamine, the surface of the perovskite nuclei was partially covered, favoring the growth of nanorod-type structure. Interestingly, researchers determined the d_{31} of nanorods to be 64 pC/N based on the lateral PFM amplitude and found that these nanorods have a longer length suitable for bending under certain stresses (Figure 6.6b). Whereas these characteristics exist, they fabricated a flexible composite film $(ATHP)_2PbBr_2Cl_2$@PDMS for PENG, with a V_{oc} of 90 V and I_{sc} of 6.5 μA. This work not only demonstrates the large transverse piezoelectricity characteristic of molecular ferroelectric $(ATHP)_2PbBr_2Cl_2$ nanorods but also opens a way for practical applications of PENG devices [87].

6.3 PFM-Aided Characterization and Design

It is urgently required that molecular ferroelectrics with excellent piezoelectric performances be precisely designed or modulated for practical applications. Consequently, several methods are invented to detect piezoelectric coefficients, including the static, quasi-static, and dynamic methods [89, 90]. These methods, though bringing precise measurement, have limitations on sample preparation, particularly in terms of the orientation, shape, and size of samples and the configuration of electrodes. In other words, respective theoretical specifications must be soundly adhered to [91, 92].

Nowadays, PFM has been demonstrated as an alternative approach for evaluating piezoelectricity and ferroelectricity [7]. This technique is believed to provide a non-destructive visualization with high-resolution imaging at micro- and nanoscale levels [93]. Naturally, molecular ferroelectrics have developed dramatically, with explosive growth in recent years. Accordingly, this section reviews the numerous contributions of PFM in assisting the characterization and design of molecular ferroelectrics with excellent piezoelectricity.

PFM allows researchers to evaluate the relative piezoelectric coefficients and further compare the piezoelectric response across various samples. PFM, with simple film manufacture, offers a feasible approach to quickly and effectively construct molecular ferroelectrics. The mechanism of PFM is based on the converse piezoelectric effect, which can induce material vibrations under an applied alternating current (AC) bias with specific frequencies [94]. Therefore, apart from the fundamental atomic force microscopy (AFM) system, PFM additionally requires a functional generator for generating an AC oscillation voltage on the conductive tip (V_{tip}) and requires a lock-in amplifier (i.e. PFM signal) for detecting the resulting motions of the cantilever [95]. The resonance enhancement is used to amplify and distinguish small PFM signals from noise. A quality factor (Q) is introduced to accurately describe and calibrate energy losses. Subsequently, the PFM amplitude signal (A), in proportion to the piezoelectric coefficient d_{33}, can be described as the following relationship: $A = d_{33}V_{tip}Q$ [96]. This is where the quality factor method of PFM comes in. The simple and quantitative method helps systematically compare the local piezoelectric response and determine the d_{33} of various samples [95–98].

A case is the construction of the molecular ferroelectric MPB. Under the quality factor method, the piezoelectric performance of the molecular solid solution $(TMFM)_x(TMCM)_{1-x}CdCl_3$ was evaluated to measure the d_{33} value within various compositions. Researchers first drove the contact resonance at selected points of each thin-film sample and then fitted the resonance curves for the Q. Finally, the relative local piezoelectric response was calculated. A distinct linearity between the intrinsic piezoelectric response and various V_{tip} was observed, excluding the effect of electrostriction. As shown in Figure 6.5a, researchers systematically compared the maximum local piezoelectric response for each composition, using the reference value of $(TMCM)CdCl_3$ ($x = 0$) as a standard (220 pC/N). It was demonstrated that the largest piezoresponse found at $x = 0.26$ in the vicinity of MPB was seven times larger than that of the standard [83]. Such a tremendous d_{33} verified by the quasi-static method on single crystals will undoubtedly be a revolutionary breakthrough in next-generation piezoelectric devices.

Besides, the quality factor method has also been performed to design a series of molecular ferroelectrics with a large local piezoelectric response through PFM, e.g. $(PFBA)_2PbBr_4$ [99], $(DMAA)CdCl_3$ [100], and $(RM3HQ)_2RbLa(NO_3)_6$ [101]. The aim is to accelerate the construction of molecular ferroelectrics and enhance their piezoelectricity.

Zhang et al. utilized PFM to investigate the piezoelectric response for a series of selectively fluorinated 2D perovskites [99]. Their thin films were all excited by increasing V_{tip} near their own resonant frequencies in vertical PFM mode. By comparing the slopes of the drive voltage versus amplitude curves, $(PFBA)_2PbBr_4$ thin films showed the strongest piezoelectric response, which was nearly two and three times larger than $(BA)_2PbBr_4$ and $(2\text{-}FBA)_2PbBr_4$, respectively. In this study, PFM was presented as an effective method for rationally improving the piezoelectricity of molecular ferroelectrics under the guidance of H/F substitution [99].

Wang et al. measured the local transverse piezoelectric response of the thin film of the molecular ferroelectric $(DMAA)CdCl_3$ (DMAA = *N*,*N*-dimethylallyl-

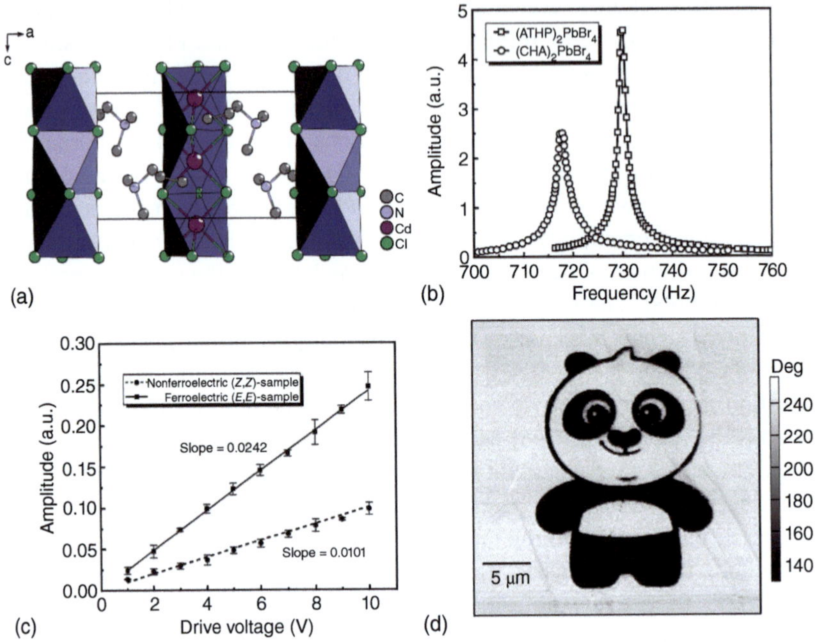

Figure 6.7 (a) Crystal structure of $(DMAA)CdCl_3$ in the ferroelectric $Pna2_1$ phase. (b) Comparison of PFM resonance peaks between $(ATHP)_2PbBr_4$ and $(CHA)_2PbBr_4$ thin films. (c) Comparison of the vertical amplitude versus the drive voltage for nonferroelectric (*Z*, *Z*) and ferroelectric (*E*, *E*) samples. (d) PFM phase image shows a panda-like domain pattern which was specially pre-defined and precisely lithographed. Source: (a) Yu-Meng You. (b) Chen et al. [88]. Reproduced with permission of American Chemical Society. (c,d) Zhang et al. [101]. Reproduced with permission from John Wiley & Sons.

ammonium) with a 1D perovskite structure (Figure 6.7a) in lateral PFM mode [100]. They first fabricated the thin film by the spinning method and demonstrated its ferroelectricity by lateral ferroelectric domain reversal. The well-confirmed lateral polarization produces the transverse piezoelectric response d_{31}. The d_{31} value can be detected by monitoring the torsional motion of the tip cantilever in lateral PFM mode. First, they drove both $(DMAA)CdCl_3$ and polyvinylidene difluoride (PVDF) thin films under different V_{tip}. Then they plotted the local effective amplitude calibrated by the quality factor versus the drive voltage curves. The slope of $(DMAA)CdCl_3$ was about two times larger than that of PVDF, indicating that the d_{31} of $(DMAA)CdCl_3$ film is about 41 pC/N. In this case, PFM provides a powerful tool for characterizing transverse piezoelectricity in practical applications [100].

Shi et al. utilized PFM imaging to demonstrate that paraelectric nonpolar regions and ferroelectric polar domains can coexist throughout a broad temperature range in the molecular ferroelectric $(RM3HQ)_2RbLa(NO_3)_6$ (RM3HQ = R-*N*-methyl-3-hydroxylquinuclidinium) [101]. This coexistence in such a rare-earth double perovskite shows that a two-phase transformation or transformation between ferroelectric domains with different orientations can likely be achieved due to low energy barriers. These factors result in the easy polarization rotation under stresses.

Accordingly, a significant $d_{33} = 106$ pC/N was confirmed by the quasi-static method on crystals. Besides, they qualitatively measured the local piezoelectric response with PFM and derived a linear series of the resonance peak values attributed to the intrinsic piezoelectric response. This discovery shows the great potential of PFM in piezoelectric imaging [101].

Similarly, Chen et al. utilized the PFM technique to compare the piezoelectric response between two 2D organic–inorganic hybrid perovskite molecular ferroelectrics by the quality factor method [88]. With the reported $(CHA)_2PbBr_4$ as a benchmark, they evaluated the d_{33} value of $(ATHP)_2PbBr_4$ as 76 pC/N (Figure 6.7b), and a giant g_{33} of 660.3 V m/N. The findings are critical to the high-performance sensor because of the relatively small dielectric constant of $(ATHP)_2PbBr_4$ [88].

Meanwhile, Zhang et al. introduced homochirality and Z/E photoisomerization and constructed the organic enantiomorphic ferroelectrics di(benzylamino)-substituted derivative of muconic acids. A lithographic process was conducted via PFM [102]. They applied a local V_{tip} to precisely lithograph desired domains with predefined patterns (Figure 6.7d). Additionally, they obtained the linear intrinsic piezoelectric response versus the driving voltage and demonstrated that the piezoelectric response of the ferroelectric (E, E)-sample is 2.4 times larger than that of the nonferroelectric (Z, Z)-sample (Figure 6.7c). This indicates why the majority of high-performance piezoelectrics are ferroelectrics by PFM [102].

6.4 Conclusions

This chapter reviewed the progress of molecular ferroelectrics with excellent piezoelectricity in terms of their performance enhancement and application exploration. Their applications are widespread, including single crystals, crystalline thin films, multiaxial plastic polycrystalline forms, composite devices, MPBs, and nanoparticles. Some recent advances achieved by PFM have also been discussed. The piezoelectric research of molecular ferroelectrics experiences a curiosity-to-application process. We anticipate that these findings can have a significant impact on the rational design and boost the widespread deployment of molecular ferroelectrics with good piezoelectricity. However, for such a novel interdisciplinary research, there is much space for discovery, and there are many hidden challenges to overcome.

First, a more clarified study is required to understand the mechanisms of how organic cations, as well as metal and halogen elements of molecular ferroelectrics, affect piezoelectricity. How much these factors contribute to the piezoelectric coefficient is still unknown. This urgently calls for a theory that can be used to quantitatively predict piezoelectric properties.

Most piezoelectric applications based on molecular ferroelectrics mentioned above are still far from industrial production and practical application because of their piezoelectric response and power output, comparably underperforming those of inorganic piezoelectric ceramics. As a result, it is first and foremost expected that novel molecule ferroelectrics with much greater piezoelectric coefficients and more polar axes can be developed, thereby helping improve the device's performance.

The problem is that a sufficiently reliable strategy that can guide us in designing such materials is still unavailable. A series of excellently reported piezoelectrics have not yet been evaluated for their application performances due to many factors, such as the time-consuming growth of single crystals. The abovementioned contradictions should be properly resolved.

In composite devices, it is necessary for us to understand the interfacial chemistry of polymer composites because the existence of interfacial polarization between the polymer matrix and piezoelectric particles can significantly affect the output performance. Besides, given that this heterogeneous contact and interface exist, the output source of a device has not been clarified in many application research studies. It still remains a puzzle for us to determine whether the intrinsic piezoelectricity of materials or external triboelectricity between materials is the output source.

Some sensitive molecular ferroelectrics may decompose when exposed to moisture, oxygen, light, and thermal fluctuations because of the active hygroscopic amine groups. To address this, polymer packaging devices were constructed, but the stability of the devices was largely compromised over time due to the absence of surface charge regeneration. Long-term durability and stability remain crucial concerns for the effective deployment of these novel molecular ferroelectrics in devices.

Last but not least, Long et al. recently carried out PFM measurements and investigated magnetoelectric coupling in a molecular ferroelectric complex [103, 104]. This work reveals a new discovery in molecular ferroelectrics, that is, the in situ visualized PFM imaging under the conditions of electric, magnetic, or optical fields can easily obtain the coupling between piezoelectricity and other multifunctionality. This motivates researchers to utilize PFM as a powerful technique for detecting and further designing multifunctional molecular ferroelectrics.

To sum up, we believe these advancements will promote the revitalization of the piezoelectric industry and bring molecular ferroelectrics considerably closer to appealing applications in piezoelectric, micromechanical, and biomechanical devices.

References

1 Valasek, J. (1921). Piezo-electric and allied phenomena in Rochelle salt. *Phys. Rev.* 17 (4): 475–481.

2 Jaffe, H. (1958). Piezoelectric ceramics. *J. Am. Ceram. Soc.* 41 (11): 494–498.

3 Wu, J., Xiao, D., and Zhu, J. (2015). Potassium–sodium niobate lead-free piezoelectric materials: past, present, and future of phase boundaries. *Chem. Rev.* 115 (7): 2559–2595.

4 Qiu, C., Wang, B., Zhang, N. et al. (2020). Transparent ferroelectric crystals with ultrahigh piezoelectricity. *Nature* 577 (7790): 350–354.

5 Zhang, W. and Xiong, R.-G. (2012). Ferroelectric metal–organic frameworks. *Chem. Rev.* 112 (2): 1163–1195.

6 Shi, P.-P., Tang, Y.-Y., Li, P.-F. et al. (2016). Symmetry breaking in molecular ferroelectrics. *Chem. Soc. Rev.* 45 (14): 3811–3827.

7 Zhang, H.-Y., Chen, X.-G., Tang, Y.-Y. et al. (2021). PFM (piezoresponse force microscopy)-aided design for molecular ferroelectrics. *Chem. Soc. Rev.* 50 (14): 8248–8278.

8 Yao, J., Pan, Q., Feng, Z.-J. et al. (2021). Hybrid organic–inorganic perovskite ferroelectrics bring light to semiconducting applications: bandgap engineering as a starting point. *APL Mater.* 9 (4): 040901.

9 Zhang, H.-Y., Tang, Y.-Y., Shi, P.-P. et al. (2019). Toward the targeted design of molecular ferroelectrics: modifying molecular symmetries and homochirality. *Acc. Chem. Res.* 52 (7): 1928–1938.

10 Liu, H.-Y., Zhang, H.-Y., Chen, X.-G. et al. (2020). Molecular design principles for ferroelectrics: ferroelectrochemistry. *J. Am. Chem. Soc.* 142 (36): 15205–15218.

11 Horiuchi, S., Tokunaga, Y., Giovannetti, G. et al. (2010). Above-room-temperature ferroelectricity in a single-component molecular crystal. *Nature* 463 (7282): 789–792.

12 Fu, D.-W., Cai, H.-L., Liu, Y. et al. (2013). Diisopropylammonium bromide is a high-temperature molecular ferroelectric crystal. *Science* 339 (6118): 425–428.

13 You, Y.-M., Liao, W.-Q., Zhao, D. et al. (2017). An organic-inorganic perovskite ferroelectric with large piezoelectric response. *Science* 357 (6348): 306–309.

14 Lv, H.-P., Liao, W.-Q., You, Y.-M. et al. (2022). Inch-size molecular ferroelectric crystal with a large electromechanical coupling factor on par with barium titanate. *J. Am. Chem. Soc.* 144 (48): 22325–22331.

15 Mu, X., Zhang, H.-Y., Xu, L. et al. (2021). Ferroelectrochemistry. *APL Mater.* 9 (5): 051112.

16 Xiong, Y.-A., Sha, T.-T., Pan, Q. et al. (2019). A nickel(II) nitrite based molecular perovskite ferroelectric. *Angew. Chem. Int. Ed.* 58 (26): 8857–8861.

17 Li, P.-F., Liao, W.-Q., Tang, Y.-Y. et al. (2019). Organic enantiomeric high-Tc ferroelectrics. *Proc. Natl. Acad. Sci.* 116 (13): 5878–5885.

18 Sha, T.-T., Xiong, Y.-A., Pan, Q. et al. (2019). Fluorinated 2D lead iodide perovskite ferroelectrics. *Adv. Mater.* 31 (30): 1901843.

19 Ai, Y., Lv, H.-P., Wang, Z.-X. et al. (2021). H/F substitution for advanced molecular ferroelectrics. *Trends Chem.* 3 (12): 1088–1099.

20 Tayi, A.S., Shveyd, A.K., Sue, A.C.H. et al. (2012). Room-temperature ferroelectricity in supramolecular networks of charge-transfer complexes. *Nature* 488 (7412): 485–489.

21 Tang, Y.-Y., Liu, J.-C., Zeng, Y.-L. et al. (2021). Optical control of polarization switching in a single-component organic ferroelectric crystal. *J. Am. Chem. Soc.* 143 (34): 13816–13823.

22 Liao, W.-Q., Zeng, Y.-L., Tang, Y.-Y. et al. (2021). Multichannel control of multiferroicity in single-component homochiral organic crystals. *J. Am. Chem. Soc.* 143 (51): 21685–21693.

23 Li, F., Lin, D., Chen, Z. et al. (2018). Ultrahigh piezoelectricity in ferroelectric ceramics by design. *Nat. Mater.* 17 (4): 349–354.

24 Li, F., Cabral, M.J., Xu, B. et al. (2019). Giant piezoelectricity of Sm-doped Pb $(Mg_{1/3}Nb_{2/3})$ O_3-$PbTiO_3$ single crystals. *Science* 364 (6437): 264–268.

25 Park, S.-E. and Shrout, T.R. (1997). Ultrahigh strain and piezoelectric behavior in relaxor based ferroelectric single crystals. *J. Appl. Phys.* 82 (4): 1804–1811.

26 Saito, Y., Takao, H., Tani, T. et al. (2004). Lead-free piezoceramics. *Nature* 432 (7013): 84–87.

27 Zheng, T., Wu, J., Xiao, D. et al. (2018). Recent development in lead-free perovskite piezoelectric bulk materials. *Prog. Mater. Sci.* 98: 552–624.

28 Wu, J. (2020). Perovskite lead-free piezoelectric ceramics. *J. Appl. Phys.* 127 (19): 190901.

29 Pan, Q., Xiong, Y.-A., Sha, T.-T. et al. (2021). Recent progress in the piezoelectricity of molecular ferroelectrics. *Mater. Chem. Front.* 5 (1): 44–59.

30 Leng, K., Li, R., Lau, S.P. et al. (2021). Ferroelectricity and Rashba effect in 2D organic–inorganic hybrid perovskites. *Trends Chem.* 3 (9): 716–732.

31 Jaffe, B., Cook, W.R., and Jaffe, H. (1971). *Piezoelectric Ceramics*. London: Academic Press.

32 Morosin, B. and Graber, E. (1967). Crystal structure of tetramethylammonium manganese (II) chloride. *Acta Crystallogr.* 23 (5): 766–770.

33 Hu, Y., You, L., Xu, B. et al. (2021). Ferroelastic-switching-driven large shear strain and piezoelectricity in a hybrid ferroelectric. *Nat. Mater.* 20 (5): 612–617.

34 Yang, D., Luo, L., Gao, Y. et al. (2019). Rational design of one-dimensional hybrid organic–inorganic perovskites with room-temperature ferroelectricity and strong piezoelectricity. *Mater. Horiz.* 6 (7): 1463–1473.

35 Ghosh, P.S., Lisenkov, S., and Ponomareva, I. (2020). Phase switching as the origin of large piezoelectric response in organic-inorganic perovskites: a first-principles study. *Phys. Rev. Lett.* 125 (20): 207601.

36 Liao, W.-Q., Tang, Y.-Y., Li, P.-F. et al. (2018). Competitive halogen bond in the molecular ferroelectric with large piezoelectric response. *J. Am. Chem. Soc.* 140 (11): 3975–3980.

37 Liao, W.-Q., Tang, Y.-Y., Li, P.-F. et al. (2017). Large piezoelectric effect in a lead-free molecular ferroelectric thin film. *J. Am. Chem. Soc.* 139 (49): 18071–18077.

38 Hu, Y., Parida, K., Zhang, H. et al. (2022). Bond engineering of molecular ferroelectrics renders soft and high-performance piezoelectric energy harvesting materials. *Nat. Commun.* 13 (1): 5607.

39 Damjanovic, D. (1998). Ferroelectric, dielectric and piezoelectric properties of ferroelectric thin films and ceramics. *Rep. Prog. Phys.* 61 (9): 1267–1324.

40 Hwang, K.S., Manabe, T., Nagahama, T. et al. (1999). Effect of substrate material on the crystallinity and epitaxy of $Pb(Zr,Ti)O_3$ thin films. *Thin Solid Films* 347 (1): 106–111.

41 Mansingh, A. (1990). Fabrication and applications of piezo-and ferroelectric films. *Ferroelectrics* 102 (1): 69–84.

42 Zhang, Z., Li, P.-F., Tang, Y.-Y. et al. (2017). Tunable electroresistance and electro-optic effects of transparent molecular ferroelectrics. *Sci. Adv.* 3 (8): e1701008.

43 Wu, H.-S., Wei, S.-M., Chen, S.-W. et al. (2022). Metal-free perovskite piezoelectric nanogenerators for human–machine interfaces and self-powered electrical stimulation applications. *Adv. Sci.* 9 (18): 2105974.

44 Zhang, Y., Liu, Y., Ye, H.-Y. et al. (2014). A molecular ferroelectric thin film of imidazolium perchlorate that shows superior electromechanical coupling. *Angew. Chem. Int. Ed.* 53 (20): 5064–5068.

45 Gao, Y., Meshkat, S., Johnston, A. et al. (2021). Electro-optic modulation using metal-free perovskites. *ACS Appl. Mater. Interfaces* 13 (16): 19042–19047.

46 Wang, J.-J., Fortino, D., Wang, B. et al. (2020). Extraordinarily large electrocaloric strength of metal-free perovskites. *Adv. Mater.* 32 (7): 1906224.

47 Song, X., Hodes, G., Zhao, K. et al. (2021). Metal-free organic halide perovskite: a new class for next optoelectronic generation devices. *Adv. Energy Mater.* 11 (11): 2003331.

48 Bie, J., Yang, D.-B., Ju, M.-G. et al. (2021). Molecular design of three-dimensional metal-free A(NH4)X3 perovskites for photovoltaic applications. *JACS Au* 1 (4): 475–483.

49 Ye, H.-Y., Tang, Y.-Y., Li, P.-F. et al. (2018). Metal-free three-dimensional perovskite ferroelectrics. *Science* 361 (6398): 151–155.

50 Wang, H., Liu, H., Zhang, Z. et al. (2019). Large piezoelectric response in a family of metal-free perovskite ferroelectric compounds from first-principles calculations. *npj Comput. Mater.* 5 (1): 17.

51 Tang, Y.-Y., Li, P.-F., Liao, W.-Q. et al. (2018). Multiaxial molecular ferroelectric thin films bring light to practical applications. *J. Am. Chem. Soc.* 140 (26): 8051–8059.

52 Das, S., Mondal, A., and Reddy, C.M. (2020). Harnessing molecular rotations in plastic crystals: a holistic view for crystal engineering of adaptive soft materials. *Chem. Soc. Rev.* 49 (24): 8878–8896.

53 Li, B., Kawakita, Y., Ohira-Kawamura, S. et al. (2019). Colossal barocaloric effects in plastic crystals. *Nature* 567 (7749): 506–510.

54 Lloveras, P., Aznar, A., Barrio, M. et al. (2019). Colossal barocaloric effects near room temperature in plastic crystals of neopentylglycol. *Nat. Commun.* 10 (1): 1803.

55 Harada, J., Shimojo, T., Oyamaguchi, H. et al. (2016). Directionally tunable and mechanically deformable ferroelectric crystals from rotating polar globular ionic molecules. *Nat. Chem.* 8 (10): 946–952.

56 Gao, H., Xue, L., Xin, S. et al. (2017). A plastic–crystal electrolyte interphase for all-solid-state sodium batteries. *Angew. Chem. Int. Ed.* 56 (20): 5541–5545.

57 Harada, J., Kawamura, Y., Takahashi, Y. et al. (2019). Plastic/ferroelectric crystals with easily switchable polarization: low-voltage operation, unprecedentedly high pyroelectric performance, and large piezoelectric effect in polycrystalline forms. *J. Am. Chem. Soc.* 141 (23): 9349–9357.

58 Harada, J. (2021). Plastic/ferroelectric molecular crystals: ferroelectric performance in bulk polycrystalline forms. *APL Mater.* 9 (2): 020901.

59 Ye, H.-Y., Ge, J.-Z., Tang, Y.-Y. et al. (2016). Molecular ferroelectric with most equivalent polarization directions induced by the plastic phase transition. *J. Am. Chem. Soc.* 138 (40): 13175–13178.

60 Harada, J., Yoneyama, N., Yokokura, S. et al. (2018). Ferroelectricity and piezoelectricity in free-standing polycrystalline films of plastic crystals. *J. Am. Chem. Soc.* 140 (1): 346–354.

61 Szafrański, M. (2011). Simple guanidinium salts revisited: room–temperature ferroelectricity in hydrogen-bonded supramolecular structures. *J. Phys. Chem. B* 115 (27): 8755–8762.

62 Pan, Q., Liu, Z.-B., Zhang, H.-Y. et al. (2017). A molecular polycrystalline ferroelectric with record-high phase transition temperature. *Adv. Mater.* 29 (29): 1700831.

63 Li, D., Zhao, X.-M., Zhao, H.-X. et al. (2018). Construction of magnetoelectric composites with a large room-temperature magnetoelectric response through molecular–ionic ferroelectrics. *Adv. Mater.* 30 (52): 1803716.

64 Wang, B., Hong, J., Yang, Y. et al. (2022). Achievement of a giant piezoelectric coefficient and piezoelectric voltage coefficient through plastic molecular-based ferroelectric materials. *Matter* 5 (4): 1296–1304.

65 You, Y.-M., Tang, Y.-Y., Li, P.-F. et al. (2017). Quinuclidinium salt ferroelectric thin-film with duodecuple-rotational polarization-directions. *Nat. Commun.* 8 (1): 14934.

66 Tang, Y.-Y., Li, P.-F., Shi, P.-P. et al. (2017). Visualization of room-temperature ferroelectricity and polarization rotation in the thin film of quinuclidinium perrhenate. *Phys. Rev. Lett.* 119 (20): 207602.

67 Vijayakanth, T., Liptrot, D.J., Gazit, E. et al. (2022). Recent advances in organic and organic–inorganic hybrid materials for piezoelectric mechanical energy harvesting. *Adv. Funct. Mater.* 32 (17): 2109492.

68 Hwang, G.-T., Byun, M., Jeong, C.K. et al. (2015). Flexible piezoelectric thin-film energy harvesters and nanosensors for biomedical applications. *Adv. Healthc. Mater.* 4 (5): 646–658.

69 Park, K.-I., Son, J.H., Hwang, G.-T. et al. (2014). Highly-efficient, flexible piezoelectric pzt thin film nanogenerator on plastic substrates. *Adv. Mater.* 26 (16): 2514–2520.

70 Sezer, N. and Koç, M. (2021). A comprehensive review on the state-of-the-art of piezoelectric energy harvesting. *Nano Energy* 80: 105567.

71 Mata, A., Fleischman, A.J., and Roy, S. (2005). Characterization of polydimethylsiloxane (pdms) properties for biomedical micro/nanosystems. *Biomed. Microdevices* 7 (4): 281–293.

72 Qi, D., Zhang, K., Tian, G. et al. (2021). Stretchable electronics based on PDMS substrates. *Adv. Mater.* 33 (6): 2003155.

73 Liu, H., Zhong, J., Lee, C. et al. (2018). A comprehensive review on piezoelectric energy harvesting technology: materials, mechanisms, and applications. *Appl. Phys. Rev.* 5 (4): 041306.

74 Huang, G., Khan, A.A., Rana, M.M. et al. (2021). Achieving ultrahigh piezoelectricity in organic–inorganic vacancy-ordered halide double perovskites for mechanical energy harvesting. *ACS Energy Lett.* 6 (1): 16–23.

75 Li, W., Li, C., Zhang, G. et al. (2021). Molecular ferroelectric-based flexible sensors exhibiting supersensitivity and multimodal capability for detection. *Adv. Mater.* 33: 2104107.

76 Lu, J., Hu, S., Li, W. et al. (2022). A biodegradable and recyclable piezoelectric sensor based on a molecular ferroelectric embedded in a bacterial cellulose hydrogel. *ACS Nano* 16 (3): 3744–3755.

77 Deswal, S., Singh, S.K., Pandey, R. et al. (2020). Neutral 1D perovskite-type abx3 ferroelectrics with high mechanical energy harvesting performance. *Chem. Mater.* 32 (19): 8333–8341.

78 Ying, W.B., Yu, Z., Kim, D.H. et al. (2020). Waterproof, highly tough, and fast self-healing polyurethane for durable electronic skin. *ACS Appl. Mater. Interfaces* 12 (9): 11072–11083.

79 Li, Z., Zhang, R., Moon, K.-S. et al. (2013). Highly conductive, flexible, polyurethane-based adhesives for flexible and printed electronics. *Adv. Funct. Mater.* 23 (11): 1459–1465.

80 Vijayakanth, T., Ram, F., Praveenkumar, B. et al. (2020). Piezoelectric energy harvesting from a ferroelectric hybrid salt [Ph3MeP]4[Ni(NCS)6] embedded in a polymer matrix. *Angew. Chem. Int. Ed.* 59 (26): 10368–10373.

81 Vijayakanth, T., Ram, F., Praveenkumar, B. et al. (2019). All-organic composites of ferro- and piezoelectric phosphonium salts for mechanical energy harvesting application. *Chem. Mater.* 31 (15): 5964–5972.

82 Jaffe, B., Roth, R.S., and Marzullo, S. (1954). Piezoelectric properties of lead zirconate-lead titanate solid-solution ceramics. *J. Appl. Phys.* 25 (6): 809–810.

83 Liao, W.-Q., Zhao, D., Tang, Y.-Y. et al. (2019). A molecular perovskite solid solution with piezoelectricity stronger than lead zirconate titanate. *Science* 363 (6432): 1206.

84 Wang, Z.L. and Song, J. (2006). Piezoelectric nanogenerators based on zinc oxide nanowire arrays. *Science* 312 (5771): 242–246.

85 Lu, L., Ding, W., Liu, J. et al. (2020). Flexible PVDF based piezoelectric nanogenerators. *Nano Energy* 78: 105251.

86 Hu, D., Yao, M., Fan, Y. et al. (2019). Strategies to achieve high performance piezoelectric nanogenerators. *Nano Energy* 55: 288–304.

87 Khan, A.A., Huang, G., Rana, M.M. et al. (2021). Superior transverse piezoelectricity in organic-inorganic hybrid perovskite nanorods for mechanical energy harvesting. *Nano Energy* 86: 106039.

88 Chen, X.-G., Song, X.-J., Zhang, Z.-X. et al. (2020). Two-dimensional layered perovskite ferroelectric with giant piezoelectric voltage coefficient. *J. Am. Chem. Soc.* 142 (2): 1077–1082.

89 Ok, K.M., Chi, E.O., and Halasyamani, P.S. (2006). Bulk characterization methods for non-centrosymmetric materials: second-harmonic generation, piezoelectricity, pyroelectricity, and ferroelectricity. *Chem. Soc. Rev.* 35 (8): 710–717.

90 Lefki, K. and Dormans, G.J.M. (1994). Measurement of piezoelectric coefficients of ferroelectric thin films. *J. Appl. Phys.* 76 (3): 1764–1767.

91 Mason, W.P. and Jaffe, H. (1954). Methods for measuring piezoelectric, elastic, and dielectric coefficients of crystals and ceramics. *Proc. IRE* 42 (6): 921–930.

92 Liu, J.M., Pan, B., Chan, H.L.W. et al. (2002). Piezoelectric coefficient measurement of piezoelectric thin films: an overview. *Mater. Chem. Phys.* 75 (1): 12–18.

93 Xiong, Y.-A., Gu, Z.-X., Song, X.-J. et al. (2022). Rational design of molecular ferroelectrics with negatively charged domain walls. *J. Am. Chem. Soc.* 144 (30): 13806–13814.

94 Gruverman, A., Alexe, M., and Meier, D. (2019). Piezoresponse force microscopy and nanoferroic phenomena. *Nat. Commun.* 10 (1): 1661.

95 Kalinin, S.V., Karapetian, E., and Kachanov, M. (2004). Nanoelectromechanics of piezoresponse force microscopy. *Phys. Rev. B* 70 (18): 184101.

96 Buragohain, P., Lu, H., Richter, C. et al. (2022). Quantification of the electromechanical measurements by piezoresponse force microscopy. *Adv. Mater.* 34: 2206237.

97 Kalinin, S.V. and Bonnell, D.A. (2002). Imaging mechanism of piezoresponse force microscopy of ferroelectric surfaces. *Phys. Rev. B* 65 (12): 125408.

98 Soergel, E. (2011). Piezoresponse force microscopy (PFM). *J. Phys. D. Appl. Phys.* 44 (46): 464003.

99 Zhang, H.-Y., Zhang, Z.-X., Song, X.-J. et al. (2020). Two-dimensional hybrid perovskite ferroelectric induced by perfluorinated substitution. *J. Am. Chem. Soc.* 142 (47): 20208–20215.

100 Wang, Z.-X., Zhang, H., Wang, F. et al. (2020). Superior transverse piezoelectricity in a halide perovskite molecular ferroelectric thin film. *J. Am. Chem. Soc.* 142 (29): 12857–12864.

101 Shi, C., Ma, J.-J., Jiang, J.-Y. et al. (2020). Large piezoelectric response in hybrid rare-earth double perovskite relaxor ferroelectrics. *J. Am. Chem. Soc.* 142 (21): 9634–9641.

102 Zhang, H.-Y., Jiang, H.-H., Zhang, Y. et al. (2022). Ferroelectric lithography in single-component organic enantiomorphic ferroelectrics. *Angew. Chem. Int. Ed.* 61 (22): e202200135.

103 Long, J., Ivanov, M.S., Khomchenko, V.A. et al. (2020). Room temperature magnetoelectric coupling in a molecular ferroelectric ytterbium(III) complex. *Science* 367 (6478): 671–676.

104 Long, J., Rouquette, J., Thibaud, J.-M. et al. (2015). A high-temperature molecular ferroelectric Zn/Dy complex exhibiting single-ion-magnet behavior and lanthanide luminescence. *Angew. Chem. Int. Ed.* 54 (7): 2236–2240.

7

High-Temperature $BiScO_3$–$PbTiO_3$ Piezoelectric Ceramics

Yazhu Dong[1,2] *and Zhiyong Zhou*[1]

[1] *Shanghai Institute of Ceramics, Chinese Academy of Sciences, 588 Heshuo Road, Jiading District, Shanghai 201899, China*
[2] *University of Chinese Academy of Sciences, Center of Materials Science and Optoelectronics Engineering, Shijingshan District, Beijing 100049, China*

7.1 Introduction

Piezoelectric materials can transform mechanical energy into electrical energy, and vice versa. They are widely used in many fields, including aerospace, automotive communications, medical health, and so on [1, 2]. Pb(Zr, Ti)O_3 (PZT)-based piezoelectric ceramics are the main commercial piezoelectric materials in practical application since their discovery in 1955 [1, 2]. PZT ceramics have extraordinary performances: piezoelectric constant (d_{33}) of 600–700 pC/N and high Curie temperature (T_C) of 350 °C near the morphotropic phase boundary (MPB), combined with the tetragonal phase and rhombohedral phase [1–4]. While many studies and applications experience showed the highest operating temperature is around 250 °C [1–4], restricting the development of the field in high-temperature (above 300 °C) working environments. In 2001, Eitel et al. combined $BiScO_3$ and $PbTiO_3$, preparing a pure perovskite structure $BiScO_3$–$PbTiO_3$ (BS–PT) binary piezoelectric ceramics system based on the relationship between the tolerance factor (t) and T_C [5]. By regulating PT content, BS–PT ceramics can obtain a high T_C of 450 °C (nearly 100 °C higher than PZT) and a high d_{33} of 450 pC/N near MPB at 0.64PT [5–7].

BS–PT ceramic is an important piezoelectric system with both high T_C and high d_{33}, which makes it a promising candidate material for high-temperature applications above 300 °C [5, 7–9]. Much attention has been focused on this binary system, followed by the breadth and depth of research work, such as phase diagrams, modification, preparation methods, and application verification [10, 11]. Meanwhile, high-performance origin studies and high-end equipment design have also been sprinted up since the discovery of the BS–PT system, which stimulated the development of a high-temperature field [12, 13]. Fruitful achievements have been obtained through multidimensional and multi-angle research on the BS–PT system.

The research diagram of BS–PT ceramics is shown in Figure 7.1, which describes the perspectives of BS–PT studies. In the BS–PT system research process, the

Piezoelectric Materials: From Fundamentals to Emerging Applications, First Edition. Edited by Jiagang Wu.

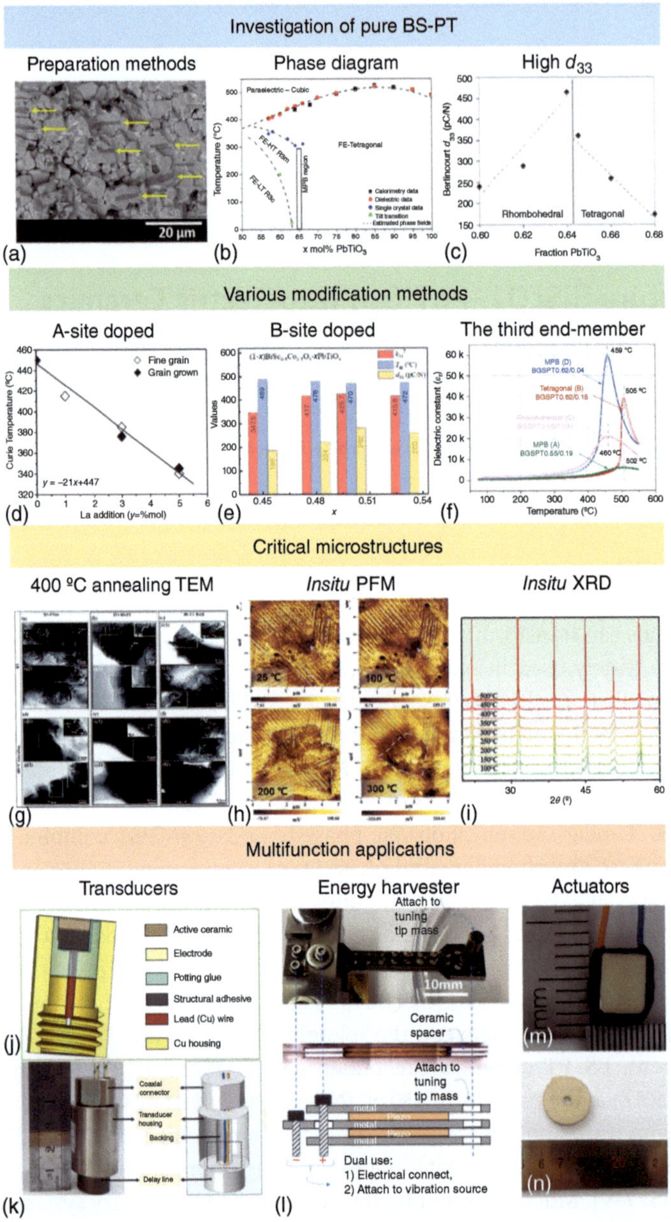

Figure 7.1 (a–n) The perspectives for BS–PT studies. Source: Adapted from Refs. [6, 13–24]. Reproduced with permission from Taylor & Francis, AIP Publishing, The Japan Society of Applied Physics, Wiley, Royal Society of Chemistry. (a) From Akça et al. [20]/ Reproduced with permission from Taylor and Francis/CC BY 4.0. (g) From Zhao et al. [17]/Reproduced with permission from Royal Society of Chemistry. (h) Refs [6, 13–24]. Reproduced with permission from AIP Publishing, IOP Publishing, John Wiley and Sons, Royal Society of Chemistry, Elsevier. (k) Dong et al. [16]/Reproduced with permission from Elsevier. (l) From Chen et al. [24]/Reproduced with permission from AIP Publishing LLC. (m) From Qaiser et al. [13]/Reproduced with permission from John Wiley & Sons. (n) [6, 13–24]. Reproduced with permission from AIP Publishing.

understanding of pure BS–PT system was dominated before 2010. The main studies focused on optimizing and revising the BS–PT binary phase diagram, the mechanism of high piezoelectric properties, and the exploring preparation methods. Afterward, more and more research shifted toward the material design and various modification methods to meet the requirements of high performance. The component modification can be divided into A-site doped, B-site doped, Pb-based end-member, Bi-based modified, and other solid solutions [3, 14, 15, 25–27]. Each modification method has a different function in BS–PT ceramics, and many achievements are obtained in piezoelectricity, temperature stability, high power, and other aspects. These studies provide a very important path for the commercialization of BS–PT ceramics. Recently, more studies to appreciate key research aspects and controversial viewpoints, which also have achieved many breakthrough results in the BS–PT-based device design and the relationship between T_C and d_{33}. These BS–PT-based high-temperature devices include transducers, actuators, and energy harvesters, which identified that BS–PT ceramics as a promising candidate for high-temperature work environments above 250 °C.

To systematically summarize these achievements is useful for future work. This chapter will focus on the development of BS–PT ceramics, the preparation method, the modification of BS–PT ceramics component design, and the potential applications at high temperatures. The grain size of BS–PT prepared by the traditional solid-state reaction method is about 3 μm, which is away from the requirements of miniaturization equipment. Therefore, scientists used the sol–gel method, high-energy ball milling, and other synthesis methods, matching with spark plasma sintering (SPS), two-step sintering (TSS), and other sintering methods, and prepared nano-sized BS–PT ceramics with high performance [28–31]. However, these chemical methods have poor repeatability and are not suitable for mass production, which indicates that the commercial application of BS–PT ceramics in miniaturization is still a long way. After this, we investigated the performance modification of BS–PT ceramics from different viewpoints and explored the values of BS–PT ceramics as a core in applications [16, 29, 32–35]. In all the studies, most of the modification studies were carried out based on BS–PT ceramics near the MPB composition for obtaining higher d_{33} and T_C, which have acquired much success. And then, the application value of BS–PT-based ceramics in high-temperature transducers, actuators, and energy harvesters was further verified by analyzing the different high-temperature devices of BS–PT-based ceramics [16, 17].

Based on our work, this chapter provides a comprehensive and specific summary of BS–PT ceramics, especially for the design ideas for different ceramic systems and high-temperature application devices, which provides a convenient and effective way to exploit the full advantages of BS–PT ceramics.

Therefore, this chapter will be arranged as follows: first, the development of the BS–PT system will be summarized from the viewpoints of the phase diagram revision and the crystal/phase/domain structure analysis. Besides, various preparation methods for BS–PT ceramics will be discussed comprehensively. Second, the modification methods for the improvement of electrical properties from the material design will be described in detail. These methods include A-site doping, B-site doping, A/B-site codoping, Pb-based doping, Bi-based doping, and so on. Third, the applications of BS–PT ceramics will be summed up from the operations of transducers, actuators, and energy harvesters. Finally, some suggestions on the problems and challenges of BS–PT ceramics will be put forward.

7.2 The Characteristics of BS–PT Piezoelectric Ceramics

BS–PT binary ceramics system was discovered in 2001 by Eitel et al. according to the crystallography relations between the Curie temperature and tolerance factors (t), in which T_C increases when the tolerance factors reduce for perovskite ceramics system [5]. $BiScO_3$ ($t = 0.91$) with monoclinic phase ($C2/c$) presents pure perovskite structure at high temperature (1413 K) and high pressure (6 GPa) [36]. $PbTiO_3$ ($t = 1.06$) with a tetragonal phase ($P4mm$) displays a classical ferroelectric perovskite structure of high T_C (490 °C), but difficulty in sintering [37]. While after $BiScO_3$ incorporating into $PbTiO_3$, the BS–PT system could be obtained by the traditional solid-phase synthesis method; meanwhile, large d_{33} of 450 pC/N and high T_C of 450 °C (higher 100 °C than PZT) appear near MPB obtained by component regulation [6, 37, 38], and a detailed introduction of BS–PT ceramics was developed in 2002 [6]. Therefore, the BS–PT system is considered a promising candidate for high-temperature applications.

7.2.1 The Revision of Phase Diagram

The phase diagram of BS–PT system was studied since 2002, and revised in 2004 [6, 18]. The phase diagram of BS–PT in Figure 7.2 shows that the BS–PT system exhibits tetragonal phase structure (T) at $x > 0.66$. When $x < 0.64$, it shows rhombohedral

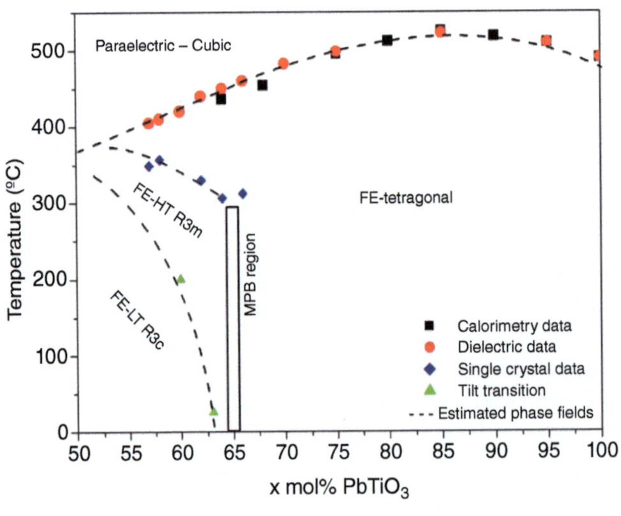

Figure 7.2 Phase diagram for the perovskite BS-PT system. Source: Eitel et al. [18]. Reproduced with permission from AIP Publishing.

phase structure (R). MPB region is located in the range, $0.64 \leq x \leq 0.66$. In addition, T_C increases first and then decreases as PT content enhances, exhibiting a severely curved line [18]. This can be contributed to the high T_C of $PbTiO_3$ and the thermodynamic instability of $BiScO_3$.

The domain structure and phase structure of the BS–PT system have been widely investigated according to this phase diagram. The secondary phases and small grains appear at high $BiScO_3$ content, which presents non-180° domain structures (with 71°, 109°, and 180° domain, and the spontaneous polarization is along the <111> direction) [18, 39]. Moreover, it exists as a superlattice at the 1/2<*hkl*> location at the high $BiScO_3$ region because of antiphase rotations of oxygen octahedral [40]. The superlattice reflections show R3c rather than R3m symmetry on the basis of (111) splitting in X-ray diffraction (XRD) patterns near $x = 0.5$, which is similar to the PZT binary system [40, 41]. Besides, in the R phase area, the crystal structure turns from R3m (regular octahedral ferroelectric phase) to R3c (inclined octahedral ferroelectric phase) as the temperature decreases due to the octahedral rotation of $[TiO_6]$. When x is around 0.60, the phase transition temperature is about 200 °C at $x = 0.60$, while it decreases by 20 °C at $x = 0.62$.

In the MPB region of BS–PT, lamellar domains and wedge-shaped domains are frequently observed [7]. MPB region is parallel to the temperature axis and perpendicular to the horizontal axis (the PT content curve), such as PZT [18, 42, 43]. The phase boundary is dependent on the composition greatly, which indicates that the optimal electrical properties may be obtained by adjusting components [19, 33, 44, 45]. Neutron powder diffraction, Rietveld refinements, XRD, and *in situ* synchrotron XRD show that the MPB region coexisted with tetragonal phase (space group: *P4mm*) and monoclinic phase (space group: *Cm*) in BS–PT ceramics [46]. Monoclinic phase in the MPB region was confirmed in <100>-oriented BS–PT single crystal with lattice parameters a = 4.0751, b = 3.9765, c = 3.9893 Å, and $\beta = 90.166°$ [47].

When the components escape from MPB, for example, PT content above 0.66, the phase structure in BS–PT ceramics displays the pure ferroelectric T phase, where 90° and 180° domains are dominated. The T_C value can reach up to 500 °C at $x > 0.75$, higher than that of PT (490 °C) [18, 37]. Random field effect, which occurs when polyvalent cations occupy equivalent lattice positions, maybe a possible reason for the enhanced T_C [48]. Besides, the valence mismatch may result in a nonlinear relationship between the Curie temperature and composition. Other PT-based systems also exhibit a similar relation, such as $Bi(Mg_{1/2}Ti_{1/2})O_3$–$PbTiO_3$, $Bi(Zn_{2/3}Nb_{1/3})O_3$–$PbTiO_3$, and $BiFeO_3$–$PbTiO_3$ [8, 49–51].

The system framework of BS–PT ceramics has been preliminarily established after the revision of the BS–PT system phase diagram. Then, scientists have gradually deepened and expanded further research on the BS–PT system. Detailed information will be illustrated in the following sections.

7.2.2 Preparation Methods of BS–PT Ceramics

Comprehensive studies on the preparation methods of BS–PT ceramics have been carried out to meet different application requirements (low sintering temperature,

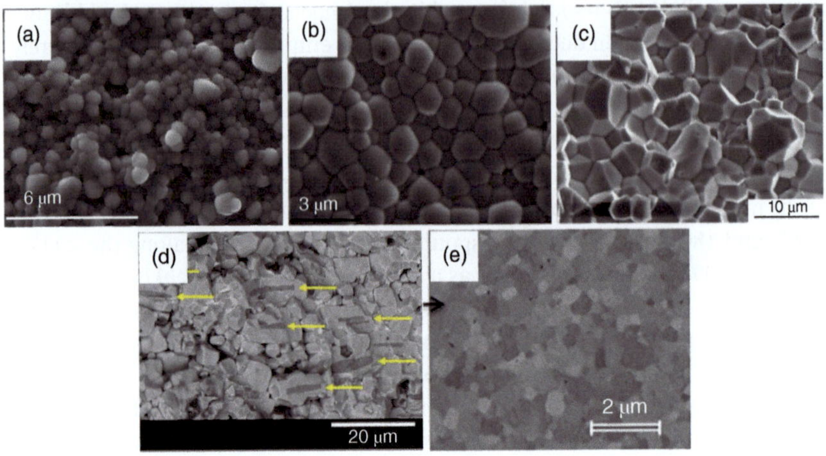

Figure 7.3 SEM images of BS–PT ceramics prepared by different methods. (a) Sol–gel method [53]. (b) Two-step sintering [54]. (c) High-energy ball milling process [55]. (d) Templates grain growth [20]. (e) SPS and two-step sintering [56]. Source: Refs. [20, 52–56]. Reproduced with permission from IEEE, Wiley, Springer, National Institute for Materials Science, Taylor & Francis. (a) From Zhao et al. [53]/Reproduced with permission from John Wiley & Sons. (b) From Zou et al. [54]/Reproduced with permission from John Wiley & Sons. (c) From Zhang et al. [55]/Reproduced with permission from Springer Nature. (d) From Akça et al. [20]/Reproduced with permission from Taylor and Francis/CC BY 4.0. (e) From Amorin et al. [56]/Reproduced with permission from Taylor and Francis/CC BY 3.0.

appropriate grain size, and suitable crystal orientation [13, 20, 29, 30]). The microstructures were shown in Figure 7.3. In this chapter, we focus on describing BS–PT ceramics obtained by different preparation methods for meeting different requirements [13, 20, 29, 30].

7.2.2.1 Synthetic Method

A mechanical method was applied to prepare BS–PT nano-powder with an average particle size of about 10 nm [57]. By mechanosynthesis to synthesize nano-powder, BS–PT ceramics can be gained with high density above 99% and high $d_{33}{}^{*}$= 400 pm/V, as well as high k_p of 0.52 by hot-pressing sintering under 1000 °C [58]. While after using SPS at 650 °C for three minutes, BS–PT ceramics with nanoscale grains were obtained with high T_C of 427 °C and d_{33} of 40 pC/N [57]. These results show that the applied pressure was a vital factor to restrict grain growth, especially for the temperature above 600 °C, and the grain size hardly varies with increased temperature at a pressure of 120 MPa [57].

It has been found that the ferroelectric transition might have disappeared for BS–PT ceramics with the nano grain size of 30 nm [57, 59]. Algueró and coworkers settled the contradiction between the polar states and the disappeared ferroelectric properties by Raman spectroscopy and impedance spectroscopy for coarse-grained and nanostructure grains [34, 59]. Raman results exhibit that the ferroelectric effect is locally preserved, and lower permittivity appears in BS–PT nanoceramics because of the dilution effect [60]. The thermodynamic theory indicates that this anomaly diminution is predicted, even when ignoring dead layers as a surface effect associated with the polarization gradient at grain boundaries [61]. Impedance spectroscopy shows that grain boundaries govern nanostructured ceramics due to the boundary effect [59].

The sol–gel method can be applied to prepare fine powders easily with a pure phase structure [62]. BS–PT nanosized powders were prepared by sol–gel method with ethylene glycol (EG) to stabilize the sol and regulate pH values to avoid precipitates. These nanosized powders were sintered by a TSS method, and grain size of the ceramics was about 1.4 μm [29]. The maximum d_{33}, k_p, and remnant polarization P_r were about 700 pC/N, 0.632, and 47.3 μC/cm^2 at room temperature and $x = 0.635$, respectively [29], which are superior to other BS–PT-based ceramics [18, 19, 33]. The nano-powder can decrease the sintering temperature from 1100 to 1000 °C, and enhance the electric properties at the same time [53]. Compared with the ceramics prepared by the conventional methods, however, the ceramics prepared by nano-powder have lower mechanical strength, lower density, and higher porosity because of the existing organic matter residue after calcination [53].

High-energy ball milling is one of the mechanical activation synthesis (MAS) techniques in the preparation of ceramics, which has lots of advantages over the solid-state reaction and wet chemical methods [63], such as the simple operation reducing costs and the lower calcined and sintering temperature decreasing the volatilization of Bi and Pb. The reaction process is not easy to control because of many uncertain factors, for example, local stress, heating, high-pressure influence, nano boundary, and inner defects [55, 63]. The process of preparing BS–PT ceramics by high-energy ball milling can be divided into several stages. At the original stage, which lasts about two hours, the refinement and amorphization of the raw material mixture oxide are dominant, and the impure perovskite phase has obviously been obtained from the measurement of XRD [55]. After about two to four hours, the BS–PT matrix was activated, followed by the crystal growth and increase in the grain size [55]. A stable perovskite structure of BS–PT nanosized powders of 18–25 nm can well be integrated by high-energy ball milling for four hours. The grain size decreased and the collision friction dominated among four to six hours, resulting in the fine grain size of BS–PT ceramics [55]. At last, the milling nanopowders were sintered in sealed crucibles at 1000–1150 °C for two hours with a grain size of 4–7 μm. However, the measurement results show that this method cannot improve the electrical properties of BS–PT ceramics compared with Eitel et al.'s work in 2001 [6, 55].

7.2.2.2 Sintering Process

In addition to the improvement of the synthesis methods discussed above, there are many studies based on the exploration of sintering methods for enhancing the performance of BS–PT ceramics. In this section, the content will focus on sintering processing, which can also obtain ceramics with both high density and fine grain size [54, 64].

TSS involves first sintering the green pellets for several minutes at a higher temperature (above normal sintering temperature) for obtaining an intermediate density and then held at a lower temperature (below 1000 °C) to make them fully dense.

In general, the bulk density of ceramics obtained by TSS is more than 97%, which is higher than that of the traditional solid-state reaction sintering [64]. The prepared BS–PT green pellets were sintered by the TSS method (sintered at 1150 °C for five minutes and then at 1040 °C for three hours), and it was found that the performance of ceramics was greatly improved. The values of k_p, d_{33}, and d_{33}^* are about 0.57, 545 pC/N, and 920 pm/V, respectively [64]. The grain size of BS–PT ceramics prepared through nanosized powders by TSS is about 200–800 nm [54]. TSS is significantly better than one-step sintering in terms of d_{33}, ferroelectric properties, and strain [64]. Besides, compared with normal sintering with the same density, TSS not only limits the growth of grain but also reduces the sintering temperature, accelerates the sintering process, and restricts the volatilization of Pb and Bi [54].

Template grain growth (TGG) is an effective method to improve the electrical performance of piezoelectric ceramics comparable to those of single crystals by aligning the original randomly oriented grains along with a certain growth orientation [20, 65]. Textured $2.5Bi(Zn_{0.5}Zr_{0.5})O_3$–$37.5BiScO_3$–$60PbTiO_3$ (BZZ–BS–PT) ceramics with a texture fraction of 91% and high relative density of 98% were achieved by TGG methods utilizing 5 vol% [001]-oriented $BaTiO_3$ (BT) plate-like templates [20]. The [001] direction BT template promotes highly oriented polycrystalline structure, followed by changing phase structure compared with random ceramic samples. Figure 7.4a shows that it is rhombohedral phase in the random orientation ceramic, while it is the dominant tetragonal phase together with rhombohedral phase in textured ceramics because the template has strong tetragonal structure [20]. The comparison result between random ceramics and textured ceramics is shown in Figure 7.4b. The textured ceramics have improved dielectric constant from 762 to 980, increased unipolar strain (@ 40 kV/cm) from 0.21% to 0.25%, and enhanced maximum piezoelectric constant (max d_{33}^*: 526 to 930 pm/V; d_{33}: 216 to 353 pC/N) [20].

Liquid phase sintering (LPS) is another technique that there are wetting liquid phases among ceramic grains in the processing of sintering conditions [66–68]. Generally, this method can result in higher density and low sintering temperature because of its special sintering theory, as shown in Figure 7.5 [66–68].

$Bi_{12}PbO_{19}$ (BP) compound exists in the BS–PT system, promoting grain growth at temperatures below 750 °C as sintering aid [66]. Besides, BP can lead to abnormal grain growth and high density above 99% of BS–PT ceramics sintered at 1000 °C because of LPS. Scanning electron microscope (SEM) and XRD results show that the calcination powder with BP can be sintered compactly at 1000 °C, while the ceramics without liquid phases are very loose, meaning more energy is needed to promote the grain growth [66, 68]. Excessive secondary phases (liquid phase) formed in the process of calcination will pin the phase boundary, resulting in the reduction in grain size [66–68]. Meanwhile, the optimal properties in BS–PT ceramics with LPS were achieved of high d_{33} (531 pC/N), large g_{33} (41×10^{-3} V m/N), high k_p value (61.8%), and high T_C (426 °C) [66]. In BS–PT ceramics, the excess Bi_2O_3 phase can usually form a liquid additive and assemble in the three-grain junctions without wetting the grain surface. This liquid phase Bi_2O_3 can compensate for the volatile Bi and Pb during the calcination and sintering processes and result in enhanced the AC resistivity five times, increased high field d_{33} by 15%, and declined $\tan\delta$ for three times [67].

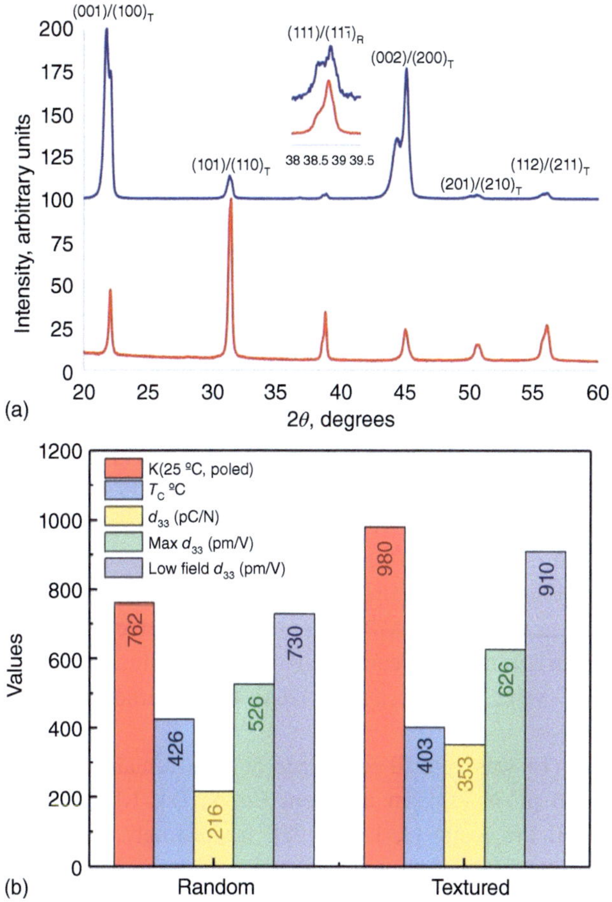

Figure 7.4 (a) XRD patterns of random (red line) and textured (blue line) samples. (b) The properties comparison of the random and textured BZZ–BS–PT samples sintered in air. Source: Akça et al. [20]. Reproduced with permission from Taylor & Francis.

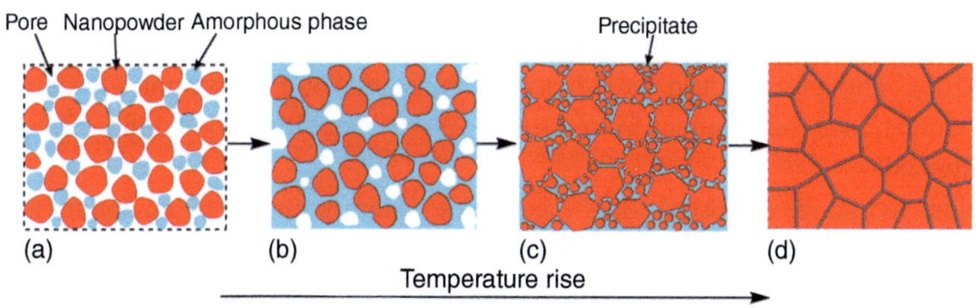

Figure 7.5 The liquid sintering processing of BS–PT ceramics: (a) Green body. (b) Liquid phase formation. (c) Densification. (d) Grain growth. Source: Yue et al. [68]. Reproduced with permission from Wiley.

7.3 The Modifications of BS–PT-Based Ceramics Near MPB

7.3.1 A-Site and B-Site Doping

In the perovskite structure, A-site is occupied by larger ion radius elements to stabilize the structure, such as rare earth and alkali earth metal elements. The B-site is occupied by smaller ion radius elements (transition metal), deciding the performance of the perovskite structure [14, 39, 40, 49]. In BS–PT ceramics, A-site La doping can decline T_C to 365 °C with unchanged d_{33} of 465 pC/N, which was completely different from PZT doping La at A-site [5, 14, 69]. However, $0.38(Bi_{1-x}La_x)ScO_3$–$0.62PbTiO_3$ (BLSPTx) ceramics away from MPB show that the grain size was much larger than other BS–PT ceramics, T_C reduced from 450 to 400 °C, and d_{33} values enhanced from 200 to 250 pC/N [70].

B-site doping is investigated through the substitution for Sc or Ti [3, 71–73]. Mn replaced B-site at Sc/Ti in 0.36BS–0.64PT ceramics enhanced T_C (from 445 to 468 °C) but inclined d_{33} value to 270 pC/N [71]. The depolarization temperature of d_{33} is 440 °C, and the high temperature (@450 °C) resistivity of BS–PT–Mn is $10^7\ \Omega \cdot cm$, which can meet the demand for high-temperature sensors and transducers in theory [71]. Wu et al. discovered that the 0.367BS–0.633PTFe ceramics with Fe (replaced for Ti) had a similar result with Mn doping. As Fe content increased, d_{33} values declined from 455 to 301 pC/N while T_C values enhanced from 426 to 466 °C without changing the phase structure [72].

0.55BSFe–0.45PT ceramics (Fe was applied to replace Sc) can remain at both T_C of 440 °C and d_{33} of 300 pC/N [33]. In addition, Co, Y, Ga, Nb, and Zr were applied to replace Sc at the B-site [3, 15, 25–27]. Interestingly, after Co, Ga, and Y substituting Sc in BS–PT-based ceramics, the performance changes exhibited a similar tendency, which was that T_C increased to 465–510 °C and d_{33} decreased to 147–426 pC/N compared with pure BS–PT components [3, 5, 15, 27]. However, the doping of Ga in BS–PT ceramics kept a high d_{33} of 510 pC/N and T_C of 438 °C [3]. The addition of Nb in BS–PT ceramics decreased the grain size without improved d_{33} and T_C, which showed a similar result to that of A-site La substitution [14, 26]. $0.36Bi(Sc_{1-x}Zr_x)O_3$–$0.64PbTiO_3$ ceramics exhibited high T_C of 420 °C and d_{33} of 476 pC/N when Zr content was 0.01 (x = 0.01). The piezoelectric and dielectric constants of this ceramics system were temperature independent from −100 to 250 °C, exhibiting good temperature stability [25].

Besides the above discussed, BS–PT ceramics system prepared by co-doping at A- and B-sites, such as $0.15BiScO_3$–$0.85(Pb_{1-3x/2}Bi_x)(Ti_{0.98}Mn_{0.02})O_3$, $0.365BiScO_3$–$0.635(Pb_{1-3x/2}Bi_x)(Ti_{0.99}Zn_{0.01})O_3$ [16, 19], had higher T_C near $PbTiO_3$ content and higher d_{33} near MPB. In summary, A/B-site doping usually increases by one property at the sacrifice of another. It is very important and difficult to find a perfect method to maintain both high d_{33} and high T_C. The research in this area still has a long way to go.

7.3.2 Bi-Based End-Member Mixing

Bi-based compounds with $6s^2$ lone pair of electrons could influence the distribution of other electron pairs. Besides, Bi-based compounds with a low tolerance factor are likely to cause a decrease in lattice symmetry and the distortion of the structure due to the smaller ionic radius, which can be applied to build MPB and increase T_C [49]. Many Bi-based additives, such as $BiMO_3$, $Bi(M_1M_2)O_3$, and $(Bi,M)TiO_3$ are solubilized in BS–PT ceramics, and they have special influence on the properties and microstructures. $BiMO_3$ includes compounds such as $BiFeO_3$ [74] and $BiGaO_3$ [21, 75]. $Bi(M_1,M_2)O_3$ includes $Bi(Zn_{0.5}Ti_{0.5})O_3$ [76], $Bi(Zn_{0.5}Hf_{0.5})O_3$ [17], and $Bi(Sn_{1/3}Nb_{2/3})O_3$ [16, 77]. $(Bi,M)TiO_3$ incorporates $(K_{0.5}Bi_{0.5})TiO_3$ [78] and $(Na_{0.5}Bi_{0.5})TiO_3$ [79].

$BiGaO_3$ has a similar structure and performance with $BiScO_3$, and they can take shape stable perovskite structure after doping in $PbTiO_3$. Similar to BS–PT ceramics, $(1-x-y)BiScO_3-yBiGaO_3-xPbTiO_3$ ceramics are also strongly composition-dependent, and the T_C and d_{33} can be improved by adjusting PT and BG content [5, 21, 75]. The phase diagram and T_C dependent on composition were exhibited in Figure 7.6 [21]. The results show that MPB shifts toward low PT content after solubilizing BG. Also, this ceramic system at $x = 0.56$ and $y = 0.19$ may achieve a high T_C of 501 °C at the sacrifice of d_{33} (151 pC/N) [75]. $BiFeO_3$ has two advantages of a low melting point of 930 °C and a high T_C of 830 °C [39]. Obviously, $0.36[(1-x)BiScO_3-xBiFeO_3]-0.64PbTiO_3$ ceramics with high density could be prepared at 1100 °C (50–100 °C lower than pure BS–PT ceramics) [5, 74]. Similar to $BiGaO_3$ doping, high T_C of 500 °C and the declined d_{33} of 125 pC/N could be achieved in BS–BF–PT ceramics. The low d_{33} can be improved to 165 pC/N after oxygen sintering by eliminating the oxygen vacancy [74].

$Bi(Zn_{1/2}Ti_{1/2})O_3$ (BZT) has been considered as a high-temperature gem due to a large c/a ratio of 1.211 and a strong spontaneous polarization of 103 μC/cm² at room temperature [80]. Besides, BZT–PT ceramics with large crystal structures display a high T_C of 710 °C, showing wide temperature stability because of zero thermal

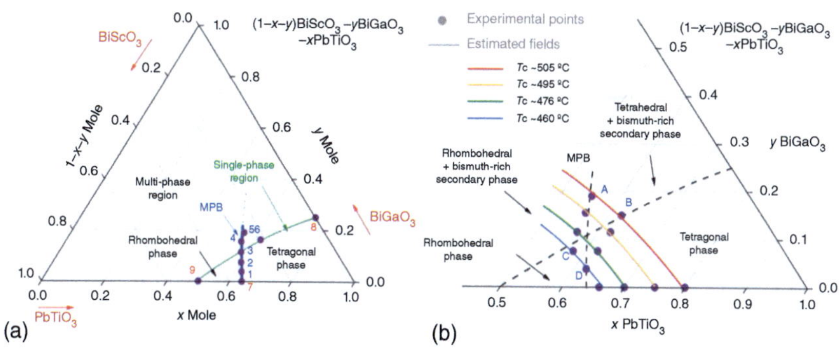

Figure 7.6 (a) Phase diagram and (b) composition dependence of T_C for $BiScO_3-PbTiO_3-BiGaO_3$ system. Source: Jiang et al. [21]. Reproduced with permission from Wiley.

expansion [80, 81]. This special characteristic can date back to Zn and Bi ions. The occupied 3s, 3p, and 3d states in Zn^{2+} can result in large tetragonal crystal distortion because of strong repulsion between Zn and O ions. While Zn 4s/4p orbitals and O 2p orbitals can shape into short covalent Zn—O bonds due to hybridization [19, 49]. Besides, Bi with a lone pair of electrons can form stronger Bi—O bonds than Pb—O [19, 82]. Dong et al. utilized the hybridization characteristics of Bi and Zn in $Bi(Zn_{1/2}Ti_{1/2})O_3$ and designed $0.365BiScO_3$–$0.635(Pb_{1-3x/2}Bi_x)(Ti_{0.99}Zn_{0.01})O_3$ ceramics [19]. MPB composition at $x = 0.01$ gained the maximum d_{33} of 495 pC/N without sacrificing T_C of 428 °C. TEM observation results show that the modification of Bi refined domain structure from the strip-like parallel domain at $x = 0$ to complex nanodomain after Bi doping. The proportion of the small grains enhanced after Bi doping, which may be contributed to more nucleation sites after Bi ions replacing Pb ions and declining crystal symmetry [19].

Figure 7.7 shows the phase diagram of $(1-y)[xBiScO_3$–$(1-x)PbTiO_3]$–$yBi(Zn_{1/2}Ti_{1/2})O_3$ ceramics, and MPB area is around $x = 0.36$ and $y = 0.01$ [76]. The ceramic system can obtain a high T_C of 425 °C, the largest ever unipolar strain of 0.45%, and a high d_{33} of 520 pC/N. PFM results and the microscopic piezoresponse hysteresis loops exhibit that the addition of BZT leads to the complexity of nanodomains, which stimulated the local piezoelectric response leading to enhanced piezoelectricity and strain [76].

$(1-x)[0.5(K_{0.5}Bi_{0.5})TiO_3$–$0.5BiScO_3]$–$xPbTiO_3$ ceramics showed the maximum d_{33} of 247pC/N and T_C of 384 °C near MPB. The increased point defects coming from the replacement of K and Bi ions disturbed the long-range coupling among the BO_6 octahedra, leading to relaxation [83]. But there was no frequency dispersion observed probably because ion displacement was dominant at the high-temperature permittivity, while the polar micro–nanodomain was minor [78, 79]. While $(Na_{0.5}Bi_{0.5})TiO_3$ ceramics displayed even worse performance, showing a d_{33} value of 141 pC/N and T_C of 350 °C in the vicinity of MPB [79].

7.3.3 Pb-Based Compound Treatment

Pb-based end-member has a key influence on the performance regulation of BS–PT ceramics. BS–PT system was developed based on PZT. BS–PT ceramics

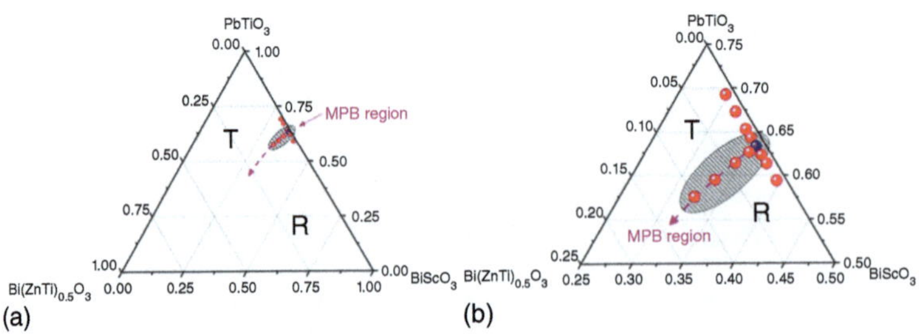

Figure 7.7 (a) Phase diagram of $BiScO_3$–$PbTiO_3$–$Bi(Zn_{1/2}Ti_{1/2})O_3$ ternary system. (b) Amplified local diagram near the MPB. Source: Liu et al. [76]. Reproduced with permission from Royal Society of Chemistry.

modified with PZT ceramics did not exhibit extraordinary characteristics. $BiScO_3$–$PbZrO_3$–$PbTiO_3$ ceramics, e.g., BS–PT doped with PZT, were studied for high-temperature actuator applications [84]. The properties, such as d_{33}, P_r, E_c, and strain, showed strong dependence on composition. The d_{33} values at room temperature were in the range of 370–430 pC/N and T_C was between 360 and 390 °C, which is lower than those of BS–PT [5, 84].

While the modification of Pb-based relax state in BS–PT ceramics can achieve relaxation ferroelectric with high d_{33} and T_C. $Pb(Mn_{1/3}Nb_{2/3})O_3$ (PMN) is considered as a classical addition for the formation of relaxation ferroelectrics. BS–PT–PMN can form a stable phase, and the phase diagram is displayed in Figure 7.8a [85]. After 10% PMN modifying BS–PT ceramics, the MPB region shifted to 0.6PT with the reduced d_{33} of 200 pC/N and the decreased T_C of 300 °C [86]. However, Segalla and coworkers designed relaxation ferroelectric (1−2*x*)BS–*x*PT–*x*PMN ceramics with MPB at $x = 0.42$ followed by d_{33} of 509 pC/N at the sacrifice of T_C (180 °C), as Figure 7.8b–d shows [85].

The investigations of relaxation behaviors for BS–PT–PMN ceramics by TEM observation are shown in Figure 7.9. Figure 7.9a,b showed that the length scales of polar micro-domain structures are at 1–50 nm and 0.1–10 mm in relaxors [28]. Domain distribution and size varied after thermal and electric processing, as Figure 7.9c–e exhibited. These macro-domain structures were surveyed by the aligned domain walls with twin habit planes close to <110> [28]. TEM observation exhibited that grains consisted of micro- and macro-domain structures. Mottled

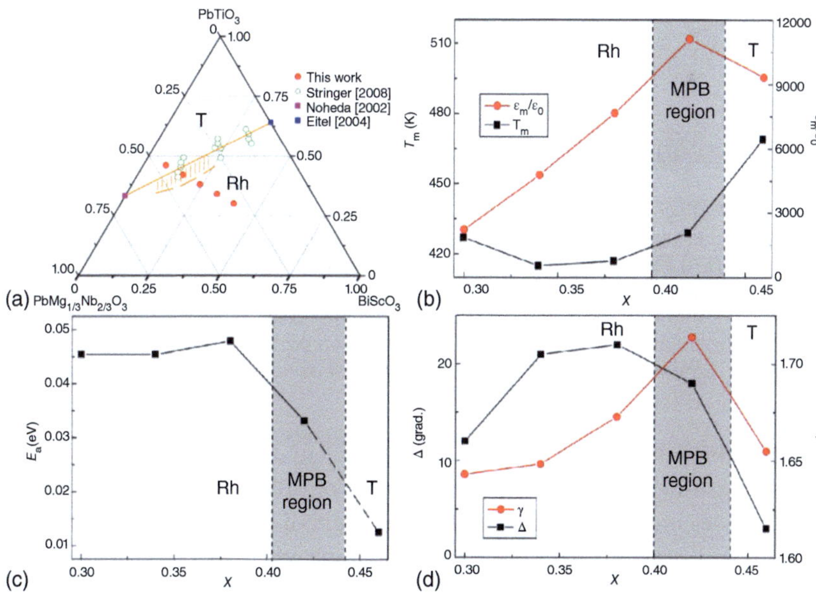

Figure 7.8 (a) Phase diagram of $BiScO_3$–$PbTiO_3$–$Pb(Mn_{1/3}Nb_{2/3})O_3$. Concentration dependencies of (b) T_m and $\varepsilon_m/\varepsilon_0$, (c) Δ and γ, (d) E_a for BS–PT–PMN ceramic. Source: Talanov et al. [85]. Reproduced with permission from Wiley.

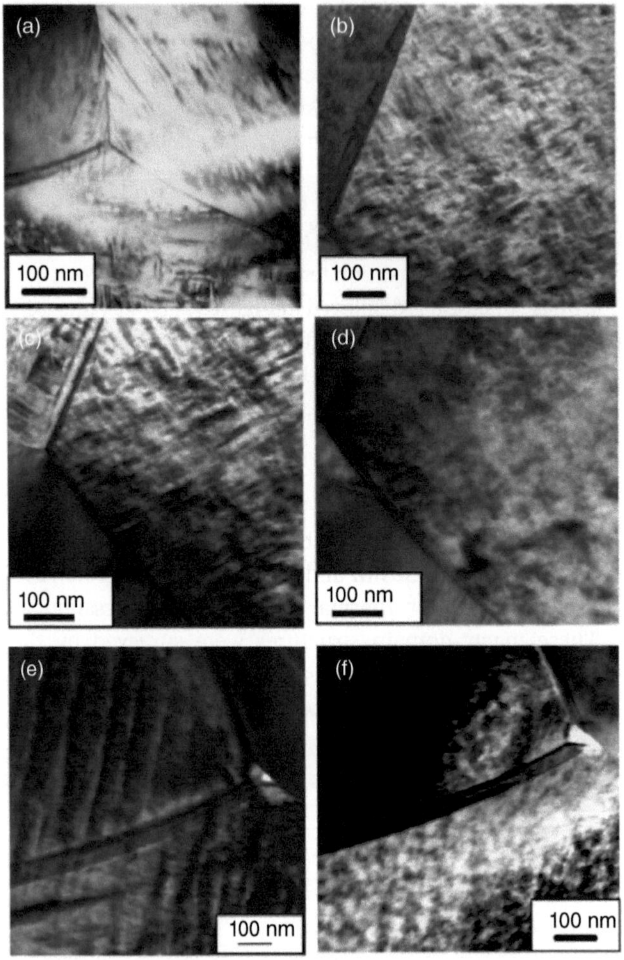

Figure 7.9 TEM images of 0.22BS–0.25PMN–0.53PT–$MgNb_2O_6$ ceramics at different measured conditions. (a) 1-mol% $MgNb_2O_6$-excess and (b) 3-mol% $MgNb_2O_6$ excess in 0.22BS–0.25PMN–0.53PT sample. Heating a zero-field cooled 3-mol% excess 0.22BS–0.25PMN–0.53PT sample at (c) 180 °C, and (d) is observed on cooling the sample after 180 °C heating. 1-mol% excess $MgNb_2O_6$ 0.22BS–0.25PMN–0.53PT sample field cooled under a 5 kV/cm ac bias at (e) 150 °C, and (f) is observed on cooling the sample after 150 °C heating. Source: Stringer et al. [28]. Reproduced with permission from John Wiley & Sons.

random domain structure relating to frozen polar micro-domains (80–100 nm) was found by comparing different diffraction contrasts of TEM techniques. A-site and oxygen vacancies breaking symmetry were the origin of relaxor behavior because the superlattice associating with B-site cation ordering was weak [28]. The freezing temperature, which could be calculated from the dielectric spectrum by the V–F equation, was 160 °C, which means that 0.22BS–0.25PMN–0.53PT ceramics showed obvious relaxor behavior above 160 °C. These mottled domains began to be activated at 160 °C.

$BiScO_3$–$PbTiO_3$–$Pb(Zn_{1/3}Nb_{2/3})O_3$ is another important ternary system [87, 88]. Where, the MPB region was located in 0.6PT with a high d_{33} of 490 pC/N, k_p of 0.57, and T_C of 417 °C [87]. Figure 7.10 shows that the temperature dielectric spectrum exhibited that the dielectric peaks displayed broader and diffused when PT content decreased, showing classical relaxor behavior, which was caused by chemical heterogeneities and stress/strain [14, 28, 87].

After that, scientists studied more characteristics on the basis of BS–PT–PZN ceramics. The multivalence Mn (Mn^{2+} and Mn^{3+}) was introduced into 0.35BS–0.60PT–0.05PZN ceramics as both hard and soft dopants. Mn replacing Ti or Sc at the B-site reduced the cell volume and promoted the movement of perovskite octahedron. The results exhibit the T_m, d_{33}, Q_m, and k_p dependent on Mn content, and this ternary ceramics with Mn can achieve a high d_{33} of 560 pC/N accompanied by high T_C of 436 °C. The grain size increased from 0.5 to 8 μm as Mn content increased because oxygen vacancies could cause mass transport during sintering [88]. Differently, the multivalence Fe in 0.35BS–0.6PT–0.05PZN ceramics near MPB expresses "hard" behavior and the properties [89]. The reasons varied in 0.35BS–0.6PT–0.05PZN ceramics with different Fe content, which was dominated by tetragonality (Fe < 0.4 mol%) and by defects (Fe > 0.4 mol%). Besides, 0.4 mol% Fe was also the boundary for the variety of grain size [89].

Zhao et al. found $Pb(Sn_{1/3}Nb_{2/3})O_3$ end-member (similar structure with PZN) and prepared $(0.95–x)BiScO_3–xPbTiO_3–0.05Pb(Sn_{1/3}Nb_{2/3})O_3$ novel ceramic system. This ceramic system could achieve high d_{33} of 555 pC/N (@ room temperature) and 2450 pm/V (@ 200 °C) with high T_C of 408 °C at MPB ($x = 0.61$) [90]. The large d_{33} and insensitive temperature characteristics could be due to the flat energy of MPB and the polarization rotation and polarization extension at the electric field stimulation [90]. *Ex-situ* d_{33} and k_p results show that the T phase of BS–PT–PSN ceramics has the highest depolarization temperature above 400 °C [90]. The theoretical calculation results of Gibbs free energy drew higher energy barrier of the tetragonal phase than the rhombohedral phase and MPB [19, 90, 91].

7.3.4 Influence of Grain Size Effect

Grain size, grain boundary, and grain distribution influence the electric properties. The grain size effect has also attracted more attention and been introduced into piezoelectric ceramics because of the requirements of miniaturization and high integration of piezoelectric devices [92]. Two famous models are applied to explain the grain size effect. One is the internal stress model, which suggests that stress (from paraelectric to ferroelectric phase transition) enhances as the grain size decreases, especially at the nanoscale [93]. Another is the domain wall model, which is an important extrinsic factor for improved dielectric and piezoelectric properties [94]. The grain size is proportional to the square of domain size, especially for the grain size of 1–10 μm [77]. The decrease in grain size means the increased grain boundary can restrict the domain wall motion and reduce extrinsic contribution [92].

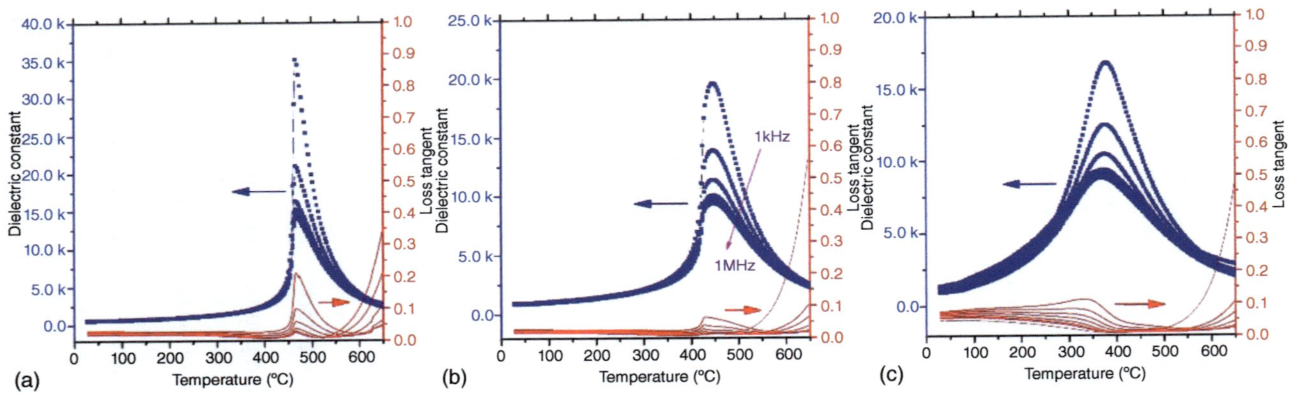

Figure 7.10 Temperature dependences of the dielectric constant and loss tangent at different frequencies from 1 kHz to 1 MHz for BS−xPT−PZN ceramics at (a) $x = 0.70$, (b) $x = 0.64$, and (c) $x = 0.54$. Source: Yao et al. [87]. Reproduced with permission from AIP Publishing.

The grain size effect of $BaTiO_3$ ceramics has been widely investigated, which could be divided into three stages. For grain size >1 μm, the domain width is smaller with a smaller grain size. The smaller domain can easily switch, leading to extraordinary piezoelectric properties [94]. For grain size < 1 μm, the reduction of domain width increases the domain wall proportion, which deteriorated d_{33} because of the "dilution" effect [95]. In BS–PT ceramics, most of the grain size effect was studied by modifying other compounds [13], while this behavior covered up the real grain size effect because of the addition. Therefore, it can be considered that the grain size effect of BS–PT ceramics has not been cleared systematically. Dong et al. synthesized BS–PT-based ceramics with different sintering temperatures from 1100 to 1180 °C, and studied the grain size effect of BS–PT–$BiSn_{1/3}Nb_{2/3}O_3$ ceramics with different grain sizes [77]. The microstructure and properties are shown in Figure 7.11 [77]. The T-phase structure was more distinct, and the domain changed from striped-like to relaxation behavior (similar to the domain structure of 0.22BS–0.25PMN–0.53PT) as the grain size increased [28, 77]. The maximum piezoelectric and ferroelectric properties could be achieved at the sintering temperature of 1120 °C with an average grain size of 2 μm [77].

Algueró et al. studied the grain size effect of BS–PT ceramics with grain sizes across scales ranging from 1.6 μm to 26 nm at 100–500 K and 100 Hz–40 GHz. Nanocrystalline powders obtained by mechanosynthesis were applied to prepare coarse-grained BS–PT ceramics by traditional sintering and to prepare submicrometer and nanostructured ceramics by SPS under tailored conditions [92]. The broadband dielectric properties in BS–PT ceramics can be contributed to polar nanoregions, which resolved the problem of broadband dielectric study [92]. The studies of impedance data provided a true size effect [59]. In detail, the imaginary part of relaxor-like peaks was narrow at small grain sizes, and the grain size was non-monotonic with the maximum permittivity (between grain size of 50 nm and 1.6 μm) at low frequencies above 250 K. They cannot be separated at microwave frequencies, which signified that the main grain size effect was a low-frequency process essentially [59, 92].

Li and coworkers discussed the grain size effect of 0.37BS–0.63PT nanoscale ceramics below 100 nm (23, 33, and 70 nm), which was obtained by sol–gel method (calcined green powder) and SPS (the first sintering step for relative density 75–80%) as well as TSS (last sintering process for relative density above 95%) [34]. The comparison of properties and microstructure shows that the ceramics possessed dense and thin grain boundaries influencing the electric performances of BS–PT nanoscale ceramics. The grain boundaries were affected by two major variations of microstructure [4, 34]. One is the domain configuration, leading to the increase in domain wall energy and lattice stress, causing the concentration of defects/charge near boundaries [34]. Another factor is the enhancement of grain boundary volume fraction, limiting the movement of domain walls sharply [34]. The strain loops and piezoelectric hysteresis loops showed that the polarizations were switchable for all ceramics of all grain sizes, which indicates that BS–PT ceramics kept a ferroelectric state even for the grain size of 23 nm. The local piezoelectric coefficient (d_{33}*) exhibited a larger value at smaller grain sizes because of intrinsic (came from grain size) and extrinsic effects (such as domains and grain boundaries) [34, 60].

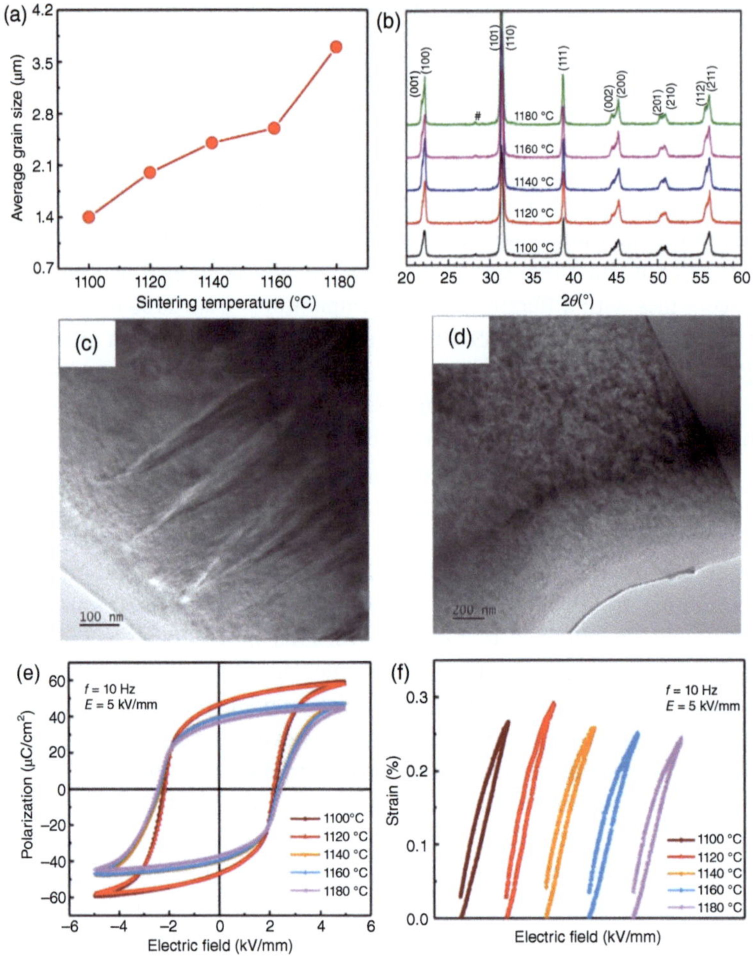

Figure 7.11 (a) The average size and (b) phase structure dependent on sintering temperature. The domain of BS–PT–BSN ceramics sintered at (c) 1100 °C and (d) 1180 °C. (e) P–E loops and (f) unipolar strain curves of BS–PT–BSN ceramics sintered at different temperatures. Source: Dong et al. [77]. Reproduced with permission from Wiley.

7.4 Applications of BS–PT-Based Ceramics

BS–PT ceramics display high T_C and high piezoelectric constant, which are promising candidates to replace PZT ceramics for high-temperature applications, such as sensors, actuators, transducers, and energy harvesters [5]. Not only are the properties of ceramics very important for high-temperature applications, but also for the design and preparation process of the devices. In this section, these devices based on BS–PT ceramics are discussed systematically.

7.4.1 Transducers Application

Ultrasonic transducers can be applied in many fields, such as ultrasonic processing, ultrasonic nondestructive testing, and medical ultrasonic imaging. Ultrasonic transducers are usually used for both receiving and transmitting sound waves. One application principle is that the driving electric field stimulates the transducer, resulting in the appearance of the displacement of piezoelectric ceramics and the ultrasonic signal. Another application principle is that the sound wave stimulates piezoelectric ceramics generating an electrical response, which can be used for detection [16, 96, 97].

Ultrasonic transducers with BS–PT-based piezoelectric ceramics are mainly utilized in high-temperature online monitoring and nondestructive testing and imaging, which are much needed in nuclear plants, well-logging, and geothermal exploration [96]. A high-temperature ultrasonic transducer was designed and prepared by pure 0.36BS–0.64PT ceramic. The results displayed that this kind of transducer can only keep the high peak–peak value of 800 mV at 200 °C, and it cannot stand higher temperatures because of the limited temperatures of isolating epoxy [97]. While the transducer based on BS–PT ceramics was prepared according to the structure in Figure 7.12a, which can keep stable electrical resonance up to 290 °C [22]. These results identified the potential of BS–PT ceramics for high-temperature applications. Dong et al. designed and prepared a kind of high-temperature ultrasonic transducer with a delay line applying BS–BPZnT ceramics, as shown in Figure 7.12b [16]. There was no epoxy in BS–BPZnT-based transducer, which may be the reason for the high-temperature application. Besides, −6 dB BW surpassed 100% and kept a little

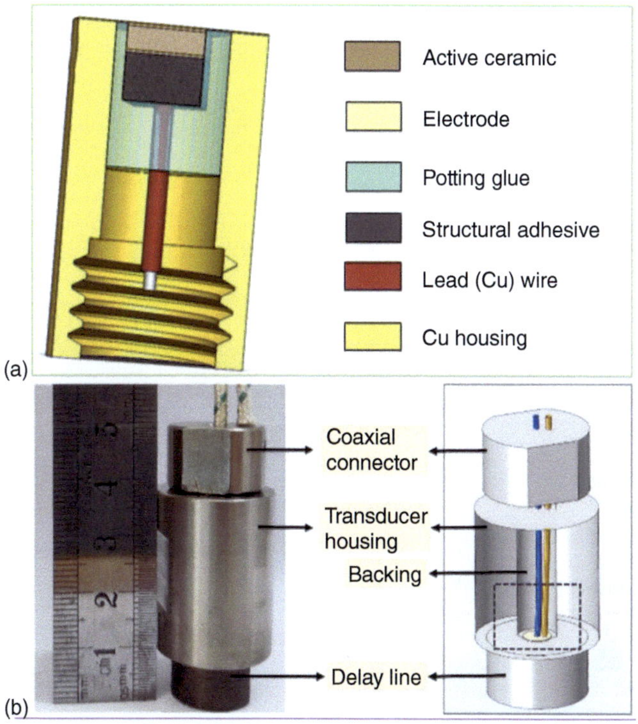

Figure 7.12 Structure diagram of (a) BS–PT-based and (b) BS–BPZnT ceramics based high-temperature ultrasonic transducer structure. Source: Refs. [16, 22]. Reproduced with permission from Elsevier, IEEE. (a) From Sun et al. [22]/Reproduced with permission from IEEE. (b) From Dong et al. [16]/Reproduced with permission from Elsevier.

enhanced tendency as the temperature increased to 450 °C, which was beneficial for high imaging resolution to get clearer images and detect finer structures [97].

Compared with piezoelectric ceramics and single crystals, piezoelectric composites have the advantage of tailored acoustic impedance, high electromechanical coupling factors, and better flexibility, which are necessary for high imaging resolution and bandwidth. While, because the polymer filler has poor thermal stability, it limits the high-temperature application of piezoelectric composites [31, 52]. Li et al. designed a 1–3 piezoelectric composite based on Mn-modified BS–PT ceramic and cyanate ester AroCy XU 371 (T_g~345 °C). The results showed that this piezoelectric composite had a high k_t of 58% and low dielectric loss (even up to 300 °C) [31]. Liu et al. prepared a transducer by 1-3 BSPT/epoxy composites, which displayed broader bandwidth (41.59%) and lower center frequency (8.93 MHz) than BS–PT-based transducer [31, 52]. The results show that 1–3 BSPT/epoxy composites have the potential in the application of high frequency and high resolution [52].

7.4.2 Actuators Application

Piezoelectric actuators, which can be activated at low frequency, are utilized in precision manipulation and miniaturized systems positioning because of their compact size and high resolution [23, 98–100].

To achieve high displacement, Chen et al. put forward a shear-bending (d_{15} mode) actuator based on BS–PT ceramics. This kind of actuator displays a circular ring with polarization along the radial direction [23]. The operating principle as well as the equipment of this shear-bending actuator were exhibited in Figure 7.13. The

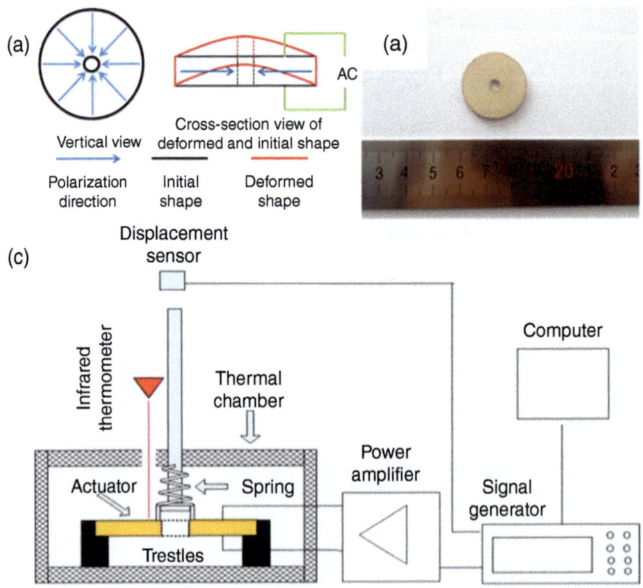

Figure 7.13 (a) Working principle; (b) photograph; (c) experimental set-up measurement system of the shear-bending actuator with ring shape. Source: Chen et al. [23]. Reproduced with permission from AIP Publishing.

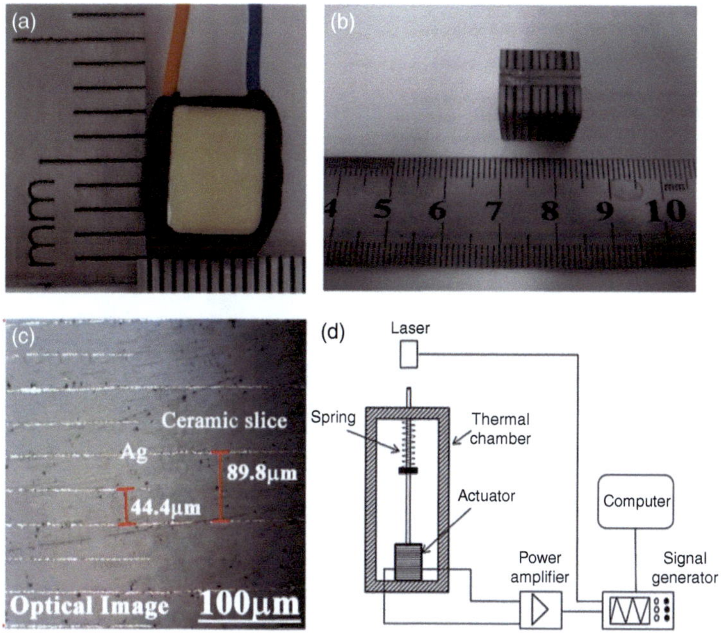

Figure 7.14 (a) Optical image of a multilayer actuator and (c) its border amplified region of BS–PT–CuO [13]. (b) Photograph and (d) its experimental setup of BS–PT multilayer actuators [12]. Source: Refs. [12, 13]. Reproduced with permission from Wiley, IEEE. (a) and (c) From Qaiser et al. [13]/Reproduced with permission from Wiley. (b) and (d) From Chen et al. [12]/Reproduced with permission from IEEE.

experiment results showed that the displacement exhibited increase at first and then a decrease with the enhancement of thermal and electric stimulation. The maximum surface displacement near the inner circle achieved 17.5 μm at 5 kV/cm and 300 °C, which identified that this actuator could work at 300 °C [23].

BS–PT piezoelectric stacks were prepared by multilayer design, which was applied to design multilayer actuators as shown in Figure 7.14 [12, 13]. BS–PT ceramics' high-temperature multilayer actuator had much better temperature stability than the PZT-5H actuator and achieved a high strain of 0.115% at 200 °C [12]. In this adhesive structure, the strain was reduced, which restricted the advantage of multilayer structure. Researchers utilized low-temperature co-firing technology, obtained Cu^{2+} doped BS–PT ceramic and designed a high-temperature multilayer actuator [13]. This multilayer actuator (4.3 mm in thick) was constituted with ceramics (44 μm in thick) and Ag electrodes (2 μm in thick). The results show that it can keep high strain above 0.13% within 20–200 °C with the driving electric field of 200 V (around 5 kV/mm), showing great potential in nanoscale positioning and precise manipulation [13].

7.4.3 Energy Harvester Application

The cantilever high-temperature energy harvester is the most widely studied for BS–PT-based ceramics. In this, the piezoelectric layer is connected to a flexible

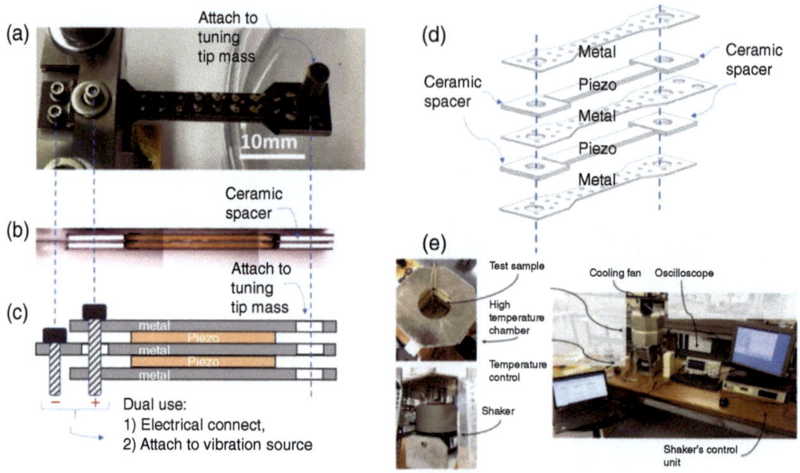

Figure 7.15 (a) Top and (b) side views of this harvester. (c) Diagram of sandwich type. (d) Explosion drawing cantilever devices. (e) The experimental setup for high-temperature measurement. Source: From Chen et al. [24]. Reproduced with permission from AIP Publishing.

cantilever, and a mass block is installed at the end of the cantilever beam for enlarging the voltage amplitude [24, 35, 45, 101].

Wu et al. designed a BS–PT-based piezoelectric harvester with a bimorph-type cantilever structure. Meanwhile, the usage of pre-tightening clamper instead of epoxy resin avoided adhesive failure at high temperatures [45]. The results exhibited that this harvester could generate electricity from room temperature to 250 °C, which displayed a maximum output power of 11 μW at 250 °C with almost no degradation (13 μW at room temperature), showing high-temperature stability [45]. Chen et al. designed a new sandwich-structured-type harvester by optimizing the mass based on BS–PT ceramics, as shown in Figure 7.15 [24]. Its maximum outpower density was 29.77 mW/g^2cm^3 at room temperature. High-temperature performance exhibited that it is a very promising strategy for intermediate-level vibrational harvesters at high temperatures up to 300 °C [24].

Zhao et al. designed a cantilever beam $(1-x)$BS–xPT piezoelectric energy harvester, and the test results are shown in Figure 7.16 [101]. This energy harvester can still generate electricity at 350 °C, and it can stabilize output power (the variation is lower than ±20%) within 100–250 °C, exhibiting great temperature stability [101]. All the design and experimental results indicate that BS–PT ceramics are promising in the application of high-temperature piezoelectric power generation. BS–PT-based energy harvesters are expected to replace batteries to realize the self-power supply of high-temperature microelectronic devices [101].

7.5 Summary and Challenge

Among high-performance Pb-based piezoelectric ceramics, BS–PT ceramics are regarded as an important candidate for high-temperature devices because of their

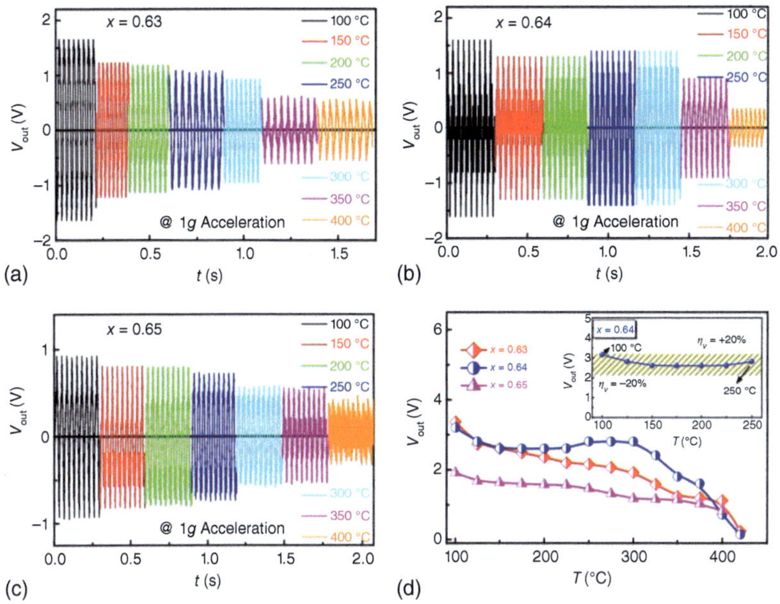

Figure 7.16 V_{out}–T curves of (a) $x = 0.63$, (b) $x = 0.64$, (c) $x = 0.65$ harvesters measured at different temperatures, (d) the output voltage as a function of temperature for three harvesters at resonate frequencies under 1g acceleration. Insert is the enlarged V_{out}–T curve for $x = 0.64$ harvesters in the temperature region of 100–250 °C. Source: Zhao et al. [101]. Reproduced with permission from Wiley.

high d_{33} and high T_C. The achievements in material design and device verification over the past two decades have been fruitful. They have been summarized and analyzed in this chapter, while some issues need to be addressed. Here, we present some existing issues and research suggestions, which may pave a path for the development and applications of BS–PT ceramics.

(1) With the development of integration and miniaturization, the grain size of BS–PT ceramics obtained by the traditional solid-state reaction method is about 3 μm, which cannot meet the requirements of miniaturized devices. The ceramic prepared by chemical method is nanoscale grain size; however, the chemical method has poor repeatability and high cost. Therefore, it is necessary to seek a low-cost and simple preparation technology to prepare fine grain size BS–PT-based ceramics.
(2) The studies on BS–PT ceramics are multidirectional, research on the stress effect, annealing atmosphere, mechanical quality factors, and conductive mechanism are still in the primary stage. The absence of these results in incomplete cognition of BS–PT ceramics. Hence, more perspectives should be considered in the following studies for better development of the BS–PT system.
(3) The superiority of BS–PT ceramics is the high T_C of 450 °C and a high d_{33} of 450 pC/N. However, the results showed that one parameter would increase at the expense of another. There is a lack of methods to increase the d_{33} and T_C at

the same time. Meanwhile, the in-depth mechanism between piezoelectricity, T_C, and thermal stability are still needed to be explored.

(4) The positioning of BS–PT ceramics is the high-temperature application, so the device design requirements are very high. Now, the adhesive applied in the device design is glue, many of which cannot stand high temperatures. Finding high temperature-resistant and suitable glue or another method to adhere all the components in the designed devices is also very challenging and important.

References

1 Jaffe, B., Roth, R., and Marzullo, S. (1954). *J. Appl. Phys.* 6 (25): 809–810.
2 Jaffe, H. (1958). *J. Am. Ceram. Soc.* 11 (41): 494–498.
3 Cheng, J., Eitel, R., Li, N., and Cross, L.E. (2003). *J. Appl. Phys.* 1 (94): 605–609.
4 Buscaglia, M.T., Buscaglia, V., Viviani, M. et al. (2004). *Nanotechnology* 9 (15): 1113–1117.
5 Eitel, R.E., Randall, C.A., Shrout, T.R. et al. (2001). *Jpn. J. Appl. Phys.* 10 (40): 5999–6002.
6 Eitel, R.E., Randall, C.A., Shrout, T.R., and Park, S.-E. (2002). *Jpn. J. Appl. Phys.* 41: 2099–2104.
7 Randall, C.A., Eitel, R.E., Shrout, T.R. et al. (2003). *J. Appl. Phys.* 11 (93): 9271–9274.
8 Randall, C.A., Eitel, R.E., Shrout, T.R. et al. (2004). *J. Appl. Phys.* 7 (95): 3633–3639.
9 Chen, S., Dong, X.L., Mao, C.L., and Cao, F. (2006). *J. Am. Ceram. Soc.* 10 (89): 3270–3272.
10 Yu, G., Dong, X.L., Wang, G.S. et al. (2010). *J. Appl. Phys.* 107: 106102.
11 Yu, G., Dong, X.L., Wang, G.S. et al. (2010). *J. Alloys Compd.* 1 (500): 56–60.
12 Chen, J.G., Liu, G.X., Li, X.T. et al. (2013). *IEEE Trans. Ultrason. Ferroelectr. Freq. Control* 3 (60): 446–450.
13 Qaiser, M.A., Ma, X.Z., Ma, R.T. et al. (2019). *J. Am. Ceram. Soc.* 9 (102): 5424–5431.
14 Eitel, R.E., Shrout, T.R., and Randall, C.A. (2004). *Jpn. J. Appl. Phys.* 12 (43): 8146–8150.
15 Yao, Z.H., Liu, H.X., Liu, Y. et al. (2008). *Appl. Phys. Lett.* 92: 142905.
16 Dong, Y.Z., Cheng, D.X., Zhou, Z.Y. et al. (2022). *Sens. Actuator A Phys.* 340: 113528.
17 Zhao, H.Y., Hou, Y.D., Yu, X.L. et al. (2021). *J. Mater. Chem. A* 4 (9): 2284–2291.
18 Eitel, R.E., Zhang, S.J., Shrout, T.R. et al. (2004). *J. Appl. Phys.* 5 (96): 2828–2831.
19 Dong, Y.Z., Zhao, K.Y., Zhou, Z.Y., and Liang, R.H. (2022). *J. Am. Ceram. Soc.* 105: 6898–6909.
20 Akça, E., Duran, C., Kowalski, B., and Sehirlioglu, A. (2021). *J. Asian Ceram. Soc.* 9: 1–8.
21 Jiang, Y., Qin, B., Zhao, Y. et al. (2008). *J. Am. Ceram. Soc.* 9 (91): 2943–2946.

22 Sun, X., Yang, X., Zhao, T.L. et al. (2021). *IEEE Trans. Ultrason. Ferroelectr. Freq. Control* 2 (69): 761–768.

23 Chen, J.G., Liu, G.X., Cheng, J.R., and Dong, S.X. (2015). *Appl. Phys. Lett.* 3 (107): 032906.

24 Chen, W.T., Gurdal, A.E., Tuncdemir, S. et al. (2020). *J. Appl. Phys.* 144102: 128.

25 Dai, Z.H., Liu, W.G., Lin, D.B., and Ren, X.B. (2018). *Mater. Lett.* 215: 46–49.

26 Chen, S., Dong, X.L., Yang, H. et al. (2007). *J. Am. Ceram. Soc.* 2 (90): 477–482.

27 Li, Q.S., Xiao, D.Q., and Zhu, J.G. (2011). *J. Inorg. Mater.* 2 (26): 185–190.

28 Stringer, C.J. and Randall, C.A. (2007). *J. Am. Ceram. Soc.* 6 (90): 1802–1807.

29 Zou, T.T., Wang, X.H., Wang, H. et al. (2008). *Appl. Phys. Lett.* 19: 93.

30 Shahzad, K., Li, L., Li, Z. et al. (2010). *Ferroelectrics* 1 (402): 142–149.

31 Li, L.L., Zhang, S.J., Xu, Z. et al. (2013). *Phys. Status Solidi A* 9 (210): 1888–1891.

32 Yu, G., Chen, X.F., Cao, F. et al. (2010). *Solid State Commun.* 23–24 (150): 1045–1047.

33 Sterianou, I., Reaney, I.M., Sinclair, D.C. et al. (2005). *Appl. Phys. Lett.* 24 (87): 242901.

34 Zhang, S., Wang, X., Wang, H., and Li, L. (2014). *J. Eur. Ceram. Soc.* 10 (34): 2317–2323.

35 Sun, Y., Chen, J., Li, X. et al. (2019). *Nano Energy* 61: 337–345.

36 Belik, A.A., Iikubo, S., Kodama, K. et al. (2006). *J. Am. Chem. Soc.* 3 (128): 706–707.

37 Belik, A.A., Azuma, M., Saito, T. et al. (2005). *Chem. Mater.* 2 (17): 269–273.

38 Shrout, T.R., Zhang, S.J., Eitel, R., Stringer, C., Randall, C.A. (2004). High performance, high temperature perovskite piezoelectrics. In: *14th IEEE International Symposium on Applications of Ferroelectrics (ISAF-04), Montreal, Canada*, 126–129.

39 Zheng, T., Wu, J., Xiao, D., and Zhu, J. (2018). *Prog. Mater. Sci.* 98: 552–624.

40 Noheda, B., Wu, L., and Zhu, Y. (2002). *Phys. Rev. B* 66 (6): 060103.

41 Zheng, H., Reaney, I., Lee, W. et al. (2002). *Ferroelectrics* 1 (268): 125–130.

42 Noheda, B., Cox, D.E., Shirane, G. et al. (1999). *Appl. Phys. Lett.* 14 (74): 2059–2061.

43 Fang, M.X., Rajput, S., Dai, Z.H. et al. (2019). *Acta Mater.* 169: 155–161.

44 Drahus, M.D., Jakes, P., Erdem, E. et al. (2011). *Phys. Rev. B* 6: 84.

45 Wu, J.G., Shi, H.D., Zhao, T.L. et al. (2016). *Adv. Funct. Mater.* 39 (26): 7186–7194.

46 Datta, K., Walker, D., and Thomas, P.A. (2010). *Phys. Rev. B* 14: 82.

47 Zhang, S.J., Randall, C.A., and Shrout, T.R. (2004). *Jpn. J. Appl. Phys.* 9A (43): 6199–6203.

48 Glinchuk, M.D. and Farhi, R. (1996). *J. Phys. Condens. Matter* 8: 6985–6996.

49 Liu, Z.H., Wu, H., Yuan, Y. et al. (2022). *Opin. Solid State Mater. Sci.* 26: 101016.

50 Shinekumar, K. and Dutta, S. (2014). *J. Electron. Mater.* 2 (44): 613–622.

51 Chen, J. and Cheng, J. (2015). *Ceram. Int.* 1 (41): 1617–1621.

52 Lin, P.F., Fei, C.M., Hou, S. et al. (2018). *IEEE Sens.* 14 (18): 5685–5690.

53 Zhao, W., Wang, X., Hao, J. et al. (2006). *J. Am. Ceram. Soc.* 4 (89): 1200–1204.

54 Zou, T.T., Wang, X.H., Zhao, W., and Li, L. (2008). *J. Am. Ceram. Soc.* 1 (91): 121–126.

55 Zhang, L., Xu, Z., Li, Z.R. et al. (2008). *J. Electroceram.* 21: 605–608.

56 Amorin, H., Alguero, M., Campo, R.D. et al. (2015). *Sci. Technol. Adv. Mater.* 1 (16): 016001.

57 Hungría, T., Amorín, H., Algueró, M., and Castro, A. (2011). *Scr. Mater.* 1 (64): 97–100.

58 Amorin, H., Hungria, T., Canovas, A.R.L. et al. (2011). *J. Nanopart. Res.* 9 (13): 4189–4200.

59 Amorín, H., Jiménez, R., Deluca, M. et al. (2014). *J. Am. Ceram. Soc.* 9 (97): 2802–2809.

60 Buscaglia, M.T., Viviani, M., Buscaglia, V. et al. (2006). *Phys. Rev. B* 6 (73): 064114.

61 Bratkovsky, A. and Levanyuk, A. (2005). *Phys. Rev. Lett.* 10 (94): 107601.

62 Wang, H., Wang, X.H., Zhang, S.P. et al. (2010). *Key Eng. Mater.* 434–435: 335–339.

63 Junmin Xue, D.W., Lee, S.-E., and Wang, J. (1999). *J. Am. Ceram. Soc.* 7 (82): 1687–1692.

64 Dai, Z., Xie, J., Liu, W. et al. (2019). *Mater. Lett.* 241: 55–59.

65 Yan, Y., Geng, L.D., Liu, H. et al. (2022). *Nat. Commun.* 1 (13): 3565.

66 Min-Seon Lee, J.-W.P. and Jeong, Y.H. (2021). *Ceram. Int.* 34405–34413.

67 Sehirlioglu, A., Sayir, A., and Dynys, F. (2008). *J. Am. Ceram. Soc.* 9 (91): 2910–2916.

68 Yue, Y., Hou, Y., Zheng, M. et al. (2019). *J. Am. Ceram. Soc.* 5 (102): 2658–2665.

69 Zhang, X., Chen, Z., Cross, L.E., and Schulze, W. (1983). *J. Mater. Sci.* 4 (18): 968–972.

70 Chen, Y., Lan, D.J., Chen, Q. et al. (2006). *Mater. Res. Soc. Symp. Proc.* 888: 1–6.

71 Zhang, S.J., Eitel, R.E., Randall, C.A. et al. (2005). *Appl. Phys. Lett.* 26: 86.

72 Wu, J.G., Yu, Y., Li, X.T. et al. (2015). *J. Am. Ceram. Soc.* 10 (98): 3145–3152.

73 Chen, S., Wang, G.S., and Dong, X.L. (2008). *Ferroelectrics* 363: 21–26.

74 Zhao, H.Y., Hou, Y.D., Yu, X.L. et al. (2018). *J. Appl. Phys.* 19 (124): 194103.

75 Jiang, Y., Qin, B., Chen, Y. et al. (2007). *IEEE Trans. Ultrason. Ferroelectr. Freq. Control* 12 (54): 2505–2509.

76 Liu, Z., Zhao, C.L., Li, J.F. et al. (2018). *J. Mater. Chem. C* 3 (6): 456–463.

77 Dong, Y.Z., Zhou, Z.Y., Liang, R.H., and Dong, X.L. (2020). *J. Am. Ceram. Soc.* 9 (103): 4785–4793.

78 Yao, Z., Liu, H., Liu, Y. et al. (2009). *Mater. Res. Bull.* 7 (44): 1511–1514.

79 Yao, Z., Song, Z., Liu, H. et al. (2013). *J. Alloys Compd.* 577: S488–S492.

80 Suchomel, M.R., Fogg, A.M., Allix, M. et al. (2006). *Chem. Mater.* 18: 4987–4989.

81 Chen, J., Xing, X.R., Sun, C. et al. (2008). *J. Am. Chem. Soc.* 130: 1144–1145.

82 Wang, F.F., Fan, L.L., Ren, Y. et al. (2014). *Appl. Phys. Lett.* 25 (104): 252901.

83 Lee, J.K., Youn, H.J., Hong, K.S., and Park, S.E. (1999). *J. Appl. Phys.* 1 (85): 368–371.

84 Hu, Q., Yang, Y., Wang, Y. et al. (2018). *Ceram. Int.* 6 (44): 6817–6822.

85 Talanov, M.V., Bush, A.A., Kamentsev, K.E. et al. (2018). *J. Am. Ceram. Soc.* 2 (101): 683–693.

86 Chen, J.G., Zhao, T.L., Cheng, J., and Dong, S.X. (2013). *J. Appl. Phys.* 14: 113.

87 Yao, Z.H., Liu, H.X., Hao, H., and Cao, M.H. (2011). *J. Appl. Phys.* 1 (109): 014105.

88 Yao, Z., Liu, H., Cao, M. et al. (2011). *Mater. Res. Bull.* 8 (46): 1257–1261.

89 Liao, Q., Chen, X., Chu, X. et al. (2013). *Sens. Actuator A Phys.* 201: 222–229.

90 Zhao, T.L., Bokov, A.A., Wu, J.G. et al. (2019). *Adv. Funct. Mater.* 12 (29): 1807920.

91 Dong, Y.Z., Zhou, Z.Y., Liang, R.H., and Dong, X.L. (2022). *J. Am. Ceram. Soc.* 2 (105): 1558–1567.

92 Balciunas, S., Ivanov, M., Grigalaitis, R. et al. (2015). *Phys. Status Solidi* 12 (252): 2727–2734.

93 Buessem, L.E.C.W.R. and Goswami, A.K. (1966). *J. Am. Ceram. Soc.* 39: 36–39.

94 Arlt, G., Hennings, D., and With, G.D. (1985). *J. Appl. Phys.* 4 (58): 1619–1625.

95 Ghayour, H. and Abdellahi, M. (2016). *Powder Technol.* 292: 84–93.

96 Kazys, R. and Vaskeliene, V. (2021). *Sensors* 21: 3200.

97 Fei, C.L., Zhao, T.L., Zhang, J.S. et al. (2018). *J. Alloys Compd.* 743: 365–371.

98 Rupitsch, S.J., Bergmann, C.P., and Brazil, P.A. (2018). *Piezoelectric Sensors and Actuators*. Heidelberg: Springer Berlin.

99 Chen, J.G., Chen, Z.J., Li, X.T., and Dong, S.X. (2013). *Appl. Phys. Lett.* 5 (102): 052902.

100 Chen, J.G., Cheng, J.R., and Dong, S.X. (2014). *J. Adv. Dielectr.* 01: 04.

101 Zhao, H.Y., Hou, Y.D., Yu, X.L. et al. (2019). *J. Am. Ceram. Soc.* 9 (102): 5316–5327.

8

$Bi_4Ti_3O_{12}$-Based High-temperature Piezoelectric Ceramics and Their Applications

Yanyan Zhang[1,2], Jianfeng Huang[1,2], Yangyang Zhou[1,2], Chun Guo[1,2], and Zhiyong Zhou[1]

[1] *Shanghai Institute of Ceramics, Chinese Academy of Sciences,, 588 Heshuo Road, Jiading District, Shanghai 201899, China*
[2] *University of Chinese Academy of Sciences, Center of Materials Science and Optoelectronics Engineering, Shijingshan District, Beijing 100049, China*

8.1 Introduction

Bismuth layer-structured ferroelectrics (BLSFs), first discovered by Aurivillius in 1949 [1], are compounds with alternating arrangements of bismuth oxide and pseudo-perovskite with the chemical formula $(Bi_2O_2)^{2+}(A_{m-1}B_mO_{3m+1})^{2-}$, where A is ions with larger radii suitable for 12 coordination, while B is ions with smaller ionic radii suitable for octahedra, and m is an integer representing the number of layers. Since their Curie temperatures (T_c, typically greater than 500 °C) are higher than those of the perovskite materials and can be used at 450 °C or even higher for sensors to meet the needs of aerospace, oil exploration, nuclear plants, and other fields, their structure, properties, and applications have been increasingly investigated in recent years.

As the typical and one of the earliest discovered BLSFs, the bismuth titanate compound ($Bi_4Ti_3O_{12}$, BIT) has gotten hundreds of reports on its high-temperature piezoelectric aspects. With a high T_c of 675 °C and large spontaneous polarization (P_s) of up to 50 μC/cm^2, BIT is a preferred material for high-temperature piezoelectric devices at 500 °C. In the twentieth century, scholars focused on the crystal structure and phase structure of BIT, while since 2000 (back in the mid-nineteenth century), research has been focused on improving the electrical properties by doping modification or optimising the preparation process, especially the former, which has been widely reported. Here, we will summarize the crystal structure, preparation process, domain structure, electrical properties, stability under external fields, and device-related reports of BIT from multiple perspectives, hoping to help readers better understand BLSFs as high-temperature piezoelectric materials.

Piezoelectric Materials: From Fundamentals to Emerging Applications, First Edition. Edited by Jiagang Wu.

8.2 Crystal Structure and Phase Transition of BIT Compounds

The crystal structure of the BIT compound reported first by Aurivillius was pseudo-tetragonal (actually orthorhombic) in 1949 [1]. The more precise crystal structure of BIT at room temperature was later determined to be monoclinic by Newnham and co-workers [2] through neutron diffraction patterns in 1972, and the phase transition process was even more thoroughly investigated in the twentieth century [3, 4]. The X-ray diffraction (XRD) data were almost consistent with the orthogonal structure (space group *B2cb*) with a = 5.448(2) Å, B = 5.411(2) Å, c = 32.83(1) Å, and Z = 4 [2].

The Bi atoms in BIT have two chemical coordination environments: the Bi atom in the perovskite layer is usually labeled Bi(1) with a coordination number of 12, while the Bi atom in the $(Bi_2O_2)^{2-}$-layer is noted as Bi(2) with an ideal coordination number of 8 and bonds to the four O(2) atoms within the $(Bi_2O_2)^{2-}$-layer and the four O(4) atoms at the top of the perovskite layer [5]. Figure 8.1 shows the external

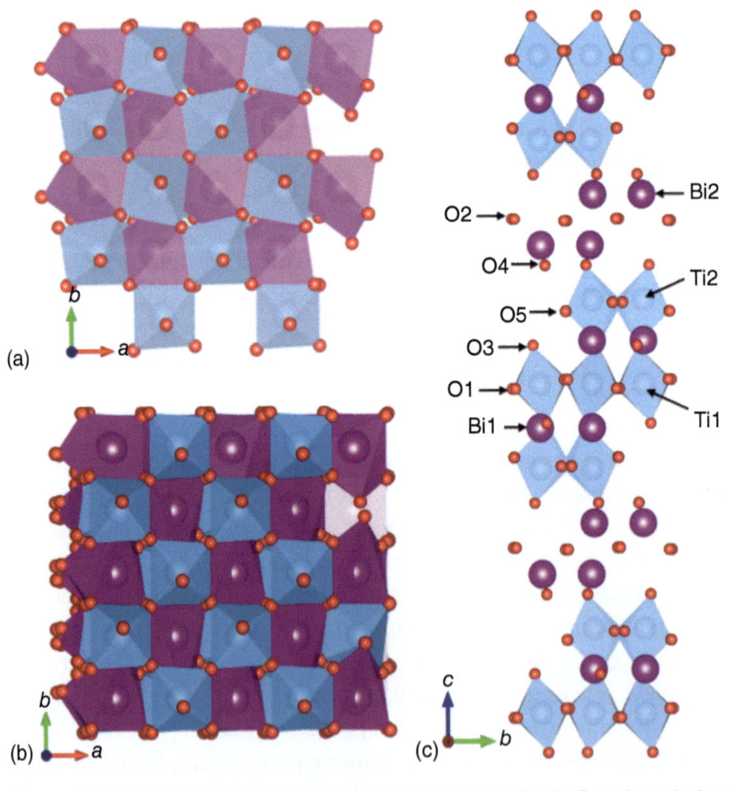

Figure 8.1 Outer (a) and inner (b) perovskite layers of $Bi_4Ti_3O_{12}$ viewed along [001], crystal structures of the aristotype (tetragonal, *I4/mmm*) and the ambient-temperature (monoclinic, *B1a1*) phases, (c) unit cell morphology and atomic positions of $Bi_4Ti_3O_{12}$. Source: (a, b) Newnham et al. [6]. Reproduced with permission from Elsevier. (c) Subbarao [3]/International Union of Crystallography/CC BY 4.0.

(a) and internal (b) perovskite layers of BIT in the [001] direction [6]. The adjacent oxygen octahedra are rotated alternately between each other with a rotation angle of about 12°. The cations of the oxygen octahedra of BIT are not in the central position and have a displacement of very large distances: 0.35 Å for the cations of the outer oxygen octahedra and 0.27 Å for the inner ones. The Ti—O bond lengths are therefore divided into two groups: three in the range 1.73–1.85 Å and three in the range 2.10–2.41 Å. The internal perovskite of the BIT does not rotate around the *c*-axis, as shown in Figure 8.1b. The Ti—O bond in the central perovskite layer of BIT has two short bonds of 1.75 Å, two bonds of 1.97 Å, and two long bonds of 2.20 Å.

In the paraelectric phase at high temperature, the Bi in the bismuth–oxygen layer is coordinated with the adjacent oxygen atoms to form four Bi—O bonds of equal length, forming a pyramid with tetrahedral symmetry. It has already been mentioned that below the T_c, the crystal structure of BIT transforms into a monoclinic phase with low symmetry, as shown in Figure 8.1c. The structure is distorted to a greater extent; the equal-length Bi—O bonds no longer exist, and the bond lengths between the bismuth and oxygen atoms are all different. In the ferroelectric phase, the Bi atom combines not only with four oxygen atoms in the Bi–O layer but also with a top oxygen atom in the perovskite layer to form a fifth Bi—O bond, the formation of which is one of the main structural changes during the phase transition. The direction of the bond is perpendicular to the direction of the polar axis [100], with an anti-parallel displacement along the *b* axes. The short Bi—O bond shears the perovskite layer, resulting in a rotation of the octahedron about 4° around the *a*-axis. Since three consecutive octahedra rotate at the same angle, the displacement of O at the external vertices is the same.

As the temperature increases, the lattice parameter of BIT shows an essentially linear rate of expansion until the phase transition occurs at 675 °C. The lattice parameter *a* expands faster than *b*, so the orthogonal distortion (*b*/*a*) decreases with the increase in temperature. At 675 °C, the lattice parameters *a* and *c* expand abruptly while *b* contracts, as shown in Figure 8.2.

A further study by Zhou and Gou et al. found that the transition from the paraelectric phase (*I4/mmm*) to the ferroelectric phase (*B2cb*) is not a single-step process [3, 4]. Γ_5^- was used to describe the shift of Ti^{4+}, and X^{3+} was used to represent the tilt

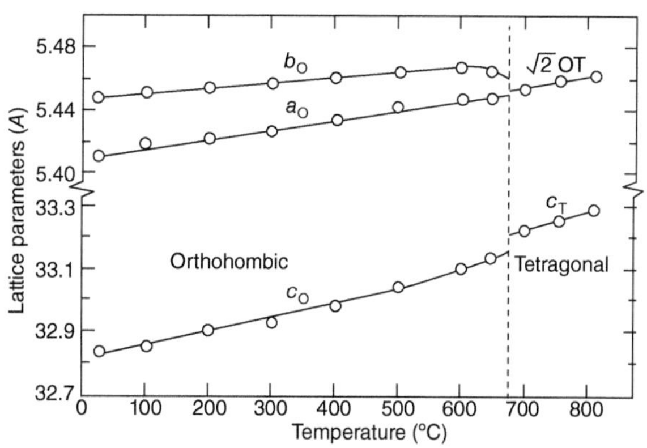

Figure 8.2 Lattice parameters of $Bi_4Ti_3O_{12}$. Source: Subbarao [3]/ International Union of Crystallography/CC BY 4.0.

of the TiO_6 octahedron. During the phase transition, the displacement of ions and the tilting of oxygen octahedra were sequential. Three independent tilt patterns of the B1a1 model were observed perpendicular to and parallel to the *c*-axis. The TiO_6 octahedra were tilted toward each other, and the tilted system could be described by Glazer notation as $a^-a^-c^0$ [7]. One possibility was that the initial distortion was a shift of Ti^{4+} leading to a distortion of the phase without any tilting of the TiO_6 octahedron, transforming to *Fmm2*, followed by a tilting of the TiO_6 octahedron through the X_3^+ mode, thereby transforming to the room temperature monoclinic phase *B2cb*. Alternatively, the initial distortion was the tilting of the TiO_6 octahedron leading to the *Cmca* phase followed by a shift of cation. Similar to $SrBi_2Ta_2O_9$, the BIT phase transition was preceded by the tilting of oxygen octahedral, followed by a shift of cation [8]. Thus, the BIT ferroelectric phase transition is from the paraelectric phase (*I4/mmm*) first to the intermediate phase (*Cmca*) and finally to the ferroelectric phase (*B2cb*).

The ferroelectric polarization of BLSF materials usually results from the synergistic contribution of the displacements of several ions [9]. In particular, the relative displacement of the entire perovskite layer with respect to the Bi_2O_2 layer (the *F2mm* mode) leads to a large P_s along the *a*-axis. Although displacements of large ions are also found along the *b*-axis, the dipole moments cancel each other due to the presence of a slip surface perpendicular to the *b*-axis. Thus, the ferroelectric polarization of BLSF materials occurs mainly along the *a*-axis. The P_s is small or zero along the *c*-axis, depending on whether m is odd (smaller P_s) or even ($P_s = 0$). In the BIT ($m = 3$) system, P_s values along the *a*-axis are 50 μC/cm², while P_s values along the *c*-axis are only 4 μC/cm² [10]. The ferroelectricity of BIT is mainly due to the displacement of Ti^{4+} and the rotation of the TiO_6 octahedral chain along the *a*-axis.

8.3 Fabrications of BIT Powders and Ceramics

The preparation process of piezoelectric ceramics is an important factor that affects the structure and properties of the materials. At present, the main synthesis methods of BIT powders are the solid-state method and the wet-chemical method. In general, the widely used fabrication of calcination and sintering for BIT is based on the solid-state reaction (SSR) method, which has been applied for years. For the improvement of the electrical properties of BIT compounds, several other techniques have been reported recently.

8.3.1 Synthesis of BIT Powders

Among all the preparation methods, the SSR is the simplest and most widely used synthesis method. However, the synthesis temperature of the powder prepared by this method is around 800 °C and has the disadvantages of coarse particle size, low purity, and inadequate powder mixing, which will have a great impact on the piezoelectric properties of the materials. The electric properties of lots of BIT-based components prepared by the SSR method will be introduced in detail in Section 8.5 of this chapter.

The sol–gel method is a wet chemical method for preparing ultrafine powders by using a colloidal dispersion system. Compared with the SSR method, the product obtained by the sol–gel method has fine particles, high purity, and good chemical uniformity. The BIT powders with high stability and good reproducibility were synthesized by the sol–gel method at a synthesis temperature of 500 °C, and with increasing the temperature to 600 °C, the grain size of BIT increased rapidly from 50 to 100 nm [11]. Besides, it was verified that the powder prepared by the sol–gel method facilitates lower sintering temperatures (T_s) while improving the densities and electrical properties of ceramics compared to SSR. The BIT ceramics prepared by the sol–gel method have a higher density (relative density: 98%), dielectric constant (ε: 249), and residual polarization (P_r: 2.699 $\mu C/cm^2$) [12].

The coprecipitation method is a preparation method in which a large amount (excess) of precipitant is added during the mixing of raw material solutions so that each component of the solution is precipitated at the same time in proportion. However, in practice, it is difficult to form precipitation at the same time as required after adding precipitant to each raw material. As a result, the formed components always have problems of uneven composition and component segregation, leading to the failure of the coprecipitation method and low purity. Marina et al. prepared W^{6+}-doped BIT-based ceramics by a chemical route based on hydroxide coprecipitation [13]. The results showed that the W^{6+}-doped BIT powder consisted of small soft agglomerates of 1 μm, which were formed by very small equiaxed particles of about 100 nm. Also, the relative density of ceramics sintered at 900 °C was found to be more than 99%, while the relative density of ceramics sintered at 1050 °C by conventional SSR method was only 94%.

The molten salt method mixes the reactants with salts that have a low melting point. After heating, the reactants undergo a chemical reaction in the molten salt, which greatly increases the diffusion rate of the dissolved reaction ions and makes it easier for the reaction to proceed in the molten salt environment. After cooling, the molten impurity salts can be cleaned to obtain the reaction products. The effects of the type of raw material and molten salt on the crystalline and microscopic properties of BIT were investigated, and it was found that larger particle sizes could be obtained using precursors of amorphous BIT as starting materials [14]. $Bi_4Ti_{2.94}W_{0.06}O_{12}$ powder with good grain dispersion and uniform grain size was prepared by the molten salt method, and then BIT-based ceramics with high grain orientation were obtained by ordinary pressureless sintering. The investigation of the source of grain orientation showed that this was mainly due to the contribution of pressure to the orientation of BIT sheet grains during molding and partly due to the contribution of grain growth during sintering [15].

The hydrothermal method is a method of synthesizing and preparing materials using an aqueous solution as the reaction solvent. This method has the advantages of uniform particle size, light agglomeration, no need for ball milling and high-temperature calcination, cheap raw materials, and good repeatability of the preparation process. However, there are many variable and uncontrollable factors in the hydrothermal method that are easily affected by the reaction conditions in the reactor. It was found that when the Bi/Ti ratio was 1.33–1.6, pure BIT powder could be synthesized by the hydrothermal method [16]. The experimental results by varying the KOH content showed that the second phase would appear when the content of KOH was less than 1.0 M or more than 5.0 M. It was also found that single-phase BIT grains were precipitated when the reaction temperature was increased to 140 °C, the reaction speed was accelerated, and the grain size became larger with the increase in the reaction temperature and time.

Solution combustion synthesis is a chemical synthesis technology that has developed rapidly in recent years. It mainly uses the redox reaction of oxidant and reducing agent to release heat and initiate combustion. In 2012 and 2014, Subohi et al. successfully prepared single-phase BIT powders by solution combustion synthesis using glycine and urea as fuel, respectively. The whole synthesis process lasted only about five minutes, but the product still needed to be calcined at 800 °C for three hours to eliminate the second phase [17, 18].

The high-energy ball milling method does not require the use of organic reagents and is synthesized in a closed reaction environment, which effectively prevents the stoichiometric mismatch of bismuth-layered compounds containing the volatile element Bi. Bi_2O_3 and TiO_2 oxides were directly passed through 15 hours of high-energy ball milling to obtain almost single-phase BIT powders, and the particle size of the mixture gradually decreased with the increasing ball milling time. After sintering at 850 °C, BIT ceramics with ε of 243, dielectric loss (tan δ) of 0.017, and P_r of 24 μC/cm^2 were obtained [19].

The characteristics of the above preparation processes are shown in Table 8.1.

At present, the synthesis of BIT powder is mainly carried out by the above methods. Each synthesis method has its own characteristics and distinctive powder morphology. The appropriate method can be selected according to the actual needs. Figure 8.3 is the scanning electron microscope (SEM) images of BIT powders synthesized by the above partial methods.

8.3.2 Fabrication of BIT Ceramics

Currently, the widely used ceramic preparation process is the traditional solid-state sintering method, but it needs a long experimental period, high sintering temperature, and poor compactness, which makes it difficult to obtain excellent electrical properties. Therefore, related texturing techniques are introduced to the preparation of BIT ceramics. The commonly used texturing methods can be briefly divided into two categories: one is thermal, force, electrical, and magnetic field-based methods such as hot forging/pressing, spark plasma sintering (SPS), and magnetic orientation, and the other is the template method based on the shape of raw materials.

8.3.2.1 Preparation Process Based on an External Field

Heat treatment technology is a kind of technology that applies a certain pressure to ceramics in a high-temperature environment to make the grains slip between grain boundaries while moving through dislocations during the growth process, and finally, obtain ceramics with preferentially oriented textures. It mainly includes hot pressing (HP), hot forging (HF), and discharge plasma sintering.

Table 8.1 Characteristics of several common preparation processes.

Method	Process temperature	Grain size	Characteristic
Solid-state reaction	~850 °C	10 nm–1 μm	Easy to operate, short preparation cycle, low cost of raw materials, and convenient promotion.
Sol–gel	400–650 °C	30–100 nm	The calcination temperature is low, the reaction controllability is strong, and the purity of the prepared material is high.
Coprecipitation	600–750 °C	100–300 nm	High failure rate, low purity, and relatively cumbersome process.
Molten salt	850–1000 °C	1–10 μm	The operation process is simple, the dispersion of the synthesized powder is good, the grain shape is regular, and the grain size difference is small.
Hydrothermal	100–180 °C	~200 nm	The particle size distribution of the powder is uniform, and the grain morphology is good, but there are many changeable and uncontrollable factors in the preparation process.
Solution combustion synthesis	~800 °C	1–10 μm	Low ignition temperature, fast heating rate, short reaction time, simple equipment, and low raw materials
High-energy ball milling	500–600 °C	50–350 nm	By rapidly refining the raw material size and enhancing the activity of the powder, the pre-sintering temperature is reduced, and the preparation efficiency is improved.

Figure 8.3 SEM images of BIT powders synthesized by some methods. (a) Solid-state reaction method, (b) sol–gel method, (c) coprecipitation method, (d) molten salt method, (e) hydrothermal method, (f) solution combustion synthesis. Source: (a) From Jiang et al. [15]/with permission from Elsevier. (b) Du et al. [11]/Reproduced with permission from John Wiley & Sons, Inc. (c) Ng et al. [20]/Reproduced with permission from Elsevier. (d) From Jiang et al. [15]/Reproduced with permission from Elsevier. (e) From Chen et al. [21]/Reproduced with permission from The Ceramic Society of Japan. (f) Subohi et al. [17]/Reproduced with permission from Elsevier.

The method of applying uniaxial pressure to a sample after pressureless sintering and holding it at high temperature for a period can yield highly oriented ceramics is called HF. In 1980, Tadashi et al. successfully prepared BIT piezoelectric textured ceramics by HF technology, as shown in Figure 8.4. The degree of grain orientation was up to 95%, and the piezoelectric properties reached 60–80% of those of single crystals [22]. The $2P_r$ of the BIT ceramics with strong anisotropy prepared by Nagata et al. was 24.6 μC/cm^2 by HF technique [27].

HP is a sintering method for preparing highly oriented ceramics in which the powder is continuously heated under uniaxial pressure while forming and sintering. The orientation and piezoelectric constant (d_{33}) of BIT ceramic materials prepared by hot-press sintering were 70–80% and 25 pC/N, respectively [28].

SPS utilizes the synergistic effect of Joule heat generated by pulse current and mechanical pressure to achieve rapid densification of materials. The $Bi_{3.15}Nd_{0.85}Ti_3O_{12}$ ceramic with significant anisotropy was obtained by the SPS sintering technique, which had two to three times more Pr and d_{33} in the vertical direction than in the parallel direction [24].

Magnetic field orientation technology is a technology to realize the directional arrangement of material grains under the action of a strong magnetic field, which requires the material to have magnetic anisotropy. In 2003, the BIT ceramics with high density and orientation distribution (orientation degree ~0.9) were first prepared by magnetic field orientation technology [26]. In the following years, the effect of powder properties on the orientation of BIT ceramics prepared by magnetic field orientation technique was intensively studied. It was shown that the Lotgering factor increased with increasing powder size, reaching saturation with an orientation of 0.96 at D50 = 0.9–1.2 μm [29].

8.3.2.2 Preparation Process Based on Templates

Template grain growth (TGG) is a technique for preparing textured and thin-layered ceramics. BIT powder prepared by the molten salt method is commonly used as a template, and it is uniformly dispersed into the substrate and then arranged by the tape casting method. The Lotgering coefficient of ceramics prepared by using BIT prepared by the molten salt method as a template and by tape casting was about 0.96 after sintering at 1000 °C [30]. The d_{33} of Nb^{5+}-modified $Bi_4Ti_{3-x/5}Nb_{x/5}O_{12}$-textured ceramics prepared by the TGG technique can be up to 30 pC/N parallel to the casting direction, 77% of that of a single crystal [25].

Through the above methods, BIT piezoelectric ceramics with a certain degree of orientation can be successfully prepared (the above SEM images of solid phase sintering are from the unpublished work of our research group). There is no significant difference in the grain morphology of the ceramics prepared by each method in the parallel and vertical directions, and they are all orderly stacked.

The characteristics of the above ceramic preparation process are shown in Table 8.2.

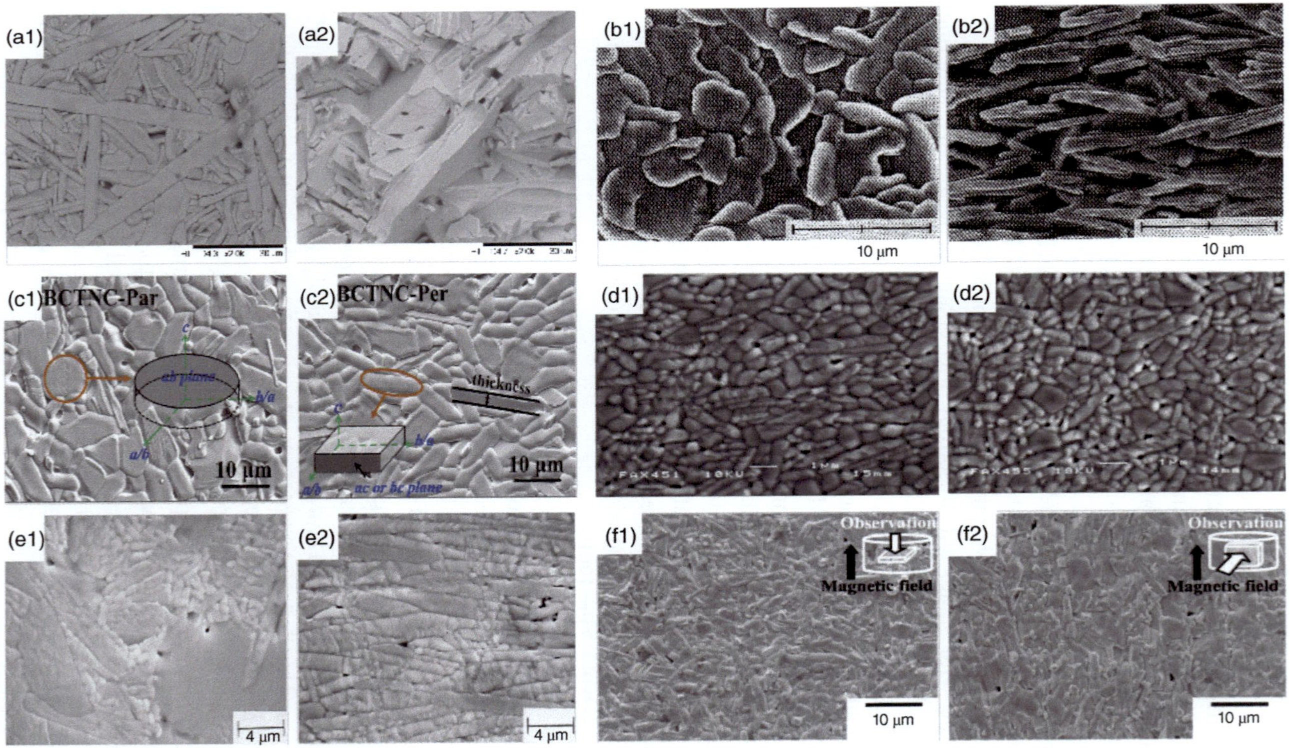

Figure 8.4 SEM images of BIT piezoelectric ceramics prepared by various methods (a) solid-phase sintering, (b) hot-forging sintering, (c) hot-pressing sintering, (d) spark plasma sintering, (e) template grain growth method, (f) magnetic field orientation method. Source: (a) From Wu (Book Author). (b) Takenaka and Sakata [22]/Reproduced with permission from IOP Publishing. (c) Shi et al. [23]/Reproduced with permission from Elsevier. (d) Liu and Shen [24]/Reproduced with permission from Elsevier. (e) Hong et al. [25]/Reproduced with permission from John Wiley & Sons, Inc. (f) Makiya et al. [26]/Reproduced with permission from The Ceramic Society of Japan.

Table 8.2 Characteristics of common ceramic preparation methods.

Method	Process parameter	Degree of orientation (%)	d_{33} (pC/N)
Solid-phase sintering	T_s: 1050–1050 °C	—	7–40
Hot-forging sintering	T_s: 930–1170 °C P: ~78.4 MPa	~95	—
Hot-pressing sintering	T_s: 750–1080 °C P: 29.4–256 MPa	70–80	16–25
Spark plasma sintering	T_s: 780–1100 °C P: 30–75 MPa	~70	[⊥] = 3.7–18.3 [//] = 7.5–20.7
Template grain growth method	T_s: 900–1150 °C Preparing templates: $Bi_4Ti_3O_{12}$	83–96	30–33
Magnetic field orientation method	T_s: 1050–1200 °C Magnetic field intensity: 10 T	~90	—

T_s, sintering temperature; P, uniaxial pressure during sintering.

8.4 Ferroelectric Domain Structure of BIT Compounds

8.4.1 Space Group Analysis of BIT Domain Boundaries

In some solid-state transitions, the crystal structures of the two phases before and after phase transitions are very similar, and there are often subgroup and supergroup relationships between their space groups. The subgroup and supergroup relationship is accompanied by a reduction in symmetry, and in the low-symmetry phase, the structure of domains is thus formed. The symmetry group of the supergroup with higher symmetry is *G*. The symmetry group of the phase with lower symmetry after the phase transition is *H*. *H* is a subgroup of *G*. The phase with lower symmetry splits into several domains, some of which are of the same type having the same crystallographic orientation, and their positions differ due to the point-like translation vector. And the domains of different types either differ in orientation or their positions differ due to something other than the point-like translation vector, or both. The supergroup of BIT in the paraelectric phase is *F4/mm* (or *14/mm*), while the ferroelectric phase becomes *B2cb*. Decomposition of the accompanying set from *F4/mm* to *B2cb*.

$$F4/mm = \{1 + 4^+ [001]\} \cdot \{1 + 2[001]\} \cdot \{1 + t(1/2,1/2,0)\} \cdot \{1 + m(001)\} \cdot B2cb$$

$$F4/mmm = \sum_{i=1}^{16} P_i \text{ B2}cb$$

By analyzing the relationship between the subgroup and supergroup in the BIT phase transition, the following independent domain boundaries exist in the BIT: antiphase domain boundary (APB), 180° domain boundary, 180° APB, 90° domain boundary, 90° APB.

8.4.2 Ferroelectric Domain Structure of BIT Crystal

In the early days, the ferroelectric domains of BIT crystals were observed by optical techniques, such as polarization microscopy. Later, with the development of microscopy, transmission electron microscopy (TEM) and Piezoresponse force microscopy (PFM) became the mainstream methods to observe the ferroelectric domains of BIT. In contrast to the large P_s of BIT, it is prone to polarization fatigue and has a smaller P_r, which is generally believed to arise from the pinning effect of domain walls (DWs). The interaction between DWs and defects (e.g. oxygen vacancies) results in a few domains that can be reversed when an external electric field is applied [31, 32]. To improve the polarization properties of BIT-based materials, it is crucial to analyze the structure of the ferroelectric domains of the BIT and their interactions with oxygen vacancies. By analyzing the ferroelectric domains with PFM, the vector information of P_s for each ferroelectric domain in the ferroelectric bulk can be obtained.

Figure 8.5a shows the in-plane PFM image of the BIT crystal in the *a–b* plane [33]. Striped domains with optical 90° DWs can be seen in the entire *a–b* plane of the crystal. To minimize the electrostatic energy, the ferroelectric DWs are usually uncharged 180° DW and 90° DW but charged 180° DWs head-to-head and tail-to-tail configurations are also seen in the BIT structure [34]. Figure 8.5b shows the out-of-plane PFM image of BIT crystal in the *a*/*b*–*c* plane. The 90° DWs observed on the *a–b* plane penetrate the crystal along the *c*-axis. It should be noted that 90° DWs is not a plane, but a strongly curved surface. BIT has a large free elastic deformation of about 0.8%. Therefore, the inhomogeneous 90° domain structure observed in BIT cannot be explained by the contribution of elastic energy alone.

The conductivity along the *a–b* plane of the BIT crystal is much higher than along the *c*-axis, which indicates more active oxygen vacancies. The movement of the 90° DWs along the *a*/*b* axis becomes viscous due to the interaction with the oxygen vacancies. When the crystal is cooled to room temperature, the complex 90° DWs structure is frozen in the *c*-axis direction before the formation of 90° DWs with the

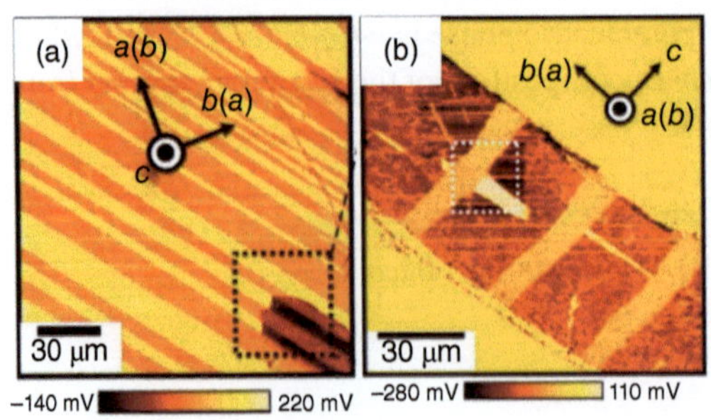

Figure 8.5 (a) In-plane PFM image, (b) out-of-plane PFM image. Source: Katayama et al. [33]. Reproduced with permission from John Wiley & Sons.

planar structure of (11a). Annealing of BIT in air at 950 °C significantly reduces the number of streaked 90° DWs, while charged 180° DWs are still present in the crystal.

8.4.3 Ferroelectric Domain Structure of BIT Ceramics

Due to the characteristics of the layered structure of the materials, the grains of BIT ceramics can be mainly divided into two types. The *a–b* plane is arranged along the *x–y* axis, marked as plane grains, and perpendicular to the *x–y* axis, marked as transverse grains. The morphology of the two grains is shown in Figure 8.6a1. It can be seen from Figure 8.6a4 that they have different PFM responses, reflecting that the polarization distributions of different grains are different [36]. In particular, the transverse grains show a strong vertical PFM (VPFM) response amplitude. The detailed analysis of the PFM signals of the planar grains shows that the structure of the ferroelectric domain of the BIT ceramics is striated. The VPFM signal shows the major type of DWs is 180°, which can be attributed to the downward and upward spontaneous polarization of the BIT ceramics. Similarly, as shown in Figure 8.6c2, 180° domains are observed in the y-lateral PFM (*y*-LPFM) signal, while there is no

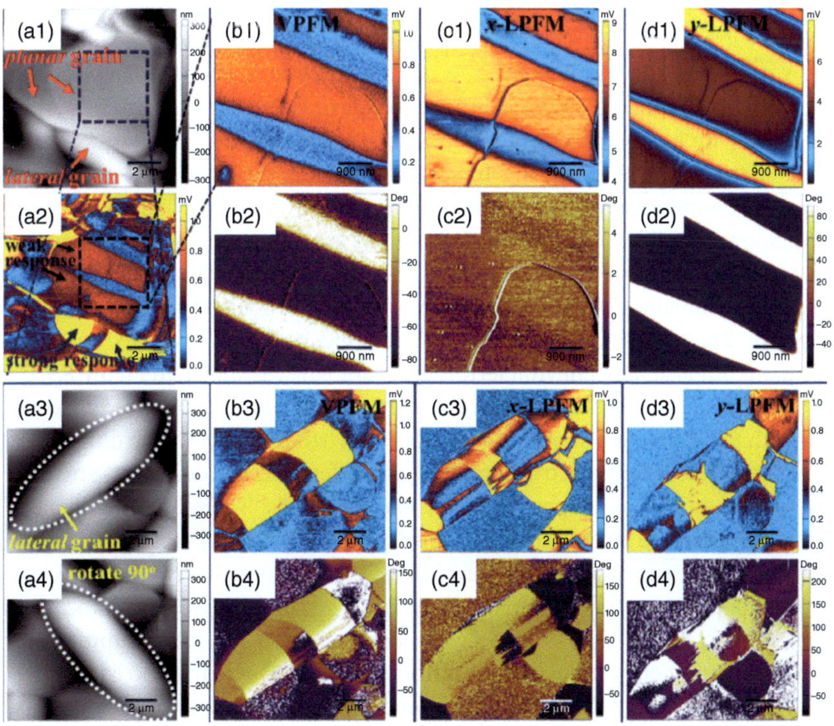

Figure 8.6 3D PFM mappings of BIT ceramics, (a1) topography, (a2) VPFM amplitude, (b1,c1,d1) amplitude, and (b2,c2,d2) phase under VPFM, *x*-LPFM, and the *y*-LPFM modes. 3D PFM mappings of a lateral grain in BIT ceramics; topography (a3) before and (a4) after 90° rotation, (b3,c3,d3) amplitude; and (b4,c4,d4) phase under VPFM and *x*-LPFM. Source: Xie et al. [35]. Reproduced with permission from Elsevier.

difference in the phase angle signal of the *x*-LPFM. Due to the difference in the polarization components of each ferroelectric domain along the *x* and *y* axes, the *x* and *y*-LPFM amplitudes are negatively correlated, and the amplitude of VPFM of the planar grains is smaller than that of LPFM [35]. Figure 8.6 shows the PFM signal of a transverse grain, which exhibits an alternating amplitude response of strong and weak signals. It is well known that the side of a BIT can be either *a*–*c* or *b*–*c* plane, or a mixture of the two. The ferroelectric inside the same grain has a large difference in piezoelectric response, indicating that the transverse grain is composed of a mixture of the *a*–*c* plane and *b*–*c* plane because the pure ac plane does not appear to have the VPFM response shown in Figure 8.6 [37]. The pure *b*–*c* plane has higher VPFM amplitude than *x*- and *y*-LPFM. According to the amplitude response and phase angle in the 3D PFM diagram, the 90° DWs formed in the transverse grains are straight, while the 180° domain nucleates on one side of the 90° DWs and grows irregularly to the other side, resulting in the formation of high local electrostatic energy, which affects the formation of domain configuration and the appearance of curved ferroelectric DWs. Due to the interaction between DWs, grain boundaries, and crystal orientation, the ferroelectric domain structure of BIT ceramics is more complex than that of their crystals.

Depending on the characteristics of BIT ceramic crystals, the morphology of the ferroelectric domains of BIT ceramics can be significantly changed by modification at the A or B sites [38]. As Figure 8.7 shows the TEM images of the ferroelectric domains of the ceramics, the width of the striped ferroelectric domains of BIT crystals is usually up to several tens of microns. The width of the striped ferroelectric domains of the doped modified BIT-based ceramics is in the range of 100–200 nm, with a clear boundary between region 1 and region 2, indicating that the spontaneous polarization directions of these two regions are different. Due to the spontaneous polarization of BIT ceramics along the *a*-axis direction, the DW between A domain and B domain in Figure 8.7b1 is 180° DW, and the boundary between region 1 and region 2 is −90° DW. The modified BIT ceramics not only have strip ferroelectric domains but also have chevron ferroelectric domains, as shown in Figure 8.7c1. The diffraction spots of the chi-shaped ferroelectric domain C in the figure are consistent with the diffraction spots of the [0*l*0] crystal band axis, while the diffraction spots of the adjacent ferroelectric domain D of the ferroelectric domain C are consistent with the diffraction spots of the crystal band axis. Therefore, the domain boundary between the ferroelectric domains C and D is 90° DW [38].

Due to the slow leap processes of solid defects such as point defects, dislocations, and grain boundaries (the intrinsic frequencies of these slow leap processes are generally in the kHz range and below, depending on the temperature). The measurement frequency range of the ferroelectric domain is from kHz to μHz, which completely covers the intrinsic frequency range of defects (e.g. point defects between adjacent equilibrium positions). The stimulus signal of the test is alternating stress, and the response signal is strain, so it applies to any object capable of propagating elastic stress waves. The height of the peak obtained is proportional to the concentration of the oxygen vacancy, and the internal dissipation peak P_2 is related to the interaction between the oxygen vacancy and the 90° domain. The increase of the P_2 peak after polarization can be observed in the experiment, indicating that the interaction between oxygen vacancy and DW is enhanced [39].

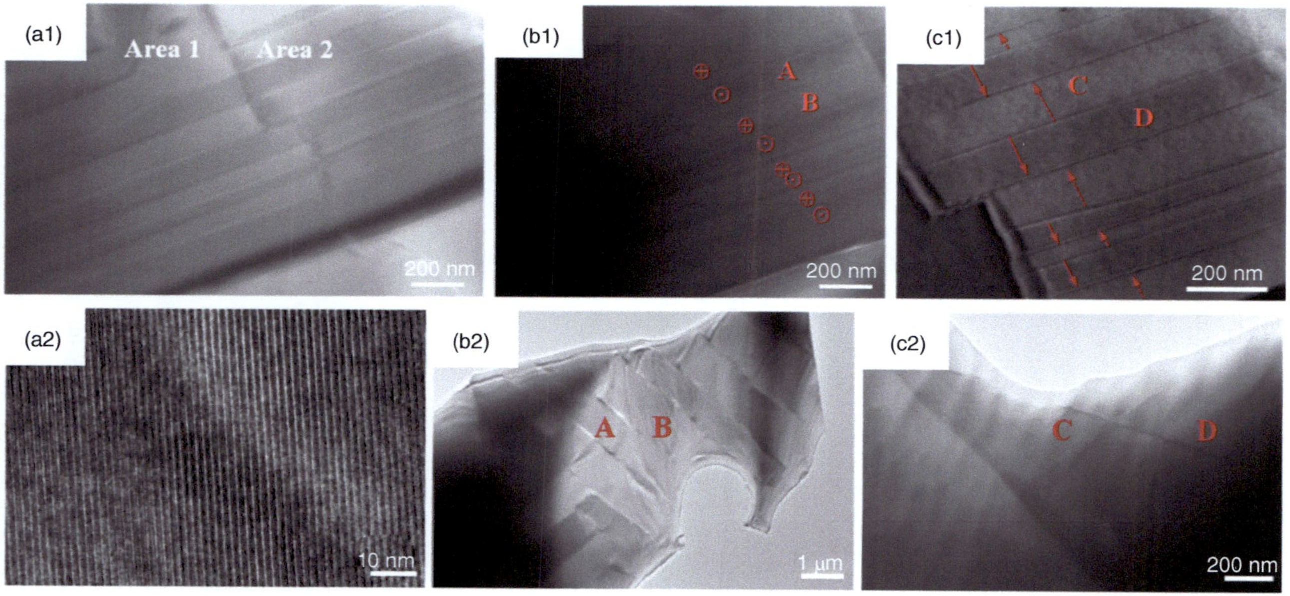

Figure 8.7 (a1) TEM image with lamellar domain structure for BIT ceramics, (b1) and (c1) TEM image of lamellar domain structure with Ps (a)-180° DWs in area 1 and area 2, (a2) TEM image with lamellar domain structure for BIT ceramics, (b2) and (c2) TEM image of lamellar domain structure with P_s(a)-180° DWs in area 1 and area 2. Source: Xie et al. [38]. Reproduced with permission from Royal Society of Chemistry.

8.5 Electrical Properties of BIT Compounds

As mentioned above, BIT has attracted much attention due to its high T_c of 675 °C and large P_s. However, its poor piezoelectric properties (d_{33}:~7 pC/N), high conductivity (~10^4 Ω·cm@500 °C), extremely high leakage current, and large coercivity field (E_c: 31.8 kV/cm) limit its practical applications. Therefore, the A or B site modification and the intrinsic relationship between doping and properties have been researched in recent years to improve the performance of the BIT. The ions commonly used to replace the A-site were rare earth ions or alkaline earth metal ions, while the ions that replaced the B-site were W^{6+}, Nb^{5+}, Ta^{5+}, Cr^{3+}, Cu^{2+}, or other ions with radii close to Ti^{4+}. And the properties of the typical BIT-based components obtained by the SSR are described in detail below.

In the early years, for the potential application of BIT in ferroelectric random access memories (FRAM), there were some studies on the modification of La^{3+} at the A-site to improve fatigue performance due to the generation of antiphase boundaries [40, 41]. As shown in Figure 8.8b, the substitution of Bi^{3+} with rare earth ions significantly reduced the T_c and dielectric loss of the BIT-based components and increased the dielectric constant at high temperatures [42]. The dielectric response is related to electron polarization, ion polarization, dipole polarization, and space charge polarization, which dominate the optical, infrared, microwave, kilohertz, or even lower frequency ranges, respectively. Increasing the disorder of the structure or the lattice distortion can increase the dielectric constant. It has also been suggested

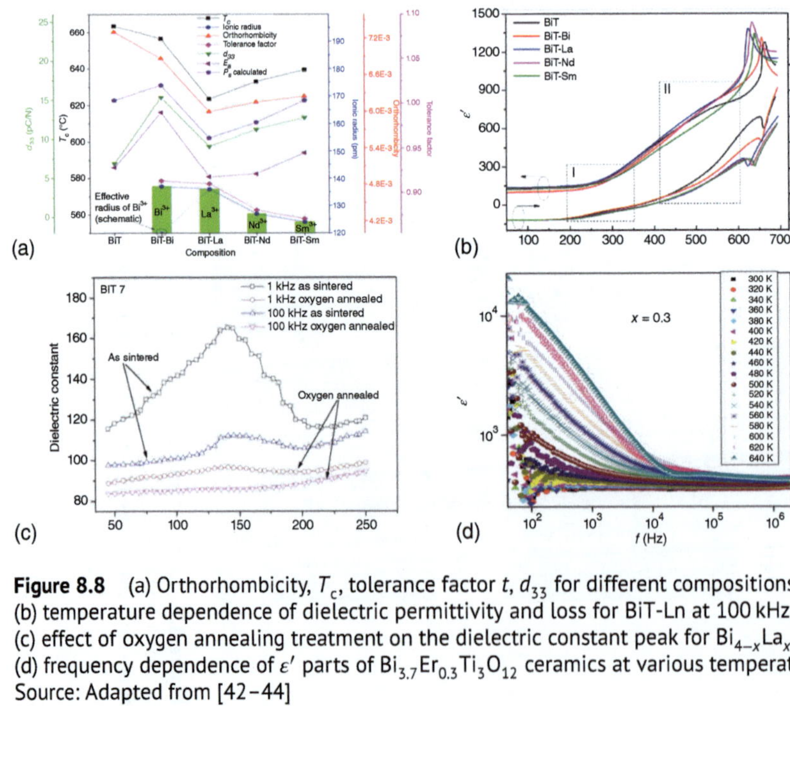

Figure 8.8 (a) Orthorhombicity, T_c, tolerance factor t, d_{33} for different compositions, (b) temperature dependence of dielectric permittivity and loss for BiT-Ln at 100 kHz, (c) effect of oxygen annealing treatment on the dielectric constant peak for $Bi_{4-x}La_xTi_3O_{12}$, (d) frequency dependence of ε' parts of $Bi_{3.7}Er_{0.3}Ti_3O_{12}$ ceramics at various temperatures. Source: Adapted from [42–44]

that the introduction of La^{3+} (1.032 Å), Nd^{3+} (0.983 Å), and Bi^{3+} (1.03 Å) of varied sizes induced random cation disorder. For the surrounding oxygen atoms, their symmetric coordination structures were required, while Bi^{3+} with 6s lone pair electrons tended to be eccentric and distorted their surrounding structures. Thus, the substitution of La^{3+} or Nd^{3+} enhanced the chemical and structural inhomogeneities, decreasing the concentration of dipole and coupling and increasing the dielectric constant. The dielectric loss of BIT comes from the contribution of multiple loss processes (ion migration loss, ion vibration loss, deformation loss, and electron polarization loss) and shows a pattern of higher in the low-frequency region and decreasing with increasing frequency. The reduction in the dielectric loss was attributed to the change in conductivity, which will be discussed in detail later. Contrary to that, the introduction of Pr^{3+} caused an increase in dielectric loss, which was speculated to be related to a new thermal activation mechanism caused by the introduction of carriers that would interact with the potential barrier [45]. The main factors that were widely accepted to affect T_c were lattice distortion (k) and tolerance factor (t), which could be calculated using Eqs. (8.1) and (8.2), respectively.

$$k = \frac{2(a-b)}{(a+b)(1)} \tag{8.1}$$

$$t = \frac{(R_A + R_O)}{\sqrt{2}(R_B + R_O)} \tag{8.2}$$

where R_A, R_B, and R_O are the ionic radii of the 12, 6, and 6 coordination sites of the A/B site ions and O^{2-}, respectively, and a and b are the lattice parameters. As shown in Figure 8.8a, the T_c of BIT-based components increases when the tolerance coefficient decreases or the orthogonality increases, which is the reason why rare earth ion modification decreases the T_c.

When a dielectric is subjected to an alternating electric field, the change in polarization lags the change in the electric field, causing the real part of the dielectric constant to change with frequency, which is called dielectric dispersion. The extremely high and anomalous dielectric constants at certain frequencies were thought to be closely related to polarization phenomena, dielectric relaxation, and space charge ions. In rare earth ions or other ion-modified components such as $Bi_{4-x}La_xTi_3O_{12}$ and $Bi_{4-x}Pr_xTi_3O_{12}$, relaxation behavior occurs when x is greater than a certain value [45, 46]. These have been attributed to effects such as the increase of the interface reaction due to the inhomogeneity of the composition caused by a large amount of doping, which hindered the carrier migration under electric field induction. Dielectric anomalies at low temperatures can be explained more concretely by the Maxwell–Wagner space charge relaxation model, which, as mentioned earlier, often occurs in dielectrics with multiple components and is thus often attributed to the presence of heterogeneity, such as grain, grain boundaries, and/or electrode effects.

Complex impedance and electrical modulus analysis are necessary and powerful techniques to understand the actual contribution of different microstructural components to the dielectric relaxation process in ceramics. Complex impedance data highlight the dominant resistive elements in the microstructural composition, based on which electrical modulus analysis can suppress the effect of electrode polarization and thus further elucidate the relaxation behavior. $M^* = j\omega C_0 Z^*$, where ω is the angular frequency and C_0 is the capacitance under vacuum. The relaxation properties of the Z''/Z''_{max} and M''/M''_{max} curves at high temperatures are also reflected by the separation of their frequency-dependent peak positions. The addition of Sm^{3+}, etc., alleviated the dielectric anomaly peak at 500 °C due to the reduction of oxygen vacancies, as shown in Figure 8.8b. In Figure 8.8c, the dielectric anomaly peak near 150 °C had been suppressed in samples of BIT-based components doped with La^{3+} after annealing at 800 °C for 48 hours in an oxygen-rich atmosphere. Therefore, this verified the possibility that the dielectric relaxation of BIT-based components was considered to be affected mainly by defects associated with oxygen vacancies [43]. Figure 8.8d showed the temperature dependence of real parts (ε') of complex permittivity of $Bi_{3.7}Er_{0.3}Ti_3O_{12}$ [44]. The introduction of rare earth ions altered the lattice distortion, thus leading to an increase in d_{33} to nearly 20 pC/N, as the trend of d_{33} was consistent with that of orthogonality, as in Figure 8.8a. The saturated polarization-electric field (P–E) hysteresis loop of the BIT is often characterized at test temperatures above 100 °C. The $2P_r$ value is related to the concentration and migration of defect ions, especially oxygen vacancies, and the switching of ferroelectric domains [47]. The defect ions accumulate at low degrees of freedom such as at DWs or electrode interfaces under the action of external electric fields, thus creating a pegging effect that hinders the polarization transition and thus affects $2P_r$. On the other hand, the local electric field of the BIT is induced by the local disorder, which has a directional effect on the surrounding dipoles and hinders the migration of defects [45, 48]. Shimakawa et al. proposed that the lattice distortion was also an influencing factor since La^{3+} equivalent doping could increase the magnitude of the remanent magnetic polarization [49–51].

Since the high conductivity of BIT leads to difficulties in polarization reversal, resulting in lower d_{33} and affecting stability at high temperatures, the most studied individually modified at B-site ions were those with higher chemical valence, such as Nb^{5+} and W^{6+} [52, 53]. As shown in Figure 8.9a, it was reported that the resistivity of BIT could be increased by nearly three orders of magnitude by replacing Ti^{4+} in the B-position with $W^{6+}Nb^{5+}$, etc. [54]. It is now widely accepted that the conductivity mechanism of BIT is a mixture of p-type conductivity and ionic conductivity because the conductivity can be significantly reduced by donor doping and the effect of oxygen partial pressure on conductivity. During high-temperature sintering, the volatilization of Bi_2O_3 generates Bi vacancies (V_{Bi}'''), while oxygen vacancies ($V_O^{\cdot\cdot}$) are created to balance the chemical valence. On cooling to room temperature, the material is gradually oxidized, and the concentration of $V_O^{\cdot\cdot}$ decreases and is accompanied by the formation of vacancies ($h\cdot$). The reaction process is shown in the following equations.

$$Bi_2O_3 = 2Bi(gas) \uparrow + \frac{3}{2}O_2(gas) \uparrow + 2V_{Bi}''' + 3V_O^{\cdot\cdot} \quad (8.3)$$

$$\frac{1}{2}O_2 + V_O^{\cdot\cdot} \rightarrow O_O^{\times} + 2h^{\cdot} \quad (8.4)$$

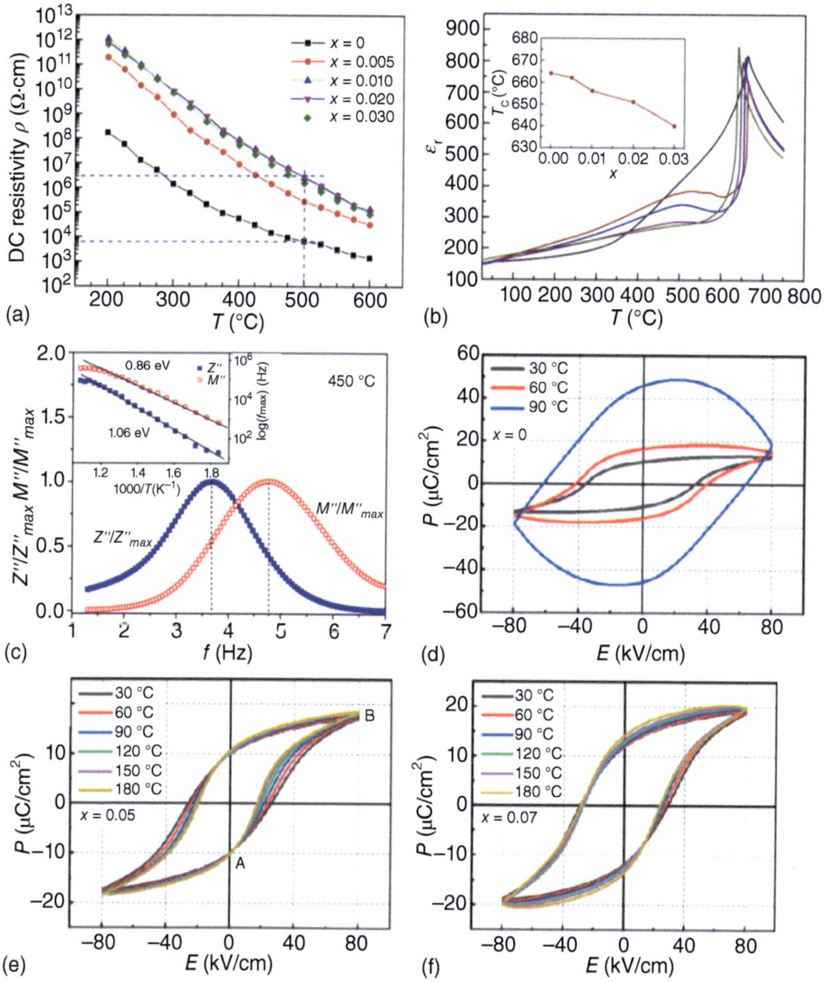

Figure 8.9 (a) Electrical conductivity as a function of temperature for the $Bi_4Ti_{3-2x}W_xNb_xO_{12}$ ceramics, (b) dielectric constant as a function of temperature for $Bi_4Ti_{3-2x}W_xNb_xO_{12}$ ceramics, (c) variation of Z''/Z''_{max} and M''/M''_{max} with frequency of $Bi_4Ti_{2.5}Fe_{0.25}Ta_{0.25}O_{12}$, (d,e) *P*–*E* loops of the $Bi_4Ti_{3-x}(Zn_{1/3}Nb_{2/3})_xO_{12}$ ceramics at 1 Hz at different temperatures, and (f) The *P*–*E* loop for BITWC. Source: Refs. [38, 54, 55]. Reproduced with permission from John Wiley and Sons, Elsevier, Royal Society of Chemistry.

The resistivity of the BIT-based component was increased because the excess electrons generated by the substitution of higher-valence ions for Ti^{4+} were bound to holes. Many reports have shown that by replacing Ti^{4+} with Nb^{5+}, W^{6+}, etc., the d_{33} of BIT increased to about 20 pC/N [53, 54, 56, 57]. Some believe this was the contribution of resistivity enhancement, while others believe it was the contribution of changes in grain morphology or domain morphology, which still needs to be further investigated. Since Ti^{4+} (0.61 Å), Nb^{5+} (0.64 Å), Ta^{5+} (0.64 Å), and W^{6+} (0.60 Å) have similar radii, their introduction did not significantly contribute to the lattice distortion relative to the rare earth elements, and thus the T_c of those was slightly decreased. The difference in the electronic configuration of the nd^0 of the transition metal ion at the B-site also influenced the T_c. Hybridization of Nb 4d orbitals with O 2p increases bonding energy and improves covalency because the extension of 4d electron orbitals is smaller than that of 5d electron orbitals. B-site with higher nd^0 electron orbitals usually decreases lattice distortion and thus T_c, so replacing the Ti^{4+}(3d) with ions having higher nd^0 configurations leads to a reduction in T_c. With high-valence ion doping, the dielectric dispersion and tan δ of the ceramics were significantly reduced. And as shown in Figure 8.9b, donor-modified BIT can effectively reduce the rate of increase of dielectric constant with temperature, especially near 400–500 °C [54]. The dielectric relaxation was evident in Figure 8.9c and could be determined by the peak frequency, whether it was due to the contribution of grain boundaries or grain bulks [55].

The effect of ion-pair modification on structure and heterogeneity is more complex, and the property changes are different compared to the individual ion modification at the B-site. The ion-pair modification of the B-site could increase the piezoelectric coefficient substantially to about 30 pC/N, and the reasons for this were also widely discussed. The increase in resistivity was one of the most recognized and important reasons for the increase in d_{33}. In addition, scholars have searched for the main reasons for the enhancement of the d_{33} mainly from the contribution of intrinsic and extrinsic to piezoelectric properties. The intrinsic contribution was judged by the change of crystal structure to the degree of tilt and rotation of TiO_6 octahedra [58], which was further judged by calculating the P_s by Xie et al. [38]. The P_s of the ferroelectric materials can be calculated by Shimakawa:

$$P_s = \sum_i \frac{m_i \times \Delta x_i Q_i e}{V} \tag{8.5}$$

where m_i, Δx_i, $Q_i e$, and V are the site multiplicity, atomic displacement, ionic charge, and volume of the unit cell, respectively. The improvement of ferroelectric properties (P_r) also contributed to the piezoelectric properties [59]. The effect of external contributions was mainly reflected through techniques such as X-ray photoelectron spectroscopy (XPS) to characterize ferroelectric domains (as detailed in the Section 8.4 above). The main ideas were that the establishment of antiphase boundaries and the increase in density of DWs, as well as the more efficient orientation of robust domains, improved the piezoelectric properties [38, 58, 60]. The variation in dielectric properties was more complex. For example, the $Bi_4Ti_{2.95}W_{0.05}O_{12.05} + 0.2\,wt\%Cr_2O_3$ ceramics had a lower dielectric constant because the Cr^{3+} pinned the DWs of ferroelectric domains [61]. The contribution of the dielectric relaxation process in $Bi_4Ti_{2.95}W_{0.05}O_{12.05} + y$ wt%Cr_2O_3 due to the directional polarization of the defect dipole related to the oxygen vacancy at high temperature proved that the interference effect of the defect dipole on the intrinsic dipole short-range interaction may also be another reason for the dielectric dispersion [62]. In the modified components, too much doping may produce a non-ferroelectric phase, as the second phase thus affecting the ferroelectric properties. For example, the secondary phase $Bi_2Fe_4O_9$ in the BIT-5$BiFeO_3$ component was not only non-ferroelectric but also contributed to the increase in leakage current density, thus reducing the ferroelectric properties of the primary phase [63]. From Figure 8.9d–f, it can be found that the P–E loops of pure BIT had a large leakage current at 90 °C with low $2P_r$. The Nb^{5+}-Zn^{2+} modification effectively reduced the leakage current so that the BIT-based component can obtain P–E loops at higher temperatures with substantially higher $2P_r$ [38] The P–E loop of the BIT base component often shows an asymmetry with unequal absolute values of $+E_c$ and $-E_c$ values, which can be represented by the internal bias field E_i.

$$E_i = \sum(e_i) = (E_{c+} + E_{c-})/2 \tag{8.6}$$

where E_i is the local microscopic electric field generated by the defect dipole, and $+E_c$ and $-E_c$ are the coercivity fields under positive and negative electric fields, respectively. Defects such as oxygen vacancies in ceramics create space charges in the electric field that can pin domains during the switching of domains, resulting in higher values of $-E_c$ than $+E_c$. In the $Bi_4Ti_{2.95}W_{0.05}O_{12.05} + y$wt%$Cr_2O_3$ (BITWC) series of components, the internal bias field of BITWC increased with y, which indicated the increase of defective dipoles. When the coupling between grain boundaries and DWs was stronger than the pinning of DWs, the E_c decreased [62, 64].

In recent years, to obtain better performance, scholars have prepared A- and B-site co-doped components by combining the doping properties of A- or B-site. The components of $Nb^{5+}/Ta^{5+}/Ce^{3+}$ co-doped BIT ceramics have significantly enhanced piezoelectric coefficients to 36.5 pC/N due to the presence of striped domains in addition to large regular bulk domains. The $Bi_{3.97}Ce_{0.03}Ti_{2.98}(WNb)_{0.01}O_{12}$ component prepared by Xie et al. had the highest piezoelectric coefficient of all the reported solid-phase sintered BIT-based components, having a d_{33} of 40.2 pC/N [65].

Resistance properties are important properties of high-temperature piezoelectric ceramics, which have been partially described earlier and are now further elaborated. The DC resistivity of BIT-based components can reach a maximum of about 10^7 Ω·cm at 500 °C. The introduction of rare earth elements, such as Ce^{3+}, Nd^{3+}, or other non-volatile ions into the A-site could reduce the concentration of volatilization of Bi, increasing the resistivity by about one order of magnitude or so. The introduction of ions with high chemical valences, such as W^{6+}, Nb^5+, and Ta^{5+}, or their combined ion pairs at the B-site effectively increased the resistivity from 10^4 Ω·cm to about 10^7 Ω·cm at 500 °C [53, 56, 57, 66, 67]. Xie et al. demonstrated that the valence change of Mn led to a decrease in oxygen vacancy concentration by the change in the ratio of lattice oxygen to vacancy oxygen area of O1s of XPS [60]. It is worth mentioning that the substitution of Ti^{4+} at the B-site by the equivalent ion pair also enhanced the resistivity due to the reduction of the carrier migration rate [38]. The magnitude of the activation energy (E_a) can reflect the change in the concentration of oxygen vacancy and its types [45, 68]. For example, the activation energy of DC conductivity of $Bi_4Ti_{2.975}W_{0.025}O_{12.025} + 0.2$ wt%Cr_2O_3 was 0.9 eV, which indicated that the external conduction mechanism dominated. The E_a of $Bi_4Ti_{2.85}W_{0.15}O_{12.15} + 0.2$ wt%Cr_2O_3 increased to 1.5 eV, close to half of the band gap, indicating the predominance of intrinsic conduction [69].

Also, microstructural components, such as grains, grain boundaries, and second equivalents, usually contribute significantly to the electrical properties of BIT-based components. Impedance analysis is a powerful technique that can distinguish the contribution of grains, grain boundaries, electrodes, and impurities to the conductive process. Conductivity measurements σac (f, T) under AC at different frequencies and temperatures are often used to characterize the jump of charges in ferroelectrics. The Z'' of $Sm_{0.75}Bi_{3.25}Ti_3O_{12}$ and $Sm_{0.75}Bi_{3.25}Ti_{2.9625}Nb_{0.03}O_{12}$ were observed to decrease with increasing temperature and merge at higher frequencies because the potential barrier of the materials decreased [70]. Several semicircles located in the Z'' and Z' relationship diagrams were considered to originate from the electrode, grain boundary, and grain bulk contributions of the ceramics, respectively. As shown in Figure 8.10a–c, typically, in some donor-doped systems,

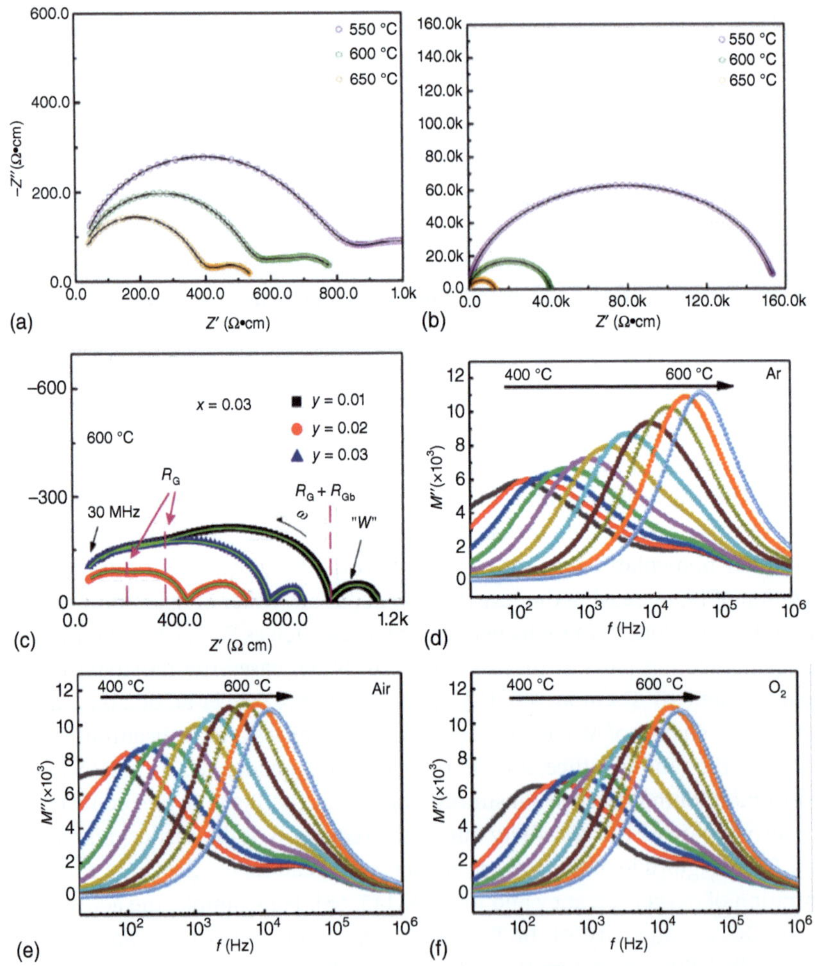

Figure 8.10 Complex Z^* plots of the compositions with (a) $Bi_4Ti_3O_{12}$, (b) $Bi_4Ti_{2.85}(Cu_{1/3}Ta_{2/3})_{0.15}O_{12}$, and (c) $Bi_{0.97}Ti_{0.75-y}Mg_y$, (d) the imaginary part of electric modulus M'' for $Bi_4Ti_{2.98}(WNb)_{0.01}O_{12}$ ceramics varied with frequency in Ar, air, and O_2 atmospheres. Source: Adapted from [59, 70–72].

Table 8.3 The electrical properties of some representative BIT-based ceramics.

Doping site	Materials	ε		$2P_r$ ($\mu C/cm^2$)	E_c (kV/cm)	T_c (°C)	d_{33} (pC/N)		References
		At RT	At T_c				At RT	Annealed[a)]	
Pure	BIT	~155[b)]	~780[b)]	8.7	38.4	675	~7	~7	[38, 42]
A Site	BIT + 3%La_2O_3	~150[c)]	~1370[c)]	—	—	~622	~9	~8.5	[42]
	BIT + 3%Nd_2O_3	~150[c)]	~1425[c)]	—	—	~631	~11	~10.5	
	BIT + 3%Sm_2O_3	~150[c)]	~1350[c)]	—	—	~635	~12.5	~12	
	$Bi_{3.6}Nd_{0.4}Ti_3O_{12}$	—	452[b)]	—	—	521	—	—	[73]
	$Bi_{3.6}Pr_{0.4}Ti_3O_{12}$	—	566[b)]	—	—	518	—	—	
	$Bi_{3.6}Gd_{0.4}Ti_3O_{12}$	—	1162[b)]	—	—	584	—	—	
	$Bi_{3.6}Dy_{0.4}Ti_3O_{12}$	—	598[b)]	—	—	588	—	—	
	$Bi_{2.9}Pr_{0.9}Ti_3O_{12}$	300[b)]	720[b)]	60	52	398	—	—	[74]
B Site	$Bi_4Ti_{2.86}Nb_{0.14}O_{12}$	—	—	—	—	655	20	—	[54]
	$Bi_4Ti_{2.95}W_{0.15}O_{12.35}$	~150[b)]	650[b)]	—	—	~615	20	—	[56]
	$Bi_{3.99}Ti_{2.98}V_{0.02}O_{12}$	138[b)]	—	—	—	679	26	—	[75]
	$Bi_4Ti_{2.96}Ta_{0.04}O_{12}$	—	—	22.4	34	—	~20	~16	[76]
	BIT + 4 at%Nb_2O_5	181[c)]	~780[c)]	~26	~55	652	19	—	[53, 57]
	BIT + 4 at%WO_3	188[c)]	~820[c)]	~35	~55	643	20	—	
	$Bi_4Ti_{2.98}W_{0.01}Nb_{0.01}O_{12}$	~150[b)]	~800[b)]	—	—	655	32	~31	[54]
	$Bi_4Ti_{2.985}(Cu_{1/3}Ta_{2/3})_{0.015}O_{12}$	~120[c)]	~580[c)]	22.52	~30	~675	34	~33	[59]
	$Bi_4Ti_{2.95}(Mn_{1/3}Nb_{2/3})_{0.05}O_{12}$	~150[b)]	~700[b)]	24.36	21.04	659	32.1	~30	[60]
	$Bi_4Ti_{2.93}(Zn_{1/3}Nb_{2/3})_{0.07}O_{12}$	~150[b)]	~800[b)]	28.4	23.5	657	30.5	~30	[38]

(Continued)

Table 8.3 (Continued)

Doping site	Materials	ε		$2P_r$ ($\mu C/cm^2$)	E_c (kV/cm)	T_c (°C)	d_{33} (pC/N)		References
		At RT	At T_c				At RT	Annealed[a)]	
	$Bi_4Ti_{2.96}W_{0.02}Cr_{0.02}O_{12}$	~150[b)]	~580[b)]	24.4	~30	651	31.1	~28	[58]
	$Bi_4Ti_{2.92}W_{0.04}Ta_{0.04}O_{12}$	—	~750[b)]	—	—	648	21.4	~20	[77]
	$Bi_4Ti_{2.965}(Cu_{1/3}Sb_{2/3})_{0.035}O_{12}$	~120[c)]	~750[c)]	~10	~25	671	24.5	24.5	[78]
	$Bi_4T_{2.95}(Cu_{1/3}Nb_{2/3})_{0.05}O_{12}$	~180[c)]	~900[c)]	18.4	~37	678	17	~17	[79]
	$Bi_4Ti_2Fe_{0.5}Nb_{0.5}O_{12}$	—	~600[b)]	—	—	~630	—	—	[55]
	$Bi_4Ti_{2.9}W_{0.1}O_{12}+4\%MnO_2$	147[c)]	950[c)]	11	42	655	21	/	[80]
	$Bi_4Ti_{2.95}W_{0.05}O_{12}+0.2\%Cr_2O_3$	146[c)]	~700[c)]	34	~50	638	28	25	[61, 81, 82]
	$Bi_4Ti_{2.98}Nb_{0.01}Ta_{0.01}O_{12}$	~168[c)]	~1200[c)]	—	—	670	26	~26	[83]
	$Bi_4Ti_{2.985}(Cu_{1/3}Nb_{2/3})_{0.015}O_{12}$	~120[c)]	550[c)]	24.2	27.32	~678	38	~35	[84]
	$Bi_4Ti_{2.9}Nb_{0.05}Ta_{0.05}O_{12}$	~180[c)]	~800[c)]	~12	~25	~690	39	~35	[85]
Co doping	$Bi_4Ti_{2.94}Ta_{0.03}Nb_{0.03}O_{12}+0.03\%CeO_2$	~150[c)]	~700[c)]	—	—	649‘	36.5	33	[86]
	$Bi_{3.97}Ce_{0.03}Ti_{2.98}(WNb)_{0.01}O_{12}$	~150[b)]	~600[b)]	~30	~30	657	40.2	36.7	[65]
	$Bi_{3.96}Ce_{0.04}Ti_{2.94}W_{0.03}Ta_{0.03}O_{12}$	~150[b)]	~600[b)]	~8	~5.5	~650	24.7	~20	[87]
	$Bi_{3.25}La_{0.75}Ti_{2.91}Ta_{0.09}O_{12}$	~200[b)]	~700[b)]	~21	~50	247	13	0	[88]
	$Bi_{3.5}La_{0.5}Ti_3O_{12}+4\,mol\%Nb_2O_5$	~175[c)]	—	—	—	—	16	0	[89]

a) Measured at room temperature after being annealed at 500 °C.
b) Measured at 1 MHz.
c) Measured at 100 kHz.

there was only one semicircle contributed by the grains at certain temperatures, e.g. in the Cu^{2+}/Ta^{5+} system [59], while for the Mg^{2+} substituted BIT composition, there were three semicircles [71]. The effect of different atmospheres of the test environment or annealing environment on the conductivity has also been explored in depth, and annealing in an oxygen atmosphere can reduce the concentration of oxygen vacancies. The change in conductivity was not significant under different test atmospheres. Figure 8.10d shows the relaxation behavior of $Bi_4Ti_{2.98}(WNb)_{0.01}O_{12}$ in different atmospheres, where the modulus peak shifted toward high frequencies with increasing temperature, indicating that the long-range conductivity gradually dominated. There was a significant diffusion of M'' in the Ar atmosphere [43, 71, 72].

Table 8.3 summarizes the electrical properties of some representative modified BIT components, where RT is the room temperature and d_{33} at 500 °C is measured at room temperature after annealing at 500 °C for several hours.

8.6 Endurance of BIT Under Different Applied Fields

8.6.1 Mechanical Fatigue Behavior of BIT Compounds

In addition to lattice defects such as ion vacancies, BIT piezoelectric ceramics also have a variety of structural defects, including pores, impurities, inclusions, and so on. Coupled with inherent fragility and vulnerability, these defects tend to promote the generation and growth of cracks and partial discharge inside the ceramics and ultimately lead to dielectric breakdown, bending fracture, and functional failure of the device. Given the importance of the reliability of these devices, there has been great interest in studying the fracture mechanism and failure behavior of piezoelectric ceramics.

The mechanical damage of W^{6+}/Cr^{3+} co-doped BIT-based components was deeply investigated, and found that transgranular fracture was the main fracture mechanism, which may be due to oxygen vacancies leading to a decrease in grain boundary strength. Three different propagation modes of the indentation crack, namely long straight-line extension, small angle deflection, and short-distance branching, were confirmed in the indentation experiment. Large grain size or low fracture toughness tends to result in low fracture strength of ceramics, and propagation of microcracks along grain boundaries leads to their eventual axial cracks. The hardness and fracture toughness of ceramics have an opposite trend with an increase in sintering temperature [90–92].

8.6.2 Thermal Depoling Behavior of BIT Compounds

Because of their high T_C, BIT ceramics are widely used in high-temperature fields, so their thermal damage problem is very significant. Some scholars have found that the concentration of oxygen vacancy, phase structure, grain orientation, and microcracks under various doping elements have a great influence on thermal damage.

The reduction of oxygen vacancy is beneficial to improve thermal fatigue stability. In the study of Nd^{3+}, Cu^{2+}/Sb^{5+}, Cu^{2+}/Nb^{5+}, and Cu^{2+}/Ta^{5+} doped BIT-based materials, it was found that oxygen vacancy played an important role in fatigue endurance, and the reduction of oxygen vacancy concentration was beneficial to the optimization of fatigue behavior and improved the stability of degradation induced by high temperature and electric field [78, 85, 93, 94]. Reverse boundary, cubic phase, and orientated structures show good resistance to heat loss. It is found that the presence of inverted phase boundaries (APBs) may play a key role in the fatigue-free behavior of BIT, and oriented ceramics grown parallel and perpendicular to the direction of shear flow have similar heat loss behavior. In 2005, Dong et al. found that materials with cubic structure showed good heat loss resistance at temperatures close to the T_c, while those with monoclinic structure showed poor heat loss resistance. At the same time, microcracks and volatilization of Bi_2O_3 are also related to thermal damage [24, 40, 95, 96].

8.6.3 Irradiation Effect on BIT Compounds

With the rapid development of the aerospace and nuclear industries, there is a higher demand for new materials for memory devices and actuators that can be used in extreme environments such as high temperatures and radiation environments.

Ionization and displacement effects are expected defects of BLSFs when exposed to radiation. Irradiation can produce many electron–hole pairs on the BLSFs due to the ionization effect, and the trapped charge affects the electrical properties of the BLSFs by generating a local electric field. Furthermore, when relatively strong energies are applied, atoms in the lattice structure will deviate from their original positions due to displacement effects. Atoms that are shifted out of their positions due to displacement effects may form vacancies in their original positions and simultaneously create impurities by interfering with the positions of other atoms or cause interstitial defects by moving farther than the original atomic positions. In the study on the structure and electrical properties of irradiated ceramics, it was found that the degradation of the electrical properties of BLSF caused by gamma-ray irradiation was mainly due to flow defects such as oxygen vacancies, while doping of rare earth metals such as La^{3+} and Nd^{3+} improved the stability of oxygen ions in the lattice structure, thus improving the electrical properties and radiation resistance [12]. Bismuth titanate has been shown to operate at fast neutron fluxes of 5×10^{20} n/cm^2 and gamma radiation of 7.23×10^{21} (gamma/cm^2), and the irradiated d_{33} is about half that of the original bismuth titanate. On the other hand, once it reaches the flux threshold of 5×10^{20} n/cm^2, the piezoelectric properties will decrease significantly [97].

8.7 High-Temperature Applications Based on BIT Ceramics

With the rapid development of automotive manufacturing, energy exploration, and aerospace, high-temperature ultrasonic nondestructive inspection (UNDE) and structural health monitoring (SHM) have received unprecedented attention and importance. As the key component of high-temperature ultrasonic inspection, the development of high-temperature piezoelectric ultrasonic transducer (HTUT) is constrained by the development of high-temperature piezoelectric materials. The research on transducers with BIT piezoelectric ceramics with high T_c as the key has received more attention.

The piezoelectric material must be properly coupled to the housing, waveguide, matching layer, etc., and bonded to the leads to produce the most consistent acoustic signal and data for its specific application. The high-temperature resistance of bismuth titanate (K15) was tested and its pulse-echo amplitude versus the time curve is shown in Figure 8.11a. The maximum recommended operating temperature of the K15 sensor was approximately 450 °C, and its ability to perform functional data acquisition at a maximum operating temperature of 546 °C reflected its trustworthy ability to operate as a high-temperature transducer. A fast Fourier transform analysis of the final received signal of the K15 sensor is shown in Figure 8.11b [98]. A problem with many high-temperature transducers is their low conversion efficiency, with high-temperature capability coming at the cost of electromechanical

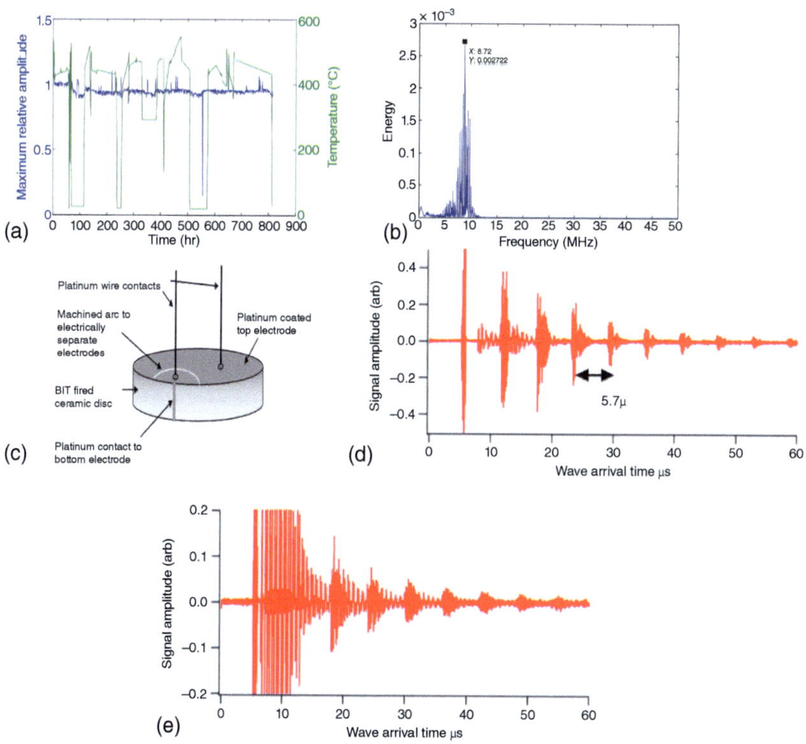

Figure 8.11 (a) Relative amplitude signals from the K15 BIT high-temperature endurance test, (b) Fast Fourier transform (FFT) analysis of the final signal of the K15 transducer, (c) transducer for bismuth titanate; back wall reflections of aluminum measured using a bismuth titanate disc transducer at (d) room temperature, (e) 190 °C. Source: Adapted from [98, 99].

couplings, such as the d_{33} of BIT-based, which was about 12–14 pC/N and about 650 pC/N for the $Pb(Zr,Ti)O_3$ (PZT)-5H. The BIT-based composite transducers prepared by the spray deposition technique BIT–PZT have improved electromechanical coupling while maintaining high-temperature performance [100]. Burrows et al. attached platinum wires to their upper surface with platinum paste and contacted the rear surface through a painted strip. The two wires were electrically separated by scribing an arc on the top electrode using a laser micromachining system, as shown in Figure 8.11c. Thickness measurements were made on aluminum and mild steel specimens, and it was found that the amplitude of the peak decreased at 190 °C, as shown in Figure 8.11d,e, and that higher temperature measurements require a coupling agent that can withstand higher temperatures [99].

Piezoelectric sensors are widely used in vibration monitoring of important equipment in the fields of aerospace, defense, and power because of their small size, lightweight, good sensitivity, and thermal stability. The operating temperature of sensors in these fields is usually higher than 500 °C, requiring good high-temperature adaptability and stability of sensors. High-temperature piezoelectric sensors with BIT piezoelectric ceramics as sensing elements have great application prospects.

Xie et al. evaluated the in-situ temperature stability of the compression mode sensor by testing the charge sensitivity to fully understand the high-temperature performance of W^{6+}/Nb^{5+} modified BIT-based ceramics. Figure 8.12a shows the charge

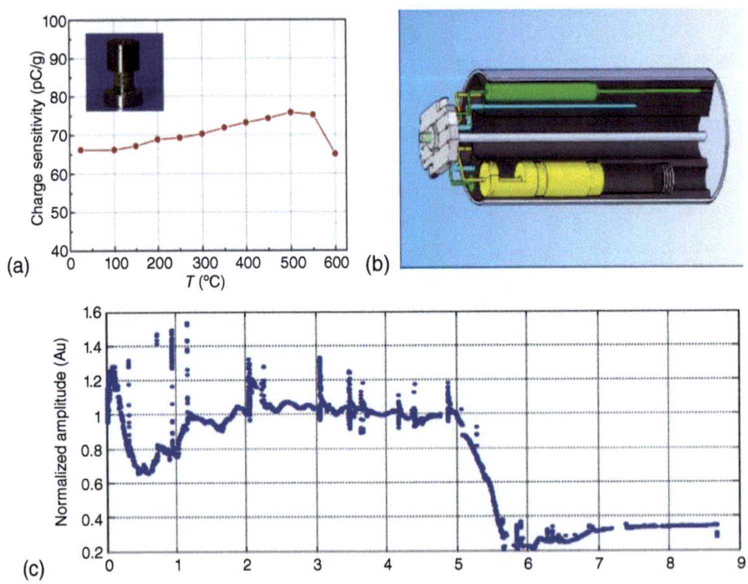

Figure 8.12 (a) Charge sensitivity of compression-mode accelerometer as a function of temperature, (b) 3D renderings of the ultra-capsule as constructed for the irradiation, (c) bismuth titanate sensor pulse-echo amplitude of fundamental harmonic as a function of accumulated fluence. Source: (a) From Xie et al. [54]/Reproduced with permission from Elsevier. (b, c) Reinhardt and Tittmann [97]. MDPI/CC BY 4.0.

sensitivity of the sensor at 160 Hz (~1000 rad/s) in the range of 25–600 °C. The sensor consists of three parts: the piezoelectric element (11 layers of $Bi_4Ti_{2.98}W_{0.01}Nb_{0.01}O_{12}$ (BTWN-1) ceramic connected in parallel), the seismic mass block, and the substrate. The piezoelectric element was located between the seismic mass block and the substrate and is fixed by screws. The sensor had a charge sensitivity of 70 ± 6 pC/g and varied very little over the measurement temperature range of 25–600 °C. This good in-situ temperature stability indicated that the W^{6+}/Nb^{5+} modified BIT ceramic was expected to be used at high temperatures up to 600 °C[54], which may be related to factors such as improved resistivity by the donor ion modification [101]. The experimental details of the operation of bismuth titanate in γ and nuclear radiation accompanied by high temperatures were described by Brian T. et al. The yellow part in Figure 8.12b shows the sensor with the BIT base as the sensing element. The BIT achieved a fast neutron injection of 5×10^{20} n/cm^2 and γ radiation of 7.23×10^{21} (γ/cm^2). The results of the study provided a new perspective for the study of sensors in high-temperature harsh environments.

The performance of the BIT sensor was measured by the amplitude of the fundamental harmonics measured by the power spectral density. The variation curve of the amplitude with the accumulated injection is shown in Figure 8.12c. Over 17 days, the pulse amplitude decreased by 80%. Thereafter, it recovered somewhat and remained small in amplitude until near the end of the experiment. That showed an echo observed before 20 μs, which indicated that the sensor was still capable of transduction, but its performance decreased significantly [97].

8.8 Conclusions

The crystal structure of BIT was confirmed by XRD and neutron diffraction to be monoclinic below T_c and tetragonal above T_c. Also, the changes in coordination environment and lattice parameters of atoms with temperature changes were deeply studied. The widely used processes for powder and formation and sintering of BIT are based on SSRs. There are also powder preparation processes such as molten salt method, sol–gel, and hydrothermal, which can compensate for the high synthesis temperature and coarse particle size of the solid-phase reaction. Sintering processes such as HF, HP, SPS, and TGG methods can improve the orientation and d_{33} of BIT-based ceramics. The influence of the structure and morphology of domains on piezoelectric properties was also well understood by PFM and TEM.

The spontaneous polarization of the BIT mainly originated from the displacement of Ti^{4+} and the rotation of the TiO_6 octahedral along the a-axis. The electrical properties of BIT-based ceramics have been substantially improved by decades of doping modifications by scholars, such as the modification of rare earth ions at the A-site, single ion and ion-pair modifications at the B-site, and synergistic modifications at the A and B-site. The d_{33} could be increased from 7 to 40 pC/N by the synergistic substitution of A and B sites, and the resistivity at 500 °C could be generally increased to 10^7 Ω cm by reducing the concentration or migration rate of carriers. Annealing under oxygen atmosphere or high-valence doping could smooth out the dielectric relaxation problem. In contrast, B-site doping had little effect on the T_c, which was generally around 650 °C, while A-site doping had a greater effect on the T_c, which could be reduced from 675 to 247 °C.

In terms of stability under external fields, transgrain fracture was the main fracture mechanism and was attributed to the decrease in the strength of the grain boundary due to oxygen vacancies. Oxygen vacancy concentration, phase structure, grain orientation, and microcracking have a great influence on thermal damage. Therefore, the reduction of oxygen vacancies is beneficial to improve thermal fatigue stability. Meanwhile, it was found that the degradation of the electrical properties of BLSF induced by gamma irradiation was mainly due to flow defects such as oxygen vacancies. There were few reports related to BIT-based piezoelectric devices for high-temperature applications, where the highest recommended operating temperature of K15 was about 450 °C. The BIT-based composite transducers prepared by spray deposition technology improved electromechanical coupling while maintaining high-temperature performance. Transducers with W^{6+}/Nb^{5+} modified BIT-based ceramics were expected to be used at high temperatures up to 600 °C. Experimental details of BIT in γ and nuclear radiation accompanied by high temperatures provided a new perspective for studying sensors in high-temperature harsh environments.

Acknowledgment

This work was supported by the National Natural Science Foundation of China (Grant No. 51932010).

References

1 Aurivillius, B. (1949). *Arkiv. Kemi.* 1: 499–512.
2 Dorrian, J., Newnham, R., Smith, D., and Kay, M. (1972). *Ferroelectrics* 3 (1): 17–27.
3 Subbarao, E.C. (1961). *Phys. Rev.* 122: 804.
4 Zhou, Q.D., Kennedy, B.J., and Howard, C.J. (2003). *Chem. Mater.* 15 (26): 5025–5028.
5 Hervoches, C.H. and Lightfoot, P. (1999). *Chem. Mater.* 11 (11): 3359–3364.
6 Newnham, R., Wolfe, R., and Dorrian, J. (1971). *Mater. Res. Bull.* 6 (10): 1029–1039.
7 Glazer, A.M. (1972). *Acta Crystallogr. Sect. B: Struct. Sci.* 28 (15): 3384–3392.
8 Macquart, R., Kennedy, B.J., Hunter, B.A. et al. (2002). *Integr. Ferroelectr.* 44: 101–112.
9 Withers, R.L., Thompson, J.G., and Rae, A.D. (1991). *J. Solid State Chem.* 94 (2): 404–417.
10 Irie, H., Miyayama, M., and Kudo, T. (2001). *J. Appl. Phys.* 90 (8): 4089–4094.
11 Du, X.F., Xu, Y.L., Ma, H.X. et al. (2008). *J. Am. Ceram. Soc.* 91 (7): 2079–2082.

12 Badge, S.K. and Deshpande, A.V. (2019). *Solid State Ionics* 334: 21–28.

13 Villegas, M., Caballero, A.C., Moure, C. et al. (1999). *J. Eur. Ceram. Soc.* 19 (6–7): 1183–1186.

14 Kan, Y.M., Jin, X.H., Wang, P.L. et al. (2003). *Mater. Res. Bull.* 38 (4): 567–576.

15 Jiang, D.W., Zhou, Z.Y., Liang, R.H., and Dong, X.L. (2021). *J. Eur. Ceram. Soc.* 41 (2): 1244–1250.

16 Shi, Y.H., Cao, C.S., and Feng, S.H. (2000). *Mater. Lett.* 46 (5): 270–273.

17 Subohi, O., Kumar, G.S., Malik, M.M., and Kurchania, R. (2012). *Phys. Rev. B Condens. Matter* 407 (18): 3813–3817.

18 Subohi, O., Kumar, G.S., Malik, M.M., and Kurchani, R. (2014). *Optik* 125 (2): 820–823.

19 Kong, L.B., Ma, J., Zhu, W., and Tan, O.K. (2001). *Mater. Lett.* 51 (2): 108–114.

20 Ng, S.H., Xue, J., and Wang, J. (2010). *J. Am. Ceram. Soc.* 85 (11): 2660–2665.

21 Chen, Z.W., Yu, Y., Hu, J.Q. et al. (2009). *J. Ceram. Soc. Jpn.* 117 (1363): 264–267.

22 Takenaka, T. and Sakata, K. (1980). *Jpn. J. Appl. Phys.* 19 (1): 31–39.

23 Shi, W., Guan, S., Li, X. et al. (2022). *J. Eur. Ceram. Soc.* 42 (15): 6968–6976.

24 Liu, J. and Shen, S.J. (2007). *J. Appl. Phys.* 102: 104107.

25 Hong, S.H., Trolier-McKinstry, S., and Messing, G.L. (2000). *J. Am. Ceram. Soc.* 83 (1): 113–118.

26 Makiya, A., Kusano, D., Tanaka, S. et al. (2003). *J. Ceram. Soc. Jpn.* 111 (9): 702–704.

27 Nagata, H., Chikushi, N., and Takenaka, T. (1999). *Jpn. J. Appl. Phys.* 38 (9): 5497–5499.

28 Takenaka, T., Nagata, H., and Hiruma, Y. (2008). *Jpn. J. Appl. Phys.* 47 (5): 3787–3801.

29 Doshida, Y., Tsuzuku, K., Kishi, H. et al. (2004). *Jpn. J. Appl. Phys.* 43 (9): 6645–6648.

30 Horn, J.A., Zhang, S., Selvaraj, U. et al. (1999). *J. Am. Ceram. Soc.* 82 (4): 921–926.

31 Dawber, M. and Scott, J.F. (2000). *Appl. Phys. Lett.* 76 (8): 1060–1062.

32 Scott, J.F. and Dawber, M. (2000). *Appl. Phys. Lett.* 76 (25): 3801–3803.

33 Katayama, S., Noguchi, Y., and Miyayania, M. (2007). *Adv. Mater.* 19 (18): 2552–2555.

34 Katayama, S., Noguchi, Y., and Miyayama, M. (2006). *Key Eng. Mater.* 320: 27–30.

35 Xie, S.X., Chen, Y., Liu, W.Y. et al. (2020). *Acta Mater.* 188: 228–240.

36 Kitanaka, Y., Noguchi, Y., and Miyayama, M. (2010). *Phys. Rev. B* 81: 9.

37 Noguchi, Y., Soga, M., Takahashi, M., and Miyayama, M. (2005). *Jpn. J. Appl. Phys.* 44 (9): 6998.

38 Xie, X.C., Zhou, Z.Y., Gao, B.T. et al. (2022). *ACS Appl. Mater. Interfaces* 14 (12): 14321–14330.

39 Li, W., Ma, J., Chen, K. et al. (2005). *Europhys. Lett.* 72 (1): 131–136.

40 Ding, Y., Liu, J.S., Qin, H.X. et al. (2001). *Appl. Phys. Lett.* 78 (26): 4175–4177.

41 Park, B., Kang, B., Bu, S. et al. (1999). *Nature* 401 (6754): 682–684.

42 Chang, Q., Fan, H.Q., and Long, C.B. (2017). *J. Mater. Sci. Mater. Electron.* 28 (6): 4637–4646.

43 Bhardwaj, S., Paul, J., Chand, S. et al. (2014). *J. Mater. Sci. Mater. Electron.* 25 (10): 4568–4576.

44 Bettaibi, A., Jemai, R., Wederni, M.A. et al. (2017). *RSC Adv.* 7: 22578.

45 Munro, A., Mata, J., Martinez, E., and Siqueiros, J. (2008). *Integr. Ferroelectr.* 101: 12–21.

46 Santos, V.B., M'Peko, J.C., Mir, M. et al. (2009). *J. Eur. Ceram. Soc.* 29 (4): 751–756.

47 Kim, J.S., Jeong, J.H., Choi, B.C. et al. (2010). *Ferroelectrics* 404: 173–179.

48 Gu, S.P., Wang, W., He, J.H., and Chen, X.B. (2007). *Integr. Ferroelectr.* 94: 56–63.

49 Shimakawa, Y., Kubo, Y., Tauchi, Y. et al. (2001). *Appl. Phys. Lett.* 79 (17): 2791–2793.

50 Du, C.L., Zhang, S.T., Cheng, G.X. et al. (2005). *Phys. Rev. B Condens. Matter* 368 (1–4): 157–162.

51 Park, J.H., Bae, J.S., Jun, B.E. et al. (2007). *Appl. Phys. A* 86 (1): 63–65.

52 Lopatin, S.S., Lupeiko, T.G., Vasiltsova, T.L. et al. (1988). *Inorg. Mater.* 24 (9): 1328–1330.

53 Zhang, L., Zhao, S.C., Zheng, L.Y. et al. (2005). *Key Eng. Mater.* 280–283: 259–262.

54 Xie, X.C., Wang, T.Z., Zhou, Z.Y. et al. (2019). *J. Eur. Ceram. Soc.* 39 (4): 957–962.

55 Rehman, F., Wang, L., Jin, H.B. et al. (2017). *J. Am. Ceram. Soc.* 100 (2): 602–611.

56 Shulman, H.S., Testorf, M., Damjanovic, D., and Setter, N. (1996). *J. Am. Ceram. Soc.* 79 (12): 3124–3128.

57 Zhang, L.N., Chu, R.Q., Zhao, S.C. et al. (2005). *Mater. Sci. Eng. B Solid State Mater. Adv. Technol.* 116 (1): 99–103.

58 Zhang, F.F., Shi, W., Guan, S.Y. et al. (2022). *J. Alloys Compd.* 907: 164492.

59 Li, X.D., Chen, Z.N., Sheng, L.S. et al. (2019). *J. Am. Ceram. Soc.* 102 (12): 7366–7375.

60 Xie, X.C., Zhou, Z.Y., Liang, R.H., and Dong, X.L. (2021). *J. Materiomics* 7 (1): 59–68.

61 Chen, Y., Liang, D.Y., Wang, Q.Y., and Zhu, J.G. (2014). *J. Appl. Phys.* 116 (7): 074108.

62 Chen, Y., Xie, S.X., Wang, H.M. et al. (2017). *J. Alloys Compd.* 696: 746–753.

63 Preethi Meher, K.R.S. and Varma, K.B.R. (2009). *J. Appl. Phys.* 106 (12): 124103.

64 Chen, Y., Pen, Z.H., Wang, Q.Y., and Zhu, J.G. (2014). *J. Alloys Compd.* 612: 120–125.

65 Xie, X.C., Zhou, Z.Y., Liang, R.H., and Dong, X.L. (2022). *Adv. Electron. Mater.* 8: 7.

66 Villegas, M., Caballero, A.C., Moure, C. et al. (1999). *J. Am. Ceram. Soc.* 82 (9): 241–246.

67 Peng, Z.H., Chen, Q., Chen, Y. et al. (2014). *Mater. Res. Bull.* 59: 125–130.

68 Hou, J.G., Vaish, R., Qu, Y.F. et al. (2010). *Mater. Chem. Phys.* 121 (1–2): 32–36.

69 Hou, J.G., Qu, Y.F., Vaish, R. et al. (2010). *J. Am. Ceram. Soc.* 93 (5): 1414–1421.

70 Prasad, N.V., Babu, S.N., Siddeshwar, A. et al. (2009). *Ceram. Int.* 35 (3): 1057–1062.

71 Long, C.B., Ren, W., Li, Y.W. et al. (2019). *J. Mater. Chem. C* 7 (29): 8825–8835.

72 Xie, X.C., Zhou, Z.Y., Liang, R.H., and Dong, X.L. (2021). *Phys. Status Solidi B Basic Res.* 258 (11): 210072.

73 Pineda-Flores, J.L., Chavira, E., Reyes-Gasga, J. et al. (2003). *J. Eur. Ceram. Soc.* 23 (6): 839–850.

74 Chen, M., Liu, Z.L., Wang, Y. et al. (2004). *Solid State Commun.* 130 (11): 735–739.

75 Takenaka, T., Nagata, H., Tokutsu, T. et al. (2008). *Mater. Sci. Technol.* 204: 167–175.

76 Hong, S.H., Horn, J.A., Trolier-McKinstry, S., and Messing, G.L. (2000). *J. Mater. Sci. Lett.* 19 (18): 1661–1664.

77 Nie, R., Yuan, J., and Zhu, J.G. (2019). *J. Mater. Sci. Mater. Electron.* 30 (15): 14445–14455.

78 Li, X.D., Zhu, L.L., Huang, P.M. et al. (2020). *J. Appl. Phys.* 127: 044102.

79 Zhao, Y.W., Fan, H.Q., Dong, G.Z., and Liu, Z.Y. (2016). *Mater. Lett.* 174: 242–245.

80 Tang, Y.X., Shen, Z.Y., Du, Q.X. et al. (2018). *J. Eur. Ceram. Soc.* 38 (16): 5348–5353.

81 Chen, Y., Xie, S.X., Liang, D.Y. et al. (2015). *Mater. Res. Bull.* 70: 272–278.

82 Chen, Y., Xie, S.X., Wang, Q.Y., and Zhu, J.G. (2016). *Prog. Nat. Sci. Mater.* 26 (6): 572–578.

83 Hou, J.G., Kumar, R.V., Qu, Y.F., and Krsmanovic, D. (2009). *Scr. Mater.* 61 (6): 664–667.

84 Li, X.D., Chen, Z.N., Sheng, L.S. et al. (2019). *J. Eur. Ceram. Soc.* 39 (6): 2050–2057.

85 Jeong, Y.G., Lee, G.J., Lee, S.H. et al. (2022). *Ceram. Int.* 48 (9): 12764–12771.

86 Zhang, F.F., Xu, Y.G., Yang, H. et al. (2022). *J. Am. Ceram. Soc.* 105 (6): 4161–4170.

87 Nie, R., Yuan, J., Chen, Q. et al. (2019). *J. Am. Ceram. Soc.* 102 (9): 5432–5442.

88 Ma, S., Cheng, X.W., Hao, J.G. et al. (2017). *Ceram. Int.* 43 (16): 13193–13198.

89 Chen, T., Wang, H.L., Zhang, T. et al. (2013). *Indian J. Phys. Proc. Indian. Assoc. Cultiv. Sci.* 87 (7): 629–631.

90 Chen, Y., Miao, C., Xie, S. et al. (2016). *J. Am. Ceram. Soc.* 99 (6): 2103–2109.

91 Chen, Y., Miao, C., Xie, S. et al. (2016). *Mater. Des.* 90: 628–634.

92 Chen, Y., Xu, J., Xie, S. et al. (2018). *Mater. Des.* 152: 54–64.

93 Li, W., Wang, C., Lu, X., and Zhu, J. (2004). *Integr. Ferroelectr.* 65 (1): 225–233.

94 Li, W., Gu, J., Song, C. et al. (2005). *J. Appl. Phys.* 98 (11): 114104.

95 Yan, H., Zhang, H., Reece, M.J., and Dong, X. (2005). *Appl. Phys. Lett.* 87 (8): 082911.

96 Xu, Q., Xie, S., Chen, Y., and Wang, Q. (2021). *Prog. Nat. Sci.-Mater.* 31 (2): 248–254.

97 Reinhardt, B.T. and Tittmann, B.R. (2021). *Sensors* 21 (18): 6094.

98 Suprock, A.D. and Tittmann, B.R. (2017). *AIP Conf. Proc.* 1086: 050007.

99 Burrows, S.E., McAughey, K.L., Edwards, R.S., and Dixon, S. (2011). *AIP Conf. Proc.* 1430: 979–985.

100 Reinhardt, B., Cypher, R., Sinding, K., and Tittmann, B. (2012). *IEEE International Ultrasonics Symposium (IUS)*, 2742–2745

101 Villegas, M., Jardiel, T., and Farias, G. (2004). *J. Eur. Ceram. Soc.* 24 (6): 1025–1029.

9

Textured Piezoceramics: Processing, Properties and Applications

Yunfei Chang[1], *Yuan Sun*[1], *and Liangliang Liu*[2]

[1] *Harbin Institute of Technology, School of Instrumentation Science and Engineering, Yikuang Street 2#, Harbin 150080, China*
[2] *Taiyuan University of Technology, College of Materials Science and Engineering, West Yingze Street 79#, Taiyuan 030024, China*

9.1 Introduction

The electrical properties of piezoelectric ceramics are determined by both intrinsic crystallographic and extrinsic microstructural characteristics. For single crystals, many electrical properties depend on the direction of measurement with respect to the crystal axes. An ordinarily prepared ceramic consists of grains with random orientations, yielding directionally averaged piezoelectric properties which are usually much lower than those of single crystals. The ceramics in which the crystal axes of grains are intentionally aligned along a particular direction are called "textured," "crystallographically textured," "crystallographically oriented," and/or "grain-oriented" ceramics [1, 2]. Texturing ceramics allows exploitation of property anisotropy and application of domain engineering, potentially offering an effective means of achieving single crystal-like piezoelectric properties in polycrystalline ceramics. Meanwhile, the limitations of single crystals, such as high cost, limited sizes and shapes, and compositional segregation can be avoided, and the merits of ceramics, such as the tolerance to a wide range of chemistries and mechanical robustness, are maintained. Therefore, textured ceramics have emerged as a very promising alternative to single crystals for various high-performance electronic applications, including ultrasonic transducers, multilayer actuators, smart sensors, and energy harvesters, especially where the use of single crystals is impractical or impossible.

Texture engineering is the technology to develop texture in sintered ceramics to enhance properties. Texture engineering was first proposed in the late 1960s to prepare piezoelectric SbSI ceramics [3]. Since then and up to the first half of the 1980s, it has been extended to crystallographically anisotropic bismuth layer-structured and tungsten-bronze (TB) structured compounds [4–8], through the utilization of morphological anisotropy to induce crystallographic texture. Because of the close relationship between grain shape and crystal structure, the texture in those materials

Piezoelectric Materials: From Fundamentals to Emerging Applications, First Edition. Edited by Jiagang Wu.

was developed by the techniques via the application of pressure (e.g. hot pressing and hot forging) during sintering of compacts containing aligned anisometric particles.

As such techniques are applicable only to compounds with anisotropic crystal structures, research on texturing piezoelectric ceramics has not progressed to a great extent since the latter half of 1980s. In the late 1990s, research on the texture engineering of piezoceramics was revived, which was motivated by the invention of a new technique called templated grain growth (TGG). The TGG process for producing textured ceramics was first reported in 1997 on alumina [9], and then successfully applied to develop texture in piezoelectric ceramics with perovskite structure [1]. In TGG, anisotropic template particles are aligned in fine equiaxed matrix particles by some method (tape casting, magnetic alignment [MA], etc.). During subsequent sintering, templates seed epitaxial nucleation and growth of the desired phase via the Ostwald ripening process to produce texture in the ceramics. Since the development of the TGG process, it has become possible to achieve higher-quality texture in a great range of piezoelectric systems [1]. Consequently, extensive studies have been carried out on texturing piezoelectric ceramics including both lead-based and lead-free systems through the TGG-related techniques, and textured piezoelectric ceramics have endured sustainable development in the past ~25 years in terms of processing, properties, and device applications [10, 11].

In this chapter, we will address the fundamental basis for texture development in piezoelectric ceramics from aspects of TGG-related theories and texture analysis, and survey processing to induce grain orientations with emphases on template selection and synthesis, template alignment techniques, and relationships among densification and texture development. The fabrication and texturing process will be analyzed in order to determine the significant factors that can affect the final texture quality of ceramics. Then, we will review recent progress on utilizing TGG-related techniques to enhance electrical properties in textured piezoelectric ceramics, based on both lead-based systems (especially relaxor PT-based ones) and lead-free systems including KNN-, BNT-, BT-, TB, bismuth-layer and perovskite-layer ones, in which texture–property relationships and related mechanisms will be discussed. Finally, the applications of textured piezoelectric ceramics will be surveyed, and the merits of those applications will be evaluated. This chapter provides a historical perspective, the present status, and future directions in the field of textured piezoelectric ceramics.

9.2 Fundamentals of Texture Development in Piezoceramics

9.2.1 Thermodynamic and Kinetic Aspects of Templated Grain Growth

TGG has been proven to be a very effective approach to fabricate high-quality textured ceramics [1, 10]. At the beginning of TGG, a minority of larger anisotropically shaped template particles are dispersed and oriented in finer and equiaxed

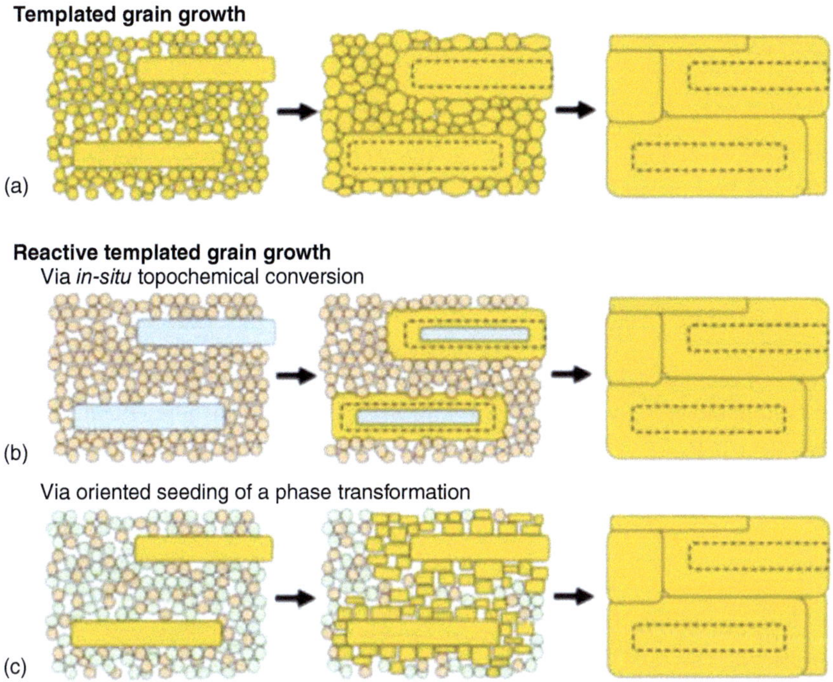

Figure 9.1 Schematics of the (a) templated grain growth and (b,c) reactive templated grain growth processes. Source: Messing et al. [10]. Reproduced with permission of Springer Nature.

matrix particles with desired phase(s) by some shear-forming methods, such as tape casting and extrusion. As shown in Figure 9.1a, during subsequent sintering, the larger oriented particles serve as templates for epitaxial nucleation of oriented grains, which grow via an Ostwald ripening process after densification. TGG takes advantage of the preferred growth of templated particles and the difference in surface-free energies between the advancing template and matrix grains. When the matrix is a chemical precursor or a mixture of precursor powders that can subsequently crystallize to the desired phase (Figure 9.1b,c), the template particles can act as nucleation sites and seed both phase transformation and oriented growth of the matrix. This process is referred to as reactive templated grain growth (RTGG). Since both TGG and RTGG processes utilize standard ceramic powder processing techniques, it is potentially less expensive to produce textured ceramics than single crystals. For the simple purpose, we refer to both TGG and RTGG processes as simply TGG in this chapter.

Here TGG was simply considered as the abnormal grain growth process owing to the obvious size advantage in a single-phase ceramic. Phenomenologically, the behavior is characterized by the fast growth of a few large templated grains into a matrix of fine grains. In the absence of the liquid phase, we assume that grain growth is controlled by a diffusion mechanism [12]. The driving force ΔG for the growth of a templated grain comes from the difference in sizes between the templated and matrix grains:

$$\Delta G = 2\gamma V\left(\frac{1}{R^*} - \frac{1}{R}\right) \tag{9.1}$$

where γ is the interfacial energy between the template and matrix, V is the molar volume, and R^* is the grain size with a zero rate of growth at the moment of concern ($\frac{dR^*}{dt} = 0$). Since R^* increases with the increase in average grain size $\langle R\rangle$, ΔG will decrease gradually. The growth rate of the templated grain is expressed as [13]:

$$R\frac{dR}{dt} = \alpha M\sigma\left(\rho\frac{R}{\langle R\rangle} - 1\right) \tag{9.2}$$

where α is a geometrical factor that takes a value of 1 for 3D and 0.5 for 2D, ρ is the mean-field parameter given as $\langle R\rangle^2/\langle R^2\rangle$ for 3D and 1 for 2D, where $\langle R\rangle^2$ denotes the average squared grain radius. M describes the grain boundary mobility, and σ represents the grain boundary energy. Herein, the size of a templated grain is defined by its volume (V)-equivalent radius, i.e., $R = \left(\frac{3V}{4\pi}\right)^{\frac{1}{3}}$. In the 2D systems, the growth rate of a templated grain could be formulated by the von Neumann–Mullins law [14]:

$$R\frac{dR}{dt} = \frac{1}{6}M\sigma(n_e - 6) \tag{9.3}$$

where n_e is the number of edges per templated grain. The edges of templated grains become rounded with an increase in annealing time, and their equilibrium shape is often a polyhedron, which is dependent on the anisotropic degree of grain boundary energy [15].

Theoretical explanations and analyses of TGG in a liquid phase-assisted environment have been attempted [1], which are based on previous experiments and theories of crystal growth from a melt. According to crystal growth theories [16–18], a templated grain of faceted crystal grows with two main modes, i.e., screw dislocation-assisted and 2D nucleation and growth. The growth rate for screw dislocation-assisted growth is expressed as:

$$v = A_1\frac{(\Delta G)}{\varepsilon}\tanh\left(\frac{\varepsilon}{\Delta G}\right) \tag{9.4}$$

For 2D nucleation and growth, the growth rate is expressed as:

$$v \propto A_2\left(\frac{\Delta G}{T}\right)^n \exp\left(-\frac{A_3\varepsilon^2}{T\Delta G}\right) \tag{9.5}$$

where $A_1, A_2,$ and A_3 are physical constants, ΔG is the driving force, ε is the step-free energy of the templated grain, and n is an exponent with value 1/2 for mononucleation and growth (MNG), and 5/6 for polynucleation and growth (PNG). Generally, the growth rate of templated grains does not show a linear function of driving force, and the interface mobility, which is a function of driving force, also varies with the crystallographic plane of the interface (Figure 9.2).

The TGG in a liquid phase-assisted matrix can be described by the model of solution-segregation [16]. A thin liquid phase layer has been observed in the crystalline interface region between the template grain and matrix grains [19].

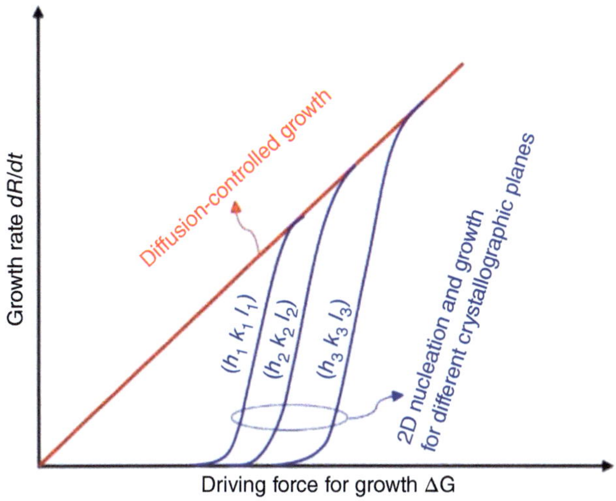

Figure 9.2 Schematic showing the variation of growth rate with the driving force for different growth mechanisms. Source: Modified from source: [12]. Reproduced with permission of Elsevier.

Epitaxial growth of templated grain occurs by dissolution of the polycrystalline matrix grains and deposition on the surface of the textured grain. Figure 9.3 shows the schematic for TGG into matrix grains with a liquid layer interface. The TGG is driven by the difference in solubility between the templated grain surface $S_{t(hkl)}$ and matrix grains S_m across a thin layer [1, 20]. For a matrix of average matrix grain size

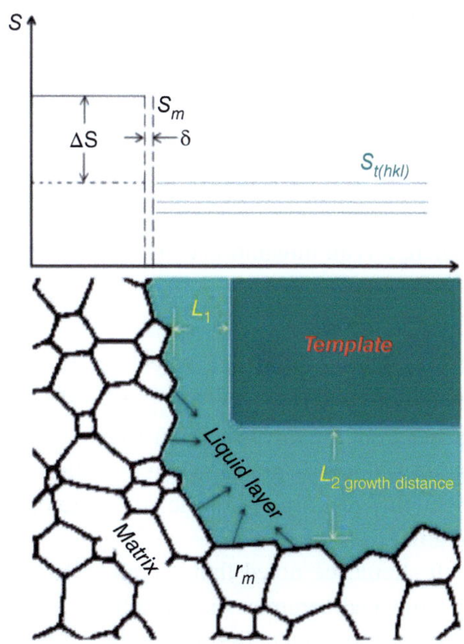

Figure 9.3 Schematic of templated grain growth of a crystal via consuming matrix grains with a liquid layer interface. Source: Modified from source: [1]. Reproduced with permission of Taylor & Francis.

r_m, the solubility of the templated grain and matrix grain can be expressed as:

$$S_{t(hkl)} = S_0^{t(hkl)} \left(1 - \frac{2\sigma_{sl}^{t(hkl)} M}{\rho R T r_m}\right) \tag{9.6}$$

and

$$S_m = S_0^m \left(1 + \frac{2\sigma_{sl}^m M}{\rho R T r_m}\right) \tag{9.7}$$

Thus, the solubility difference ΔS between templated grain and matrix grain can be expressed with the assumption of $S_0^{t(hkl)} = k_{(hkl)} S_0^m$ to be:

$$\Delta S = (1 - k_{(hkl)}) S_0^m + \frac{2M}{\rho R T r_m} \left(\sigma_{sl}^m + k_{(hkl)} \sigma_{sl}^{t(hkl)}\right) S_0^m \tag{9.8}$$

The differential equation for growth of templated grain has been used by Seabaugh et al. [16]:

$$\frac{dL}{dt} = \frac{D \Delta S}{\rho \delta} \tag{9.9}$$

where L is the average growth distance of templated grain, D is the diffusion coefficient, and δ is the average thickness of diffusion layer. With the assumption of $\delta = c r_m$ and definition $K_m = \frac{D M \sigma_{sl}^m S_0^m}{R T \rho^2 c}$, Eq. (9.9) can be rewritten as:

$$\frac{dL}{dt} = \frac{\rho R T K_m}{M \sigma_{sl}^m r_m} (1 - k_{(hkl)}) + \frac{2(K_m + K_{t(hkl)})}{r_m^2} \tag{9.10}$$

The matrix grain growth is usually governed by the following equation:

$$r_m^3 = 1.05 K_m (t + t_0) \tag{9.11}$$

Thus, the growth distance of a templated grain as a function of annealing time can be obtained:

$$L(t) = 1.48 \frac{\rho R T K_m^{2/3}}{M \sigma_{sl}^m} (1 - k_{(hkl)}) (t + t_0)^{2/3} + 5.44 \left(K_m^{1/3} + \frac{K_{t(hkl)}}{K_m^{2/3}}\right) (t + t_0)^{1/3} + C \tag{9.12}$$

where C is an integration constant. Equation (9.12) can be simplified as follows:

$$L(t) = M_{t(hkl)} (t + t_0)^{2/3} + M_m (t + t_0)^{1/3} + C \tag{9.13}$$

with definition $M_{t(hkl)} = 1.48 \frac{\rho R T K_m^{2/3}}{M \sigma_{sl}^m} (1 - k_{(hkl)})$ and $M_m = 5.44 \left(K_m^{1/3} + \frac{K_{t(hkl)}}{K_m^{2/3}}\right)$. Equation (9.13) contains $M_{t(hkl)}(t + t_0)^{2/3}$ and $M_m(t + t_0)^{1/3}$ two parts. The front part arises from the intrinsic difference in solubility between (hkl) surface of templated grain and matrix grains, while the latter part arises from the extrinsic difference in solubility resulting from the curvature of interface between templated and matrix grains.

It should be noted that the above model describes the initial stage of TGG in a templated ceramic. As the volume fraction of liquid phase decreases and average grain size increases, the growth rate of templated grain decreases. Especially, when growing templated grains impinge on each other, the lower driving force at the boundaries leads to a significant decrease in overall growth rate. The normal grain growth behavior is often observed in the microstructure of impinged templated grains. So, the growth environment must be considered when describing TGG behavior.

9.2.2 Texture Analysis

The texture quality of piezoelectric ceramics needs to be determined in order to understand the fundamental relationships between crystallographic orientation and electrical properties. Different methods have been applied to characterize orientation patterns and determine the orientation distribution, and most of them are based on diffraction techniques including X-ray, electron, and neutron diffractions [21, 22]. X-ray diffraction (XRD) with a pole-figure goniometer is a routine method. For some applications, synchrotron X-rays provide unique opportunities. Electron diffractions using the scanning electron microscope (SEM) and transmission electron microscope (TEM) are also useful because they permit one to correlate microstructures, neighbor relations, and texture. As X-ray and electron diffractions have less bulk penetrating capabilities than neutrons, neutron diffraction offers some distinct advantages for materials that impede X-rays, etc.

The Lotgering method [23] based on XRD, is the most common method because of its simplicity. In this method, the relative intensity of the diffraction lines in the 2θ scan patterns, which are related to the texture direction, is used to estimate the texture quality, i.e., Lotgering factor F. For example, to estimate the <001>-texture degree F_{001} by the Lotgering method:

$$F_{00l} = \frac{P_{(00l)} - P_o}{1 - P_o}; P_{(00l)} = \frac{\sum I_{(ool)}}{\sum I_{(hkl)}}; P_o = \frac{\sum I_{o(00l)}}{\sum I_{o(hkl)}} \tag{9.14}$$

where $\sum I_{(00l)}$ and $\sum I_{(hkl)}$ are the summations of all 00l and hkl peak intensities for a textured sample, and $\sum I_{o(00l)}$ and $\sum I_{o(hkl)}$ are the summations of all 00l and hkl peak intensities for a randomly oriented sample. F values are respectively 0 and 1 for a non-textured ceramic and a single crystal, while it scales between 0 and 1 for textured ceramics. It should be noted that the degree of orientation obtained by this method is not accurate because the intensity of diffraction lines from slightly misaligned grains is not accounted for during the $\sum I_{(00l)}$ and $\sum I_{(hkl)}$ measurement.

More quantitative texture analysis methods are needed to better understand grain orientation features in textured ceramics. Generally, the texture of a single-phase ceramic is defined by the orientation distribution function (ODF) of crystallites, and ODF is defined as a multiple of a random distribution (MRD), i.e., the density of grains oriented along a certain direction with respect to the density of all grain orientation directions in a randomly oriented material [24]. The ODF can be calculated by measuring rocking curves and (inverse) pole figures using X-ray, electron, and neutron diffraction techniques. To be specific, the rocking curve technique rocks a sample by fixing the detector at a certain 2θ angle to measure diffraction intensities from the lattice planes for the determination of the preferred orientation degree.

Pole figures show the densities of a given crystallographic direction along all sample orientation directions. Inverse pole figures represent the densities of all crystallographic directions along a given sample direction.

March–Dollase texture model [25] is generally used to quantitatively analyze grain orientation features, in which the orientation distribution function $F(f, r, \omega)$ is described as follows:

$$F(f, r, \omega) = (1 - f) + f\left(r^2\cos^2\omega + \frac{\sin^2\omega}{r}\right)^{-\frac{3}{2}} \tag{9.15}$$

where ω is the angle between scattering vector and texture axis, f is the texture fraction, which describes the volume fraction of textured grains, and r is the orientation parameter, which describes the width of grain orientation distribution. Note that f here is different from the Lotgering factor F_{001}. Besides, $r = 1$ for a random sample and $r = 0$ for a perfectly textured sample. Figure 9.4 describes texture qualities of the regularly 1200 °C-textured and liquid-phase-assisted 975 °C-textured $0.16Pb(Yb_{1/2}Nb_{1/2})O_3$–$0.52Pb(Mg_{1/3}Nb_{2/3})O_3$–$0.32PbTiO_3$ (0.16 PYN–0.52PMN–0.32PT) ceramics are characterized, respectively, by 2θ scans and rocking curves using XRD technique, and inverse pole figures and pole figures

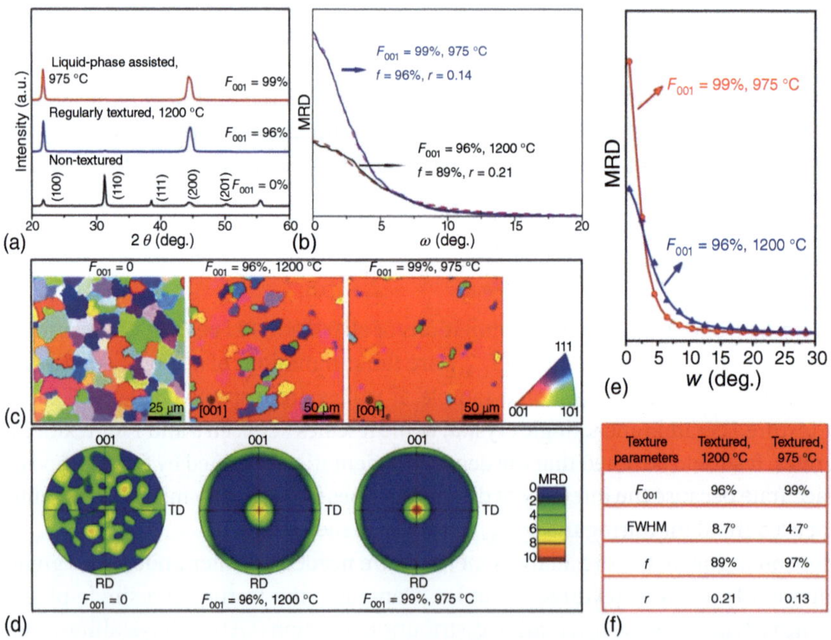

Figure 9.4 Evaluation of texture quality by (a) XRD 2θ scans using the Lotgering factor, (b) fitting XRD rocking curves to the March–Dollase (MD) texture model, and (c) EBSD inverse pole figures, and (d) pole figures for the regularly 1200 °C-textured and liquid-phase-assisted 975 °C-textured 0.16PYN–0.52PMN–0.32PT ceramics. (e) Grain orientation distributions were obtained by fitting $(001)_c$ pole figure data to the MD model and (f) the calculated texture quality parameters. Here the XRD pattern, EBSD inverse pole figure, and pole figure of non-textured counterpart were also presented. Source: Chang et al. [26]. Reproduced with permission of American Chemical Society.

using electron backscatter diffraction (EBSD) technique [26]. Although the two types of samples show similar F_{001} values, their f and r values possess considerable differences. For the same samples, XRD rocking curve and EBSD pole figure results yield similar f and r values.

9.3 Processings for Producing Grain Orientations in Piezoceramics

9.3.1 Template Selection and Synthesis

For the preparation of high-performance textured piezoelectric ceramics, proper selection of template is very important for TGG [1, 10, 11, 27]. Generally, there are some criteria for templates in order to achieve high texture degrees: (i) high-aspect-ratio morphology (e.g. platelet, needle). Such morphology is needed in order to make those templates mechanically well aligned under an applied shear force during green forming; (ii) a similar crystal structure and matched axis with the desired phase and crystallographic orientation. Those are needed to ensure epitaxial nucleation and oriented grain growth of templates via absorbing the surrounding matrix particles during heating; (iii) thermodynamic stability within the environment. Template should not react or dissolve into the matrix material before they have served their function; (iv) large size difference as compared to matrix particles. This is to enable favorable thermodynamic and kinetic conditions for oriented grain growth. The composition, orientation, size, number, and distribution of template particles seriously influence the texture quality of piezoelectric ceramics. In the following, the development of both perovskite and non-perovskite structured templates will be reviewed.

9.3.1.1 Perovskite-Structured Templates

It is difficult to directly synthesize anisometric perovskite templates under conventional conditions due to their high structural symmetry. Till now, the most successful approach to synthesizing anisotropic perovskites has been based on topochemical conversion (TMC) of low-dimensional precursors through insertion, deletion, or exchange of individual atoms, with morphological features maintained [28–42]. Low-dimensional compounds, such as Ruddlesden–Popper, Lepidocrocite, Dion–Jacobson, and Aurivillius phases, are typical precursor candidates for such conversions. Figure 9.5 shows a schematic diagram of the crystal structure transition from Aurivillius $SrBi_4Ti_4O_{15}$ to perovskite $SrTiO_3$ as an example [40]. During the conversion process, the $[Bi_2O_2]^{2+}$ sublayers decompound from the Aurivillius crystal lattice in the form of Bi_2O_3-based liquid phases. Ions from the reactants penetrated into the interior $(A_{n-1}B_nO_{3n+1})^{2-}$ pseudo-perovskite blocks and replaced Bi^{3+} ions in the A-site to form a target perovskite phase. After topochemical reaction(s), each original Aurivillius single-crystal particle is replaced by an aligned perovskite particle.

Topochemical and morphological conversions of Aurivillius phases to perovskite templates have been studied based on titanate microplatelets [39–41]. Figure 9.6

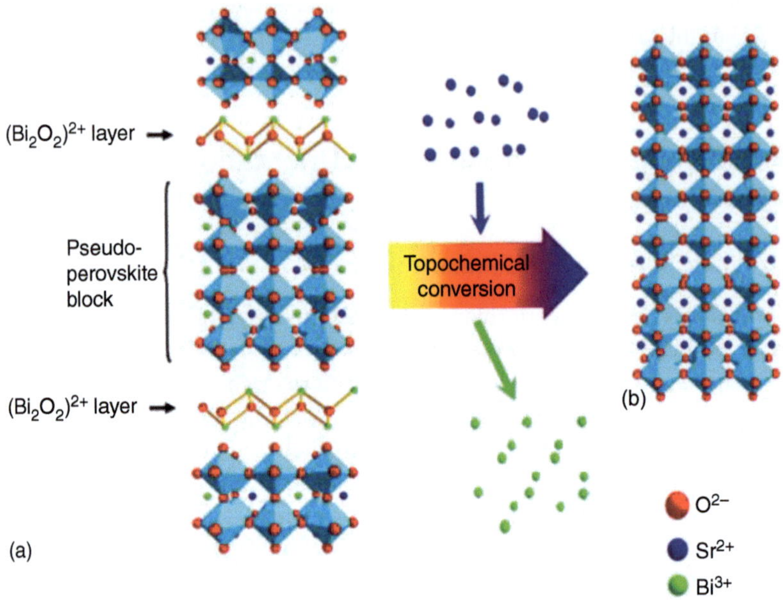

Figure 9.5 Schematic illustration of crystal structure and conversion scheme of TMC reaction: (a) Aurivillius $SrBi_4Ti_4O_{15}$; (b) perovskite $SrTiO_3$. Source: Chang et al. [40]. Reproduced with permission of American Chemical Society.

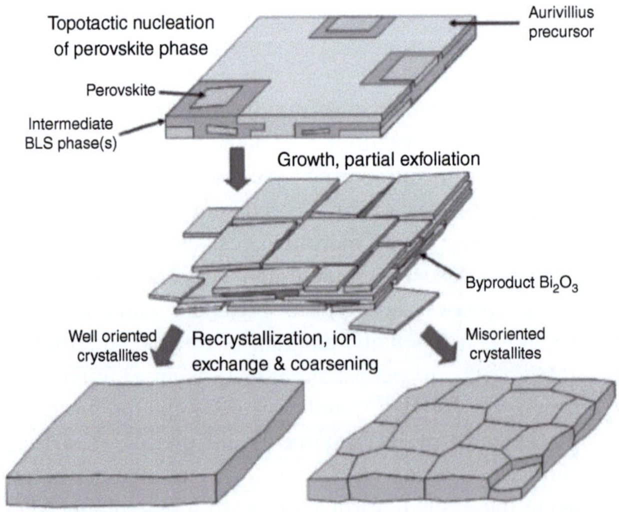

Figure 9.6 Schematic model of the Aurivillius to perovskite topochemical conversion reaction. Source: Poterala et al. [39]. Reproduced with permission of American Chemical Society.

shows the related schematic model of the Aurivillius to perovskite TMC reaction [39]. Generally, the conversion mechanism was proposed to occur in at least two sequential stages. In the first stage, the perovskite phase forms through multiple topotactic nucleation events, either directly on the precursor phase or through a related intermediate phase. This process proceeded very fast at relatively low temperatures. However, because of the loss of epitaxy and expulsion of by-product, such structural conversion caused substantial microstructure/crystallinity damages (e.g. extensive exfoliation and disintegration). As a result, the initial single-crystal precursors are replaced by aligned polycrystalline aggregates. In the second stage, the aligned crystallites recrystallize and sinter to form dense perovskite platelets via Ostwald ripening. The obtained platelets may be polycrystalline or single crystal in nature, depending on several parameters such as heating temperature and time. Higher temperatures and longer hold times may be required here to heal the microstructure/crystallinity damages caused by structural conversion.

Several perovskite templates with controlled morphology and orientation have been successfully synthesized through the TMC approach, such as $BaTiO_3$ [39, 43–48], $SrTiO_3$ [39–42, 49], $PbTiO_3$ [35, 39, 50], $Bi_{0.5}Na_{0.5}TiO_3$ [51–54], $NaNbO_3$ [55–57] and $KNbO_3$ [36]. For example, Liu et al. produced <001>-oriented $BaTiO_3$ platelets with an average size of 5–10 μm and a thickness of 0.5 μm through conversion from $BaBi_4Ti_4O_{15}$ microplatelets in 2007 [43]. Sato et al. synthesized <110>-oriented $BaTiO_3$ particles via TMC from needle-like TiO_2 particles [44]. Lv et al. synthesized <111>-oriented $BaTiO_3$ platelets via the reaction of $Ba_6Ti_{17}O_{40}$ with excess $BaCO_3$ [45]. Figure 9.7 shows the SEM images of {001}, {110}, and <111>-oriented $BaTiO_3$ platelet templates synthesized by TMC reactions. Li et al. [42] reported the synthesis of <111>-oriented $SrTiO_3$ microplatelets by using the <111>-oriented $BaTiO_3$ platelets. Figure 9.8 shows the detailed topochemical procedures, including the XRD patterns and SEM images of the precursors and targeted $SrTiO_3$ microplatelets. The length and thickness of the <111>-oriented $SrTiO_3$ microplates are in the ranges of ~8–16 and ~0.8–1.4 μm, respectively, with an aspect ratio on the order of 10.

9.3.1.2 Non-Perovskite Structured Templates

Templates with non-perovskite structures (such as bismuth-layered and TB structures) are much easier to synthesize due to the low crystal symmetry, for which

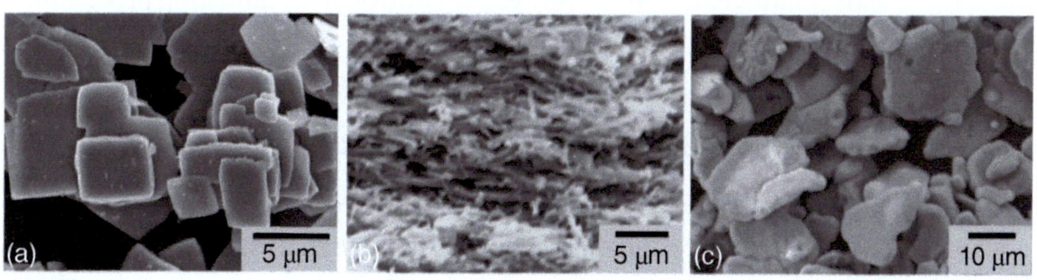

Figure 9.7 SEM images of (a) {001}, (b) {110} and (c) {111}-oriented $BaTiO_3$ platelet templates. Source: (a) From Liu et al. [43]/Reproduced with permission from John Wiley & Sons. (b) From Sato et al. [44]/Reproduced with permission from John Wiley & Sons. (c) From Lv et al. [45]/Reproduced with permission from John Wiley & Sons.

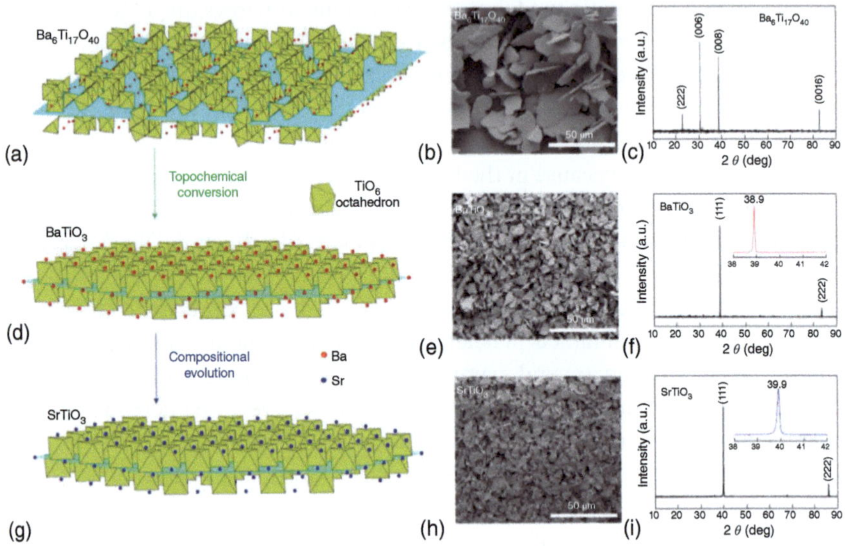

Figure 9.8 Fabrication process of <111>-oriented $SrTiO_3$ templates. (a,d,g) Schematic of the crystal structures, (b,e,h) SEM images and (c,f,i) XRD patterns of $Ba_6Ti_{17}O_{40}$ microplatelets, <111>-oriented $BaTiO_3$ microplatelets, and <111>-oriented $SrTiO_3$ microplatelets. Source: From Li et al. [42]. Reproduced with permission from Springer Nature.

molten salt synthesis (MSS) is the commonly utilized method. For bismuth-layered templates, Horn et al. fabricated $Bi_4Ti_3O_{12}$ platelets with an average thickness of 0.5 μm and an average diameter of ~5–20 μm by adding amorphous $Bi_4Ti_3O_{12}$ powder in NaCl and KCl salts [58]. $Bi_4Ti_3O_{12}$ platelets doped with different elements were also synthesized [59]. In addition, bismuth-layered $BaBi_4Ti_4O_{15}$, $SrBi_4Ti_4O_{15}$, $PbBi_4Ti_4O_{15}$, and $Bi_{2.5}Na_{3.5}Nb_5O_{18}$ plates were synthesized by MSS [35, 40, 43, 55], which were often used as precursors to fabricate perovskite templates as described above. For TB structured templates, needle-like $Sr_{0.5}Ba_{0.5}Nb_2O_6$ templates have been more desirable to be synthesized due to their relatively good piezoelectric properties. Duran et al. found that the chemical compositions of those templates were difficult to control, and the aspect ratios of templates were usually small [60]. Needle-like $KSr_2Nb_5O_{15}$ particles with well-controlled morphology were synthesized by MSS because those aspect ratios can be relatively easily controlled by adjusting processing parameters [61–63].

9.3.2 Techniques for Template Alignments and Inducing Texture

A large number of techniques have been developed for producing texture in ceramics, which are categorized by mechanical, electromagnetic, or thermal driving forces for particle alignment and preferential growth [10]. As mentioned earlier, TGG-type processes enable the achievement of ceramics with high texture quality, and the template alignment and subsequent growth via the consuming matrix directly determine the final texture quality of ceramics. Several techniques have been combined with TGG processes to generate template alignment, such as tape casting [1], magnetic slip casting [64], hot pressing [65], ice templating [66], and additive manufacturing (AM) [67]. The following sections will describe detailedly three of the representative techniques, i.e., tape casting, magnetic slip casting, and AM.

9.3.2.1 Tape Casting

As one of shear-forming techniques, tape casting is the most common technique utilized to align anisometric template particles in matrix powders [1, 10]. A schematic of the related tape casting process is illustrated in Figure 9.9. A high-viscosity dispersion (slurry) including anisometric templates, matrix particles, solvents, dispersants, binders, and plasticizers is extruded from a reservoir and spread over a moving surface under a carefully controlled blade referred to as a doctor blade, and the tape is achieved after evaporating the solvents. For the platelet templates, a regular blade is enough to achieve template alignment in the matrix (Figure 9.10a), in which the large surfaces of those platelets are parallel to the tape casting direction. For the needlelike templates, a gated doctor blade is designed to modify the local shear field to orient those particles along the tape casting direction (Figure 9.10b), which are formed by spacing needles at a small interval. The dried tapes are cut, stacked, and laminated to form green compacts. After cold isostatic pressing and binder burnout,

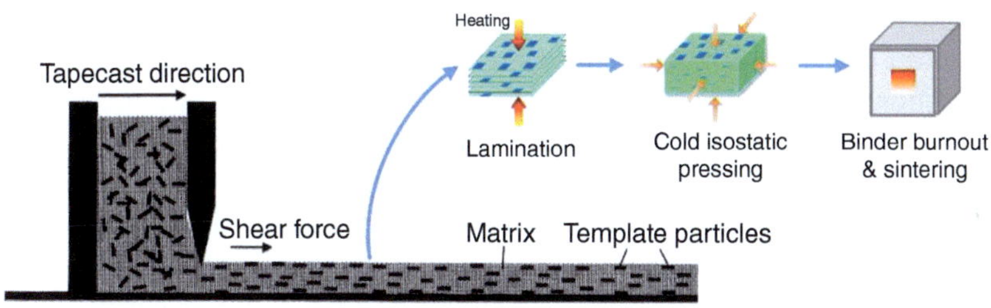

Figure 9.9 Schematics of template alignment by tape casting and fabrication of textured ceramics. Source: Modified from Messing et al. [1]. Reproduced with permission of Taylor & Francis.

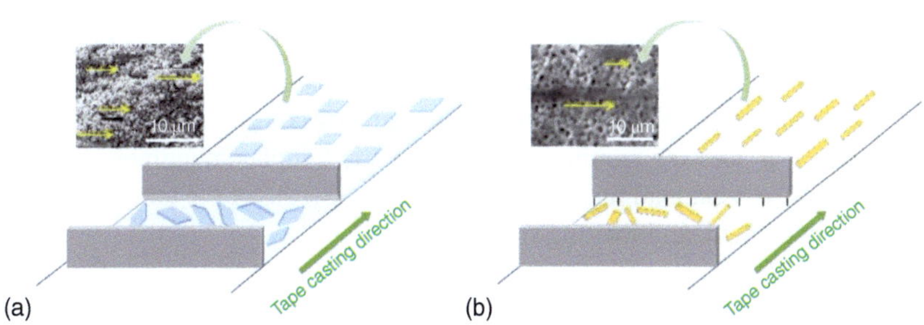

Figure 9.10 Schematics of orienting (a) platelet and (b) needlelike template particles by (a) regular or (b) gated doctor blade techniques. Source: (a) From Liu et al. [68]/Reproduced with permission from American Chemical Society. (b) From Chang et al. [63]/Reproduced with permission from Springer Nature.

the samples are sintered to form the desired texture, during which the aligned templates direct nucleation and oriented grain growth of the matrix.

For the defined types of matrix and templates, multiple variables associated with slurry preparation and casting have substantial impacts on the template alignment quality in the tapes. The slurry preparation process has been categorized in order to produce a high-quality slurry, and variables that need to be considered include matrix, template particles, solvents, binder, plasticizer, and dispersant. The conditions for shear-based alignment of large-aspect-ratio templates have been demonstrated to depend on the flow field under the doctor blade and slurry rheology during tape casting. Processing parameters, such as blade gap, casting speed, slurry viscosity, and reservoir height, can affect template alignment in the tapes [27]. Generally, for orientating templates uniformly, the slurry needs to have a higher viscosity, and be cast through a gap with higher height than the thickness of templates, and at an increased speed, etc.

9.3.2.2 Magnetic Slip Casting

Magnetic slip casting applies an external magnetic field during the casting process to orient the preferential crystalline axis of the ceramic powders. Applying an external magnetic field to a diamagnetic material induces an internal magnetic field in the other direction, causing a repulsive force. Generally, the strength of the applied field needs to be very high (≥10 T) to create such an internal magnetic field [64]. Transforming the diamagnetic powders using magnetic elements or applying a dynamic MA can lower the external magnetic field strength required. Under magnetic fields, the orientation of ceramic powder results from its crystalline or shape anisotropy. The former arises from anisotropy in the magnetic susceptibility due to electron orbits that cover different areas along different directions in the crystal. The latter is the shape of the particle that determines the anisotropy in susceptibility and the preferential orientation. The use of magnetic slip casting in conjunction with TGG has been demonstrated successfully to produce strong textures in piezoelectric ceramics such as $Bi_4Ti_3O_{12}$ [69] and $Pb(Mg_{1/3}Nb_{2/3})O_3–PbTiO_3$ [70]. During sample fabrication, the magnetic field or the sample can be rotated, which is often favorable for MA along a single direction but for biaxial orientation. The green samples with templates were unidirectionally slip-cast within the rotating magnetic field using a cylindrical sample chamber partially filled with casting plaster. Dynamic MA samples were rotated in a horizontal magnetic field during casting. Compared with tape casting, MA eliminates the requirement for morphology control of templates used for TGG. Several processing parameters influence the template alignment quality during MA, such as solid loading, casting viscosity, properties of the template, casting time, magnetic field, and rotating speed, etc.

9.3.2.3 Additive Manufacturing

Layer-wise slurry and direct writing AM can be used to fabricate textured ceramics via combining TGG with manipulation of the rheological properties of the feedstock, printhead geometry, and printing speed, etc. [67] Layer-wise slurry AM has three main methods. One method involves a slurry containing ceramic powders

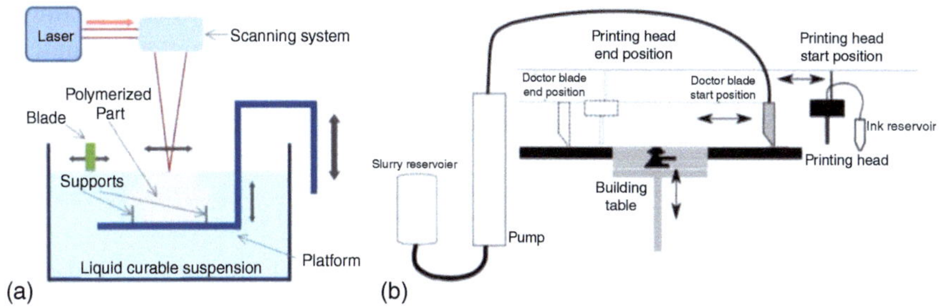

Figure 9.11 Schematic of layer-wise slurry additive manufacturing via (a) photopolymerization and (b) binder gelation. Source: Zocca et al. [73]. Reproduced with permission of Göller Verlag GmbH.

and a photoactive polymer deposited in a layer. The 3D shape can be built up utilizing those layers in which the photoactive polymer is cross-linked with light (Figure 9.11a) [71]. Another method builds the 3D ceramic shape via drying of a ceramic slurry in cross-sections using a laser [72]. The gelation method fixes the 2D layers by the drop-wise addition of a chemical cross-linking agent to the related slurry (Figure 9.11b) [73]. For all the three methods, a ceramic slurry is deposited as a layer by a moving doctor blade, which creates a shear field similar to that for tape casting. Generally, it has been demonstrated that ceramic formulations for tape casting and layer-wise AM have similar ceramic powder and solvent/polymer contents, and their rheological behavior and layer heights fall within the same ranges [67].

Direct writing AM can also provide the requisite shear conditions for template alignment and thus enable the fabrication of textured ceramics with unique crystallographic orientations. As shown in Figure 9.12, direct writing includes the extrusion of filaments of ceramic paste through a nozzle and the layering and/or stacking of filaments to form macroscopic geometries [74]. Printed filaments maintain shape by rapid drying and/or rheological recovery to produce both spanning lattices and macroscopically dense structures. More recently, research on utilizing direct writing to align high-aspect-ratio templates and then to produce textured piezoelectric ceramics has been carried out [74, 75], such as in BT-templated $Pb(In_{1/2}Nb_{1/2})O_3$–$Pb(Mg_{1/3}Nb_{2/3})O_3$–$PbTiO_3$ (PIN–PMN–PT) ceramics. It has been demonstrated that using an oval cross-section nozzle decreases the misalignment angle of template particles in relation to the surface of the printed part, thereby

Figure 9.12 Schematic of the direct writing 3D Printer where the deposition nozzle moves in x–z and the stage moves in y showing the paste extrusion mechanism and resulting flow field. Source: Walton et al. [74]. Reproduced with permission of John Wiley and Sons.

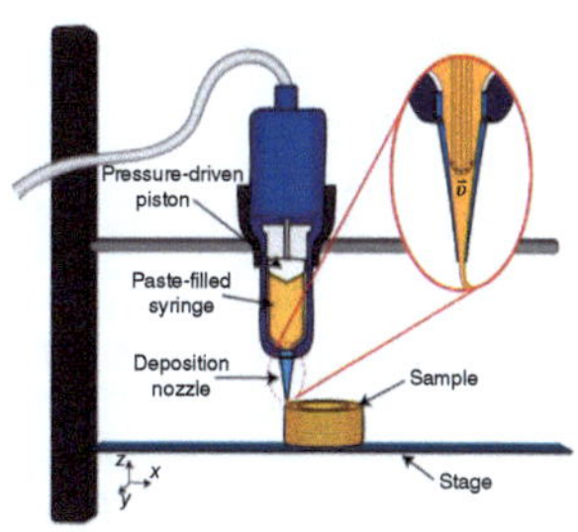

creating particle alignment closer to that of templated ceramics via tape casting. Some challenges associated with using direct writing to produce template ceramics exist, such as rheological tailoring of the printing pastes, integrity of the ceramic during printing and drying, and control of the unaligned core of deposited filaments, etc.

9.3.3 Densification and Texture Evolution During Sintering

Sintering is the final but also very important stage for fabricating textured piezoelectric ceramics, which provides the driving force to TGG and determines the final texture quality of the ceramics. During sintering, ceramic texture develops as a result of epitaxial growth of an ensemble of well-aligned templates into fine-grained matrix powder(s). Generally, texture development via TGG occurs in three main stages: (i) nucleation of templates, (ii) rapid growth of textured grains, and (iii) slow growth of textured grains after impingement. Relationships among sintering parameters (temperature, time, etc.), densification, and texture evolution have been explored in previous years. Figure 9.13 shows microstructure evolution during sintering BT-templated PMN–PT ceramics as an example [76]. The heterogeneous nucleation and epitaxial growth of oriented PMN–PT crystallites were initiated at the template-matrix interface at 900 °C (Figure 9.13a). With increasing holding time from 1 to 10 minutes, sizes of template grains and matrix all showed increased trends, but the latter remained much smaller than the former (Figure 9.13b–d), thus providing a large drive force for oriented crystalline interface migration. Therefore, textured grains grow rapidly by consuming small matrix grains. Large, textured grains began to impinge on each other at 900 °C for 20 minutes (Figure 9.13e), which slowed down the entire oriented grain growth rate as a result of the lower driving force at the boundaries between the impinged grains. Finally, most of the matrix grains were replaced by the textured grains (Figure 9.13f), producing a textured PMN–PT microstructure. Similar TGG behaviors have been observed in several other textured ceramics, such as PIN–PMN–PT (Figure 9.14a,b) [19], $Pb(Mg_{1/3}Nb_{2/3})O_3$–$PbZrO_3$–$PbTiO_3$ (PMN–PZT) [77], (Ba, Ca)(Ti, Sn)O_3 (BCTS) [78], and (K, Na)NbO_3 (KNN) [79] based ones.

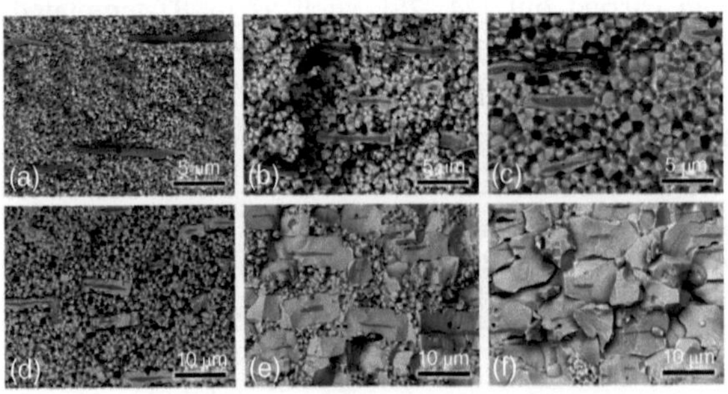

Figure 9.13 SEM images of fractured cross-sections of PMN–PT ceramics with 3 vol% BT templates sintered at 900 °C for different times: (a) 1 minutes, (b) 3 minutes, (c) 5 minutes, (d) 10 minutes, (e) 20 minutes, and (f) 40 minutes. Source: From Zhou et al. [76]. Reproduced with permission from AIP Publishing LLC.

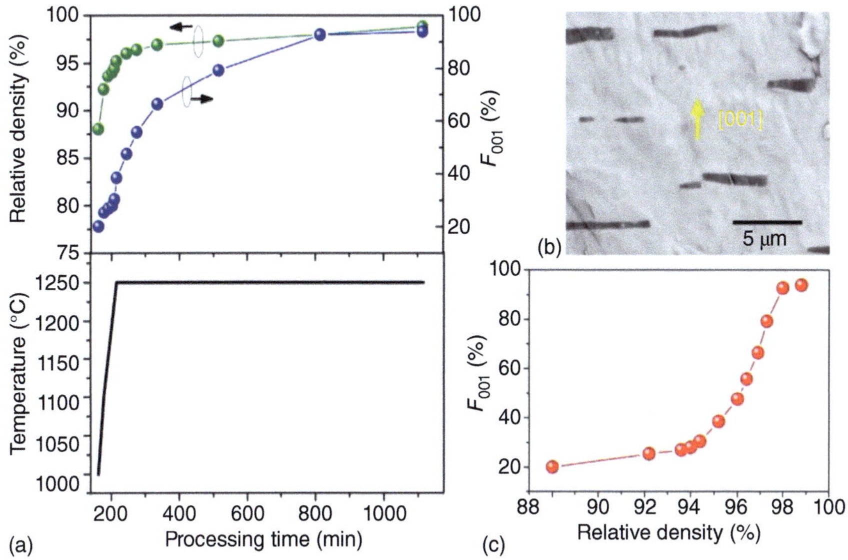

Figure 9.14 (a) Variations in relative density (RD) and texture fraction (F_{001}) as a function of processing time and temperature during sintering of BT-templated PIN–PMN–PT ceramics; (b) SEM image of the textured ceramics sintered at 1250 °C for 10 hours, and (c) F_{001} as a function of relative density. Source: Chang et al. [19]. Reproduced with permission from Elsevier.

Significant TGG was usually observed at high densities due to the dependence of grain growth on microstructure connectivity (e.g. pore geometry). The limited templated growth below the critical density could be attributed to the non-uniform wetting of anisometric particles or the pinning of grain boundaries by pores [80]. Generally, a plot of F_{001} versus density is expected to resemble a typical sintering trajectory (a plot of grain size with respect to density). Figure 9.14c shows such a plot by taking BT-templated PIN–PMN–PT ceramics as an example [19]. It can be seen that the onset of TGG occurs with $F_{001} \sim 20\%$ at ~88% RD. As RD increases to 94%, F_{001} slightly increases to 28%. Further increasing RD to 99% results in a substantial increase in F_{001} to 94%. The sharp increase in texture development at RD above 94% is consistent with established models for TGG, suggesting that a higher density promotes TGG by reducing pore drag on the oriented crystalline interface.

A detailed understanding of oriented crystalline interface phenomena is of critical importance for monitoring the TGG process. As shown in Figure 9.15a [19], both template-textured grain interface and textured grain-matrix interface were clearly observed during BT templating PIN–PMN–PT. Different reactions occurred at the template-matrix interface during TGG when using different types of templates. For example, BT, ST, BNT–PT, and PMN–PT particles have been utilized as templates to produce $[001]_c$-texture in PMN–PT ceramics via different mechanisms [81–85]. Homoepitaxial nucleation and growth occurred at the template-matrix interfaces when using PMN–PT template particles [85], while heteroepitaxial nucleation and growth occurred when using BT [81], ST [82], and BNT–PT [83] template particles.

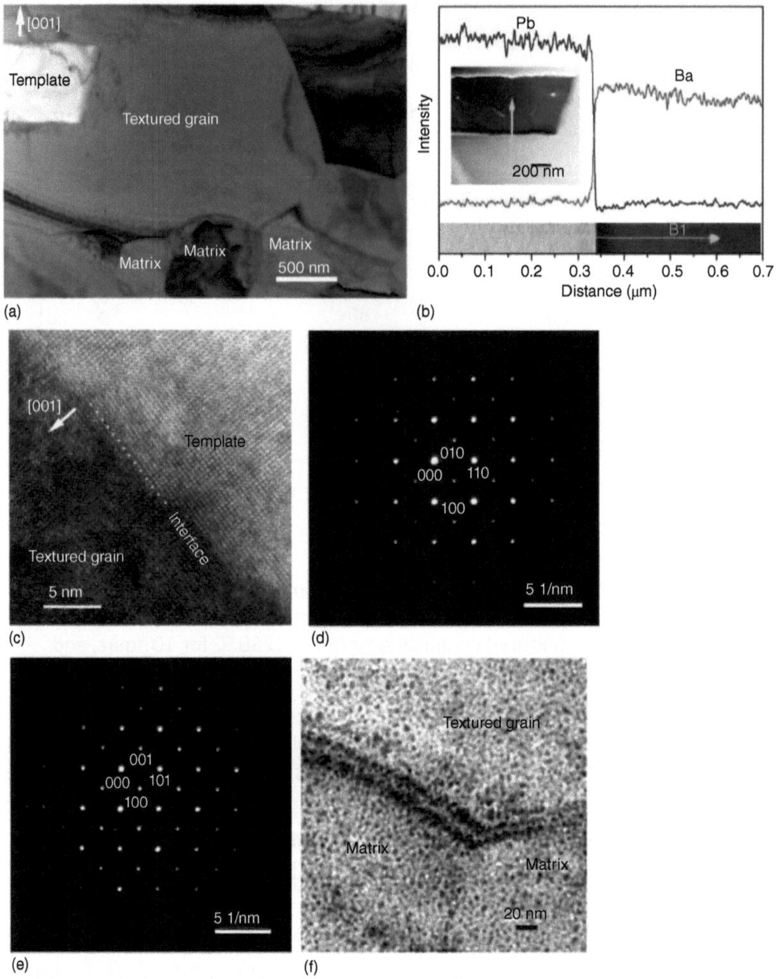

Figure 9.15 (a) TEM image of cross-section of a BT templated PIN–PMN–PT specimen heated to 1100 °C without holding; (b) EDS analysis across a BT template in a textured grain; (c) HR-TEM image of the interface between BT template and textured PIN–PMN–PT grain; (d,e) SAED patterns corresponding to the textured PIN–PMN–PT grain and BT template regions, respectively. (f) Presence of a liquid phase layer at the oriented crystalline interface. Source: From Chang et al. [19]. Reproduced with permission from Elsevier.

Furthermore, BT templates are much more stable than ST and BNT–PT ones in the PMN–PT matrix [81–85]. To be specific, ST and BNT–PT templates completely dissolved into the PMN–PT matrix during the TGG, while the BT template stayed very stable so that a composite microstructure with well-oriented templates inside the PMN–PT textured grains was observed. A similar microstructure was also observed in BT-templated PIN–PMN–PT ceramics (Figure 9.14b). Almost no barium diffusion into the matrix was observed from line scanning element analysis of EDS across such a textured grain (Figure 9.15b), and no crack was observed near the template-textured grain interface (Figure 9.15c). When the crystallization interface migrated from the template boundary into the matrix, the textured grains that have the same grain orientation as templates along the texture direction (Figure 9.15d,e) grew in a homoepitaxial way via Ostwald ripening. Generally, the kinetic of textured grain boundary migration through the matrix can be increased by the presence of a liquid phase layer (Figure 9.15f), where epitaxial growth of the textured grain occurs by dissolution of matrix grains and deposition on the lowest energy surface of the textured grains [19].

9.4 Recent Progress on Enhancing Electrical Properties of Textured Piezoceramics

Textured piezoelectric ceramics include mainly perovskites (general formula ABO_3), TB structures, Aurivillius phases (or bismuth-layered structures), and perovskite layer structures. For the perovskites, lead-based systems such as relaxor PT-based ones and lead-free-based systems such as KNN-based, BNT-based, and BT-based ones are the most commonly studied piezoelectrics. Progresses on enhancing piezoelectric properties of ceramics through crystallographic texturing have been discussed in reviews written by Messing et al. [1, 10] in 2004 and 2017. The following will mainly survey recent progress on enhancing piezoelectric properties and related mechanism exploration in textured piezoelectrics.

9.4.1 Textured Perovskite-Structured Piezoelectrics

9.4.1.1 Lead-Based Piezoelectrics

$Pb(Zr, Ti)O_3$ (PZT) ceramics are the commercially utilized piezoelectric materials. However, a very limited number of texture papers can be found on PZT ceramics because of the instability of titanate templates in matrix with higher Zr content before TGG starts. In 2015, Miwa et al. [86] reported the fabrication of $[001]_c$-oriented $Pb(Zr_{0.52}Ti_{0.48})O_3$ ceramics via MA, but piezoelectric properties did not show any improvements due to the low Lotgering factor of 25%. In recent years, the most studied lead-based systems have been focused on relaxor-PT-based ones and relaxor-PZT-based ones with low Zr contents.

Binary relaxor-PT ceramics (especially $Pb(Mg_{1/3}Nb_{2/3})O_3$–$PbTiO_3$, PMN–PT) have endured sustainable development in recent years in terms of improving piezoelectric properties. As shown in Table 9.1, the template seriously influences the texture quality, piezoelectric properties, and phase transition temperatures of the $[001]_c$-textured PMN–PT ceramics. Substantially enhanced piezoelectric constants d_{33} and d_{33}^* values have been achieved in BNT–PT and ST-textured ceramics, but dissolving those templates into bulk PMN–PT ceramics results in relatively low F_{001} values and substantially reduced rhombohedral–tetragonal phase transition temperatures T_{r-t} and/or Curie temperatures T_c [82–84]. BT templates can produce PMN–PT ceramics with high F_{001} values, but those templates do not react with the PMN–PT matrix and thus remain as residual inclusions in the textured ceramics [81]. As a result, it can reduce the piezoelectric response of the ceramics via

Table 9.1 Phase transition temperatures and piezoelectric properties of $[001]_c$-textured $(1-x)$PMN–xPT ceramics with different template types and/or amounts.

x	Template	F_{001}/f	d_{33} (pC/N)	d^*_{33} [a)] (pm/V)	T_{r-t} (°C)	T_c (°C)	References
0.285	7 vol% BNT–PT	92%	1016	—	76	129	[83]
0.280	5 vol% ST	81%	940	1200[b)]	—	103	[82]
0.325	5 vol% ST	69%	—	1310[c)]	—	120	[84]
0.325	1 vol% BT	98%	1000	—	120	160	[81]
0.325	5 vol% BT	98%	847	—	90	162	[81]
0.325	5 wt% PMN–PT	72%	—	1000[c)]	67	165	[85]
0.325	1 vol% BNT + 1 vol%BT	98%	1031	1305[c)]	45	165	[87]

The (a) d^*_{33} values were calculated using S_{max}/E_{max} at E_{max}= (b) 5 kV/cm and (c) 10 kV/cm.

mechanical clamping. Therefore, subsequent work on PMN–PT textured ceramics has been focused on developing ceramics with a higher F_{001} while using fewer templates. As shown in the table, decreasing BT amount from 5 to 1 vol% substantially increases both d_{33} from 847 to 1000 pC/N and T_{r-t} from 90 to 120 °C. Amorín et al. showed that PMN–PT particles can be used as templates to avoid the above issues [85], but the templates have a low aspect ratio and thus cannot be efficiently aligned during fabrication, restricting the property improvement in textured ceramics.

Substantially enhanced piezoelectric properties have been achieved in textured PMN–PT binary ceramics, but their low phase transition temperatures T_{r-t} and T_c seriously limit the working temperature range. Recent years, extensive effort has been focused on texturing ternary and polynary systems with higher T_{r-t} and T_c values, such as $Pb(In_{1/2}Nb_{1/2})O_3$–$Pb(Mg_{1/3}Nb_{2/3})O_3$–$PbTiO_3$ (PIN–PMN–PT) [88–91], $Pb(Yb_{1/2}Nb_{1/2})O_3$–$Pb(Mg_{1/3}Nb_{2/3})O_3$–$PbTiO_3$ (PYN–PMN–PT) [26, 92, 93], $Pb(Mg_{1/3}Nb_{2/3})O_3$–$PbZrO_3$–$PbTiO_3$ (PMN–PZ–PT) [77, 94–96] and $Pb(In_{1/2}Nb_{1/2})O_3$–$Pb(Sc_{1/2}Nb_{1/2})O_3$–$PbTiO_3$ (PIN–PSN–PT) [97, 98], etc. Generally, as shown in Table 9.2, those textured ceramics possessed comparable or even better piezoelectric properties comparable to the binary ones, together with higher phase transition temperatures. For example, textured PIN–PMN–PT based ceramics with F_{001} ~ 93–97% show d_{33} values of 725–927 pC/N, higher T_c about 200–225 °C, and larger coercive fields [88–91]. Textured PYN–PMN–PT based ceramics with F_{001} ~ 91–99% show d^*_{33} values around 990–1830 pm/V and T_c about 200–214 °C [26, 92, 93]. The d_{33} values around 750–1100 pC/N and T_c values about 204–235 °C have been achieved in textured PMN–PZ–PT ceramics with F_{001} ~ 90–98% [77, 94–96]. Other examples include, but are not limited to, textured PSN–PIN–PT ceramics with F_{001} ~ 99% show d_{33} value around 1090 pC/N and T_c about 247 °C [98]. Those ceramics show remarkably broadened temperature operating ranges, which are very promising for electromechanical devices with high performance and wide temperature usage range.

Table 9.2 Phase transition temperatures and piezoelectric properties of relaxor-PT-based ternary and polynary textured ceramics.

Composition	F_{001}	d_{33} (pC/N)	d^*_{33} [a)] (pm/V)	T_{r-t}/T_d (°C)	T_c (°C)	References
0.28PIN–0.40PMN–0.32PT	93%	824	1555[d)]	120	203	[88]
0.36PIN-0.30PMN-0.34PT	95%	780	850[f)]	—	225	[89]
0.25 wt% CuO and 2 mol% MnO_2 doped 0.24PIN–0.42PMN–0.34PT	97%	725	700[e)]	105	205	[90]
0.125 wt% CuO doped 0.28PIN–0.40PMN–0.32PT	97%	927	1510[c)]	120	200	[91]
0.155PYN–0.465PMN–0.38PT	91%	—	1340[b)]	—	214	[92]
0.16PYN–0.52PMN–0.32PT	96%	772	990[d)]	120	200	[93]
0.25 wt% CuO doped 0.16PYN–0.52PMN–0.32PT	99%	1087	1830[d)]	120	205	[26]
0.50 wt% CuO doped 0.42PMN–0.25PZ–0.33PT	98%	860	1050[d)]	120	222	[94]
0.40PMN–0.22PZ–0.38PT	98%	950	1300[d)]	—	235	[95]
0.40PMN–0.25PZ–0.35PT	90%	1100	1200[d)]	—	204	[77]
Quenched 0.40PMN–0.25PZ–0.35PT	92%	750	760[g)]	—	—	[96]
0.19PIN–0.445PSN–0.365PT	99%	770	—	172	270	[97]
CuO/B_2O_3 doped 0.19PIN–0.445PSN–0.365PT	99%	1090	1270[c)]	110	247	[98]

The (a) d^*_{33} values were calculated using S_{max}/E_{max} at E_{max} = (b) 5 kV/cm, (c) 10 kV/cm, (d) 20 kV/cm, (e) 30 kV/cm, (f) 40 kV/cm and (g) 50 kV/cm.

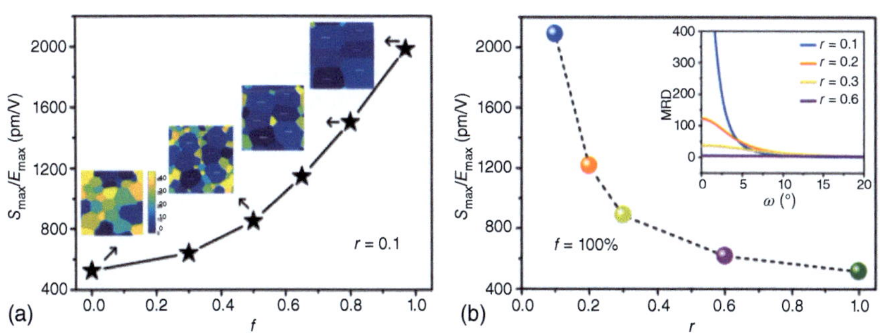

Figure 9.16 Dependence of d^*_{33} of PYN–PMN–PT ceramics on texture parameters (a) f and (b) r. Source: Chang et al. [26]. Reproduced with permission of American Chemical Society.

Despite the great success achieved in improving performance, a substantial performance gap still existed between high-F_{001} relaxor-$PbTiO_3$ ceramics and single crystals in terms of piezoelectric properties. The effects of texture parameters (f and r) on piezoelectric properties have been studied via computational simulations. As shown in Figure 9.16, the texture quality seriously influences the piezoelectric response of the textured ceramics. Simulations show that d_{33}^* (S_{max}/E_{max}) is more sensitive to grain misorientation r than to texture fraction f [26]. Perfectly textured ceramics with f = 100% and r = 0 have maximized piezoelectric response. Subsequent work on relaxor-$PbTiO_3$ textured ceramics has focused on developing ceramics with higher texture degrees by using fewer templates and/or the introduction of sintering aid(s). For example, Yang et al. fabricated <001>-textured $Pb(In_{1/2}Nb_{1/2})O_3$–$Pb(Sc_{1/2}Nb_{1/2})O_3$–$PbTiO_3$ (PIN–PSN–PT) ceramics with $F_{001} \sim 99\%$ via adding CuO and B_2O_3 liquid aids [98]. As shown in Figure 9.17a, the achieved piezoelectric coefficient d_{33} and electromechanical coupling factor k_{33} are much larger than those of classic PZT ceramics. Especially, the k_{33} values are approximately around 87–90%, being comparable to those of relaxor-PT crystals. Moreover, the operating temperature range ($T_{r\text{-}t} \sim 160$–$200\,°C$) of those textured ceramics is much wider than state-of-the-art relaxor-PT crystals (Figure 9.17b). Yan et al. also achieved single crystal-like k_{33} and k_{31} values in 0.40PMN–0.25PZ–0.35PT ceramics with F_{001} over 98% and less BT amount ≤ 3 vol% (Figure 9.17c,d), and found that the strong covalent bonding between B-cation and oxygen via *d-p* hybridization contributes most to k through a theoretical model-[99]. In addition, Chang et al. synthesized PYN–PMN–PT textured ceramics with $F_{001} \sim 99\%$ ($f \sim 97\%$ and $r \sim 0.13$) through liquid-phase-assisted TGG via adding CuO, where single-crystal properties, large $d_{33}^* \sim 1830$ pm/V, $d_{33} \times g_{33} \sim 61.3 \times 10^{-12}$ m^2/N and $T_c \sim 205\,°C$ were simultaneously achieved [26]. Especially, d_{33}^* and $d_{33} \times g_{33}$ values correspond to ~180% enhancement as compared to the regularly textured ceramics with $F_{001} \sim 96\%$.

9.4.1.2 Lead-Free Based Piezoelectrics

Due to growing environmental and human health concerns about lead toxicity, extensive attention has been moved to lead-free piezoelectric materials. Lead-free textured ceramics, especially the perovskite-structured systems, have endured sustainable development in recent years. The following sections will review progress in enhancing the electrical properties of more common textured lead-free systems.

KNN-Based Ceramics (K, Na)NbO_3 (KNN)-based ceramics have attracted considerate attention due to their outstanding piezoelectric properties and high T_c values among lead-free systems. Pure KNN ceramic with an MPB located at K : Na = 1 shows inferior piezoelectric properties ($d_{33} \sim 80$ pC/N), very low ε_r, and high phase transitional temperatures (i.e. orthorhombic-tetragonal phase transition temperatures $T_{O\text{-}T}$ and T_c). Many studies have been carried out on texturing KNN ceramics, in which platelike $NaNbO_3$ particles are the most commonly used template. Table 9.3 shows texture quality and piezoelectric properties of KNN-based ceramics with relatively high $T_{O\text{-}T}$ temperatures. Pure KNN and liquid phase (CuO and

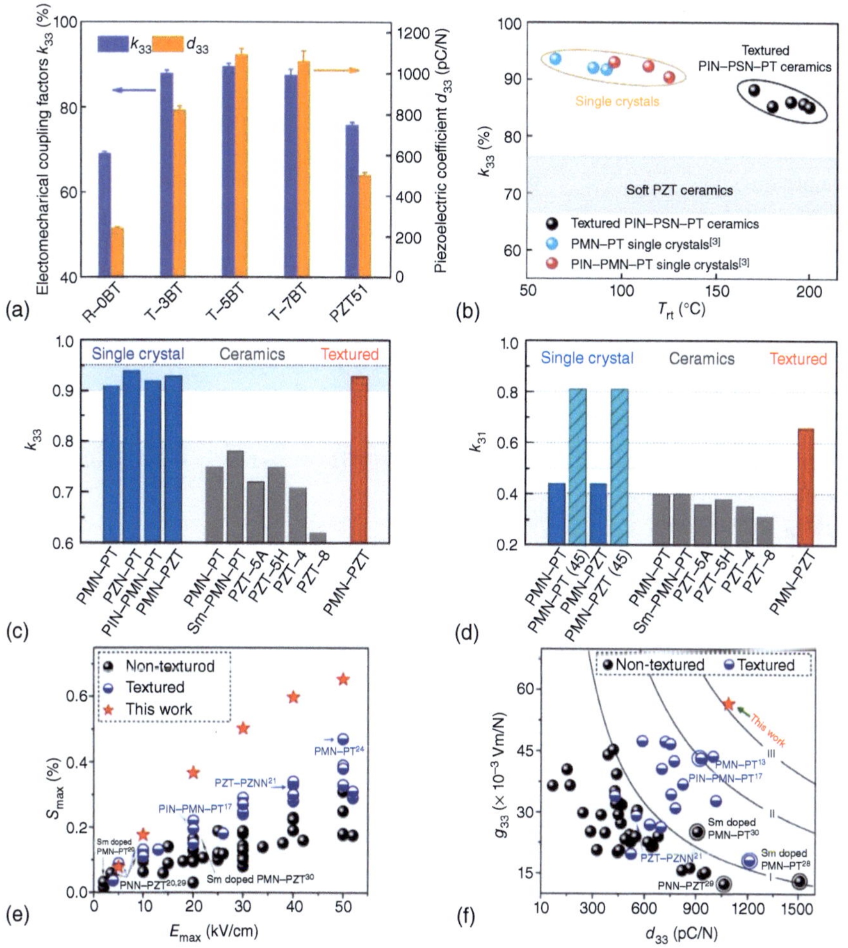

Figure 9.17 (a) The k_{33} and d_{33} values for nontextured (R-0BT), 3 vol% BT (T-3BT), 5 vol% BT (T-5BT), and 7 vol% BT-textured (T-7BT) 0.19PIN–0.445PSN–0.365PT ceramics, compared with commercially available "soft" PZT ceramics with similar Curie temperature; (b) The k_{33} as a function of $T_{r\text{-}t}$ for a series of textured PIN-PSN-PT ceramics and relaxor-PT single crystals, where the k_{33} values of state-of-the-art "soft" PZT ceramics are marked in figure by the gray color; comparisons of (c) k_{33} and (d) k_{31} among single crystals, random ceramics, and PMN–PZT textured ceramics; comparisons of (e) electric field E_{max} induced unipolar strain S_{max} and (f) piezoelectric figure of merit $d_{33} \times g_{33}$ among the textured PYN–PMN–PT ceramics with $F_{001} \sim 99\%$ and recently reported piezoelectric ceramics. The black lines I, II, and III in (f) indicate the plots of $d_{33} \times g_{33} = 20.0 \times 10^{-12}$, 40.0×10^{-12}, and 60.0×10^{-12} m^2/N, respectively. Source: Adapted from [26, 98, 99].

$K_4CuNb_8O_{23}$) doped KNN ceramics with texture degree $F_{001} \sim 80$–99% have been fabricated [100–107]. Those ceramics show improved d_{33} values of 123–225 pC/N and low ε_r of 202–810, while maintaining high $T_{O\text{-}T}$ values around 120–200 °C and T_c about 320–415 °C. Simply Li-, Ta- or Sb-substituted KNN ceramics with high $T_{O\text{-}T}$ values ~143–170 °C have also been textured, which possess enhanced d_{33} values of

Table 9.3 Texture degree and piezoelectric properties of KNN-based ceramics with relatively high orthorhombic–tetragonal phase transition temperatures.

Composition	F_{001}	d_{33} (pC/N)	ε_r	T_{O-T} (°C)	T_c (°C)	References
1.0 mol% CuO-doped $(K_{0.5}Na_{0.5})NbO_3$	96%	123	211	—	—	[100]
$(K_{0.5}Na_{0.5})NbO_3$	93%	133	~500	168	405	[101]
$(K_{0.5}Na_{0.5})NbO_3$	88%	180	286	200	395	[102]
$(K_{0.5}Na_{0.5})NbO_3$	96%	225	~500	170	383	[103]
$(K_{0.5}Na_{0.5})NbO_3$	80%	163	810	~120	320	[104]
1.0 mol% CuO-doped $(K_{0.5}Na_{0.5})NbO_3$	97%	158	—	—	—	[105]
0.5 mol% $K_4CuNb_8O_{23}$-doped $(K_{0.45}Na_{0.55})NbO_3$	95%	133	432	~170	401	[106]
1.0 mol% CuO-doped $(K_{0.476}Na_{0.524})NbO_3$	99%	146	202	183	415	[107]
$(K_{0.5}Na_{0.5})(Nb_{0.97}Sb_{0.03})O_3$	97%	208	382	160	352	[108]
$(K_{0.5}Na_{0.5})_{0.98}Li_{0.02}NbO_3$	40%	280	~550	~170	430	[109]
$(K_{0.5}Na_{0.5})_{0.98}Li_{0.02}NbO_3$	98%	192	263	155	439	[110]
$(K_{0.5}Na_{0.5})(Nb_{0.85}Ta_{0.15})O_3$	99%	184	401	143	322	[110]

around 184–280 pC/N and low ε_r of about 263–550 [108–110]. The high polymorphic phase transition (PPT) temperatures T_{O-T} of those textured ceramics can provide wide temperature usage ranges for practical device applications. For example, the textured $(K_{0.5}Na_{0.5})(Nb_{0.97}Sb_{0.03})O_3$ ceramics maintained high and stable electromechanical coupling k and d_{33} values from −70 to 160 °C [108].

There are also many texture studies on pursuing high room temperature piezoelectric properties in KNN-based systems, through texturing ceramics with PPT temperatures near room temperature [55, 111–124]. Phase boundaries such as R–O, O–T, and/or R–O–T of KNN systems were modified through compositional modifications, e.g., ion substitutions and adding new solid solutions. Especially, as shown in Figure 9.18, Saito et al. developed textured $(K_{0.44}Na_{0.52}Li_{0.04})(Nb_{0.84}Ta_{0.10}Sb_{0.06})O_3$ (LF4T) ceramics with F_{001} of 91% and achieved a high d_{33} of 416 pC/N in 2004 [55], corresponding to 1.6 times as large as those of non-textured counterparts. This pioneering work led to a resurgence of studies into KNN-based piezoelectric ceramics. As shown in Table 9.4, many other textured KNN-based ceramics with relatively complex compositions and low phase transitional temperatures (T_{O-T} and T_c) have been subsequently textured. Benefited from $[001]_c$ grain orientation ($F_{001} \sim 80–98\%$) and phase coexistence at room temperature (e.g. $T_{O-T} < 70\,°C$), the d_{33} values of these ceramics are substantially enhanced [111–124] as compared to those with high PPT temperatures in Table 9.3. Some d_{33} values reach as high as 620–805 pC/N, accompanied by the dramatic increase of

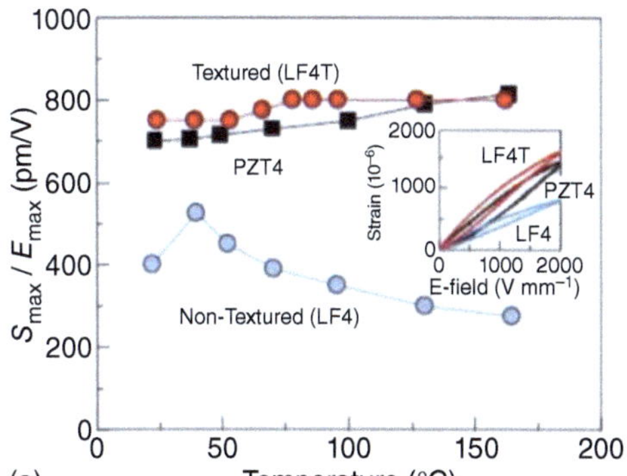

(a)

Properties	LF4T	PZT4
T_c (°C)	253	250
K_p	0.61	0.60
d_{31} pC/N	152	170
d_{33} pC/N	416	410
g_{31} (10^{-3} V m/N)	11.0	8.3
g_{33} (10^{-3} V m/N)	29.9	20.2
$\varepsilon_{33}^{T} / \varepsilon_0$	1570	2300
S_{max} / E_{max} pm/V	750	700

(b)

Figure 9.18 (a) Temperature dependences of electric-field induced strain for non-textured LF4 and the textured LF4T ceramics; (b) Piezoelectric properties of LF4T and PZT4 ceramics. Source: Saito et al. [55]. Reproduced with permission from Springer Nature.

ε_r to be ~2700–3300. For example, as shown in Figure 9.19, Li et al. fabricated $0.96(K_{0.5}Na_{0.5})(Nb_{0.965}Sb_{0.035})O_3$–$0.01CaZrO_3$–$0.03(Bi_{0.5}K_{0.5})HfO_3$ textured ceramics with F_{001} ~ 98%, which show high piezoelectric properties with d_{33} of 704 pC/N and k_p of 76% [121]. The excellent piezoelectric properties were attributed to the high texture degree, facilitated polarization rotation, and optimized engineered domain status.

BNT-Based Ceramics $(Bi_{0.5}Na_{0.5})TiO_3$ (BNT)-based systems are also considered one of the key lead-free candidates owing to their high remnant polarization (P_r), high E_c, and large electric field-induced strain (S_{max}). BNT can form a variety of piezoelectric solid solutions with other components. Among those solid solutions, $(1-x)(Bi_{0.5}Na_{0.5})TiO_3$–$x$BiTiO$_3$ (BNT–BT) ceramics have received special attention, owing to their substantially enhanced piezoelectric properties achieved near the MPB ($0.05 \leq x \leq 0.07$). Textured ceramics of 0.94BNT–0.06BT and similar compositions have been synthesized through TGG-related processes via the use of various templates such as $Bi_4Ti_3O_{12}$ (BiT) [125–127], $NaNbO_3$ (NN) [128–130], $SrTiO_3$ (ST) [131], $BaTiO_3$ (BT) [132], BNT [133, 134], and BNT–BT [135]. As shown in Table 9.5, although high F_{001} values were achieved, those textured ceramics did not show very high d_{33} values. Moreover, the addition of heterogeneous templates caused both lattice mismatches with the matrix and compositional changes, which could lower the depolarization temperature and thereby cause instability in the final textured ceramics. Using BNT platelets as templates can avoid the above problems and meanwhile, produces a relatively high d_{33} value of about 322 pC/N in the textured ceramics with F_{001} ~ 92% [133], which corresponds to 200% enhancement as compared to its randomly oriented counterpart. Besides, 0.93BNT–0.07BT ceramics with F_{001} ~ 89% were textured via utilizing 10 wt% 0.93BNT–0.07BT templates, producing single crystal-like actuating performance with up to 0.70% free strain at 70 kV/cm [135], without lowering the depolarization temperature of the ceramics.

Table 9.4 Piezoelectric properties of KNN-based textured ceramics with PPT temperatures near room temperature.

Composition	F_{001}	d_{33} (pC/N)	ε_r	T_{O-T} (°C)	T_c (°C)	References
$(K_{0.44}Na_{0.52}Li_{0.04})(Nb_{0.84}Ta_{0.10}Sb_{0.06})O_3$	91%	416	1570	—	253	[55]
$(K_{0.470}Na_{0.545})(Nb_{0.55}Ta_{0.45})O_3$	80%	390	2700	—	145	[111]
$(K_{0.5}Na_{0.5})_{0.95}Li_{0.05}Nb_{0.93}Sb_{0.07}O_3$	91%	331	1393	66	314	[112]
$0.95(K_{0.5}Na_{0.5})(Nb_{0.96}Sb_{0.04})O_3$–$0.01BaHfO_3$–$0.04(Bi_{0.5}Na_{0.5})ZrO_3$	95%	490	~3400	20[a)]	~210	[113]
$0.92(K_{0.5}Na_{0.5})NbO_3$–$0.06BaZrO_3$–$0.02(Bi_{0.5}Li_{0.5})TiO_3$	88%	271	2112	50	251	[114]
$0.98(K_{0.5}Na_{0.5})_{0.95}Li_{0.05}Nb_{0.93}Sb_{0.07}O_3$–$0.02BaZrO_3$	85%	234	~1600	48	~240	[115]
$0.95(K_{0.49}Na_{0.49}Li_{0.02})(Nb_{0.8}Ta_{0.2})O_3$–$0.05CaZrO_3$	90%	330	~1500	—	192	[116]
$0.96(K_{0.5}Na_{0.5})(Nb_{0.935}Sb_{0.065})O_3$–$0.04SrZrO_3$	97%	620	~2700	—	124	[117]
$0.915(K_{0.45}Na_{0.5}Li_{0.05})$–$0.075BaZrO_3$–$0.01Bi_{0.5}Na_{0.5}TiO_3$	93%	360	1702	—	254	[118]
$0.99(K_{0.5}Na_{0.5})_{0.95}Li_{0.05}Nb_{0.93}Sb_{0.07}O_3$–$0.01CaZrO_3$	87%	310	2032	23	280	[119]
$0.99(K_{0.49}Na_{0.49}Li_{0.02})(Nb_{0.72}Sb_{0.03}Ta_{0.25})O_3$–$0.01CaZrO_3$	82%	391	1814	32	223	[120]
$0.96(K_{0.5}Na_{0.5})(Nb_{0.965}Sb_{0.035})O_3$–$0.01CaZrO_3$–$0.03(Bi_{0.5}K_{0.5})HfO_3$	98%	700	~2100	60	242	[121]
$0.97(K_{0.5}Na_{0.5})_{0.98}$–$Ag_{0.02}(Nb_{0.96}Sb_{0.04})O_3$–$0.03(Bi_{0.5}Na_{0.5})ZrO_3$	98%	550	1475	60[a)]	265	[122]
$0.97(Na_{0.4625}Li_{0.0375}K_{0.5})(Nb_{0.89}Sb_{0.11})O_3$–$0.03CaZrO_3$	96%	625	~3100	28	127	[123]
$0.96(Na_{0.5}K_{0.5})(Nb_{0.93}Sb_{0.07})O_3$–$0.02BaZrO_3$–$0.02(Bi_{0.5}Ag_{0.5})ZrO_3$	98%	805	~3300	43	167	[124]

a) Those temperatures corresponded to the R–O–T phase transition.

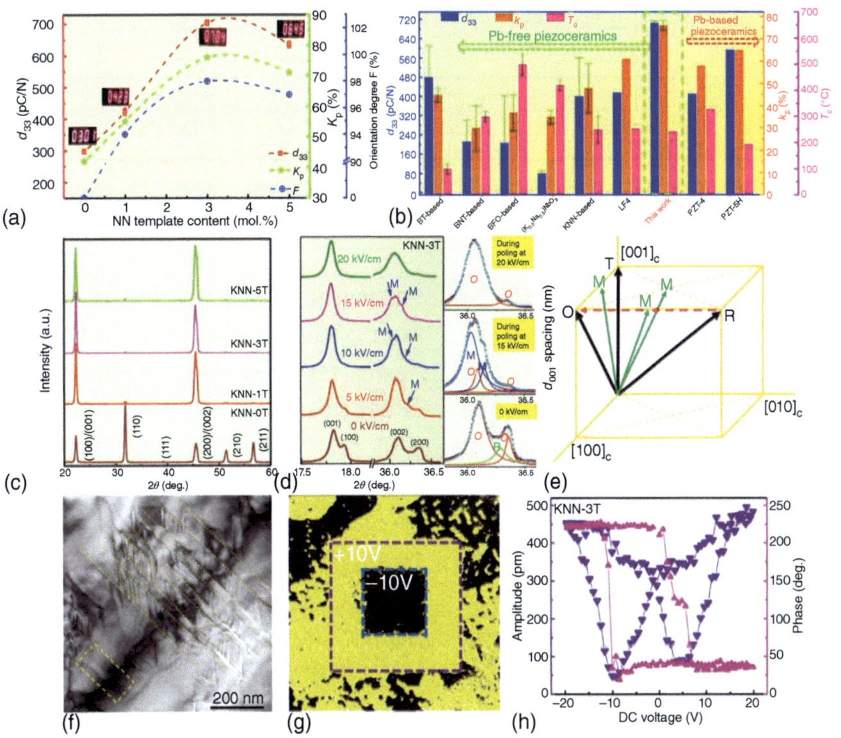

Figure 9.19 (a) Piezoelectric properties of KNN-xT non-textured and textured samples; (b) comparison of some important parameters in representative lead-based and lead-free piezoelectric ceramics; (c) XRD patterns of KNN-xT; (d) *in-situ* synchrotron XRD for KNN-textured samples; (e) schematic of the possible polarization rotation path from $[111]_R$ to $[101]_O$ in $(10\bar{1})$ plane; (f) HR-TEM images of nanodomains; (g) PFM phase images; and (h) microscopic piezoelectric response for the textured KNN-3T samples. Source: Li et al. [121]. Reproduced with permission of John Wiley and Sons.

More complex BNT-based systems have also been textured using different templates [136–146], such as $(Bi_{0.5}Na_{0.5})TiO_3$–$(Bi_{0.5}K_{0.5})TiO_3$–$BaTiO_3$ (BNT-BKT-BT) with BiT, BT, and NN templates; $(Bi_{0.5}Na_{0.5})TiO_3$–$BaTiO_3$–$(K_{0.5}Na_{0.5})NbO_3$ (BNT-BT-KNN) with BNT and BiT templates, and $(Bi_{0.5}Na_{0.5})TiO_3$–$(Bi_{0.5}K_{0.5})TiO_3$–$BiAlO_3$ with $Na_{0.5}Bi_{4.5}Ti_4O_{15}$ (NBiT) template, as shown in Table 9.6. A variety range of F_{001} values ~70–96% was achieved in those textured ceramics, producing obviously enhanced piezoelectric strain S as compared to their randomly oriented counterparts. However, other parameters such as d_{33} and k_p are still very low and did not show obvious improvement, suggesting the relative ineffectiveness of TGG in this case. Overall, the increased ratio of BNT-based textured ceramics in piezoelectric response is much lower than that in their lead-based counterparts, and their piezoelectric properties are also lower than those of KNN systems. Fundamental research focusing on nano-scale polar regions, domain engineering, grain-template interface, grain boundary, and the poling method, etc., is required for BNT-based textured ceramics to further enhance their piezoelectric properties.

Table 9.5 Electrical properties of BNT–BT textured ceramics with different types of templates.

Ceramic system	Template	F_{001}/f	d_{33} (pC/N)	S (E)	T_d (°C)	T_c (°C)	References
0.925BNT–0.075BT	BiT	74%	202	0.20% (50 kV/cm)	140	235	[125]
0.94BNT–0.06BT	BiT	93%	290	—	—	—	[126]
0.94BNT–0.06BT	BiT	96%	241	—	140	260	[127]
0.94BNT–0.06BT	NN	78%	297	0.24% (60 kV/cm)	57	290	[128]
0.94BNT–0.06BT	NN	79%	—	0.51% (120 kV/cm)	80	270	[129]
0.94BNT–0.06BT	NN	78%	—	0.46% (75 kV/cm)	—	—	[130]
0.945BNT–0.055BT	ST	94%	200	0.26% (70 kV/cm)	—	—	[131]
0.94BNT–0.06BT	BT	91%	260	—	120	280	[132]
0.93BNT–0.07BT	BNT	92%	322	—	100	300	[133]
0.94BNT–0.06BT	BNT	87%	299	—	140	294	[134]
0.93BNT–0.07BT	BNT-7BT	89%	—	0.70% (70 kV/cm)	125	325	[135]

BT-Based Ceramics $BaTiO_3$ (BT)-based ceramics are also among the most promising lead-free replacements owing to their relatively high piezoelectric properties, low losses, and high tenability. Pure $BaTiO_3$ ceramic has a tetragonal structure at room temperature and possesses three-phase transition temperatures, i.e., rhombohedral–orthorhombic temperature $T_{R\text{-}O} \sim -90\,°C$, orthorhombic–tetragonal temperature $T_{O\text{-}T} \sim 0\,°C$ and Curie temperature $T_c \sim 120\,°C$. Randomly oriented BT ceramics have a low d_{33} coefficient of up to 190 pC/N. $[001]_c$, $[110]_c$ and $[111]_c$-textured tetragonal BT ceramics have been respectively synthesized via TGG, RTGG, and MA [147–151], etc. As shown in Table 9.7, $[001]_c$-textured BT ceramics with $F_{001} \sim 97\%$ show a d_{33} value ~274 pC/N [147], $[110]_c$-textured BT ceramics with an average grain size of 75 μm and average domain size of 800 nm show a d_{33} of 788 pC/N [148], and $[111]_c$-textured BT ceramics exhibited a d_{33}^* of 627 pm/V [151]. Besides, tetragonal $(Ba_{0.92}Ca_{0.08})TiO_3$ and $(Ba_{0.85}Ca_{0.15})(Ti_{0.95}Zr_{0.05})O_3$ ceramics have been textured by TGG along $[001]_c$ and $[111]_c$ directions [48, 152]. Interestingly, both types of ceramics textured along $<001>_c$ polar axis show much higher piezoelectric properties than those textured along the nonpolar $<111>_c$ texture, indicating an "extender" ferroelectricity nature. Moreover, thermally stabilized strain responses were achieved over a broadened temperature range between RT and ~100–110 °C in those $[001]_c$-textured ceramics, owing to wider temperature ranges of tetragonal phase and stable domain structure.

A&B sites co-substituted $BaTiO_3$ ceramics have attracted considerable attention since large $d_{33} \sim 620$ pC/N and $d_{33}^* \sim 1140$ pm/V were reported in randomly oriented $(Ba_{0.85}Ca_{0.15})(Ti_{0.90}Zr_{0.10})O_3$ ceramics [153]. Both the structural instabilities and facilitated polarization rotation are believed to contribute to the high d values. As shown in Table 9.8, the $[001]_c$-textured $(Ba_{0.85}Ca_{0.15})(Ti_{0.9}Zr_{0.1})O_3$ ceramics

Table 9.6 Phase transition temperatures and piezoelectric properties of other typical BNT-based textured ceramics.

Ceramic system	Template	F_{001}/f	d_{33} (pC/N)	S (E)	T_d/T_{f-r} (°C)	T_c (°C)	References
0.83BNT–0.17BKT	BT	79%	12	0.32% (45 kv/cm)	17	300	[136]
0.80BNT–0.13BKT–0.07BT	BT	93%	190	0.48% (140 kV/cm)	165	300	[137]
0.88BNT–0.08BKT–0.04BT	NN	96%	185	0.04% (20 kV/cm)	110	270	[138]
0.852BNT–0.11BKT–0.038BT	BiT	77%	10	0.55% (65 kV/cm)	100	375	[139]
0.85BNT–0.11BKT–0.04BT	BiT	81%	314	0.41% (40 kV/cm)	100	310	[140]
0.91BNT–0.06BT–0.03KNN	BNT	83%	—	0.23% (60 kV/cm)	95	314	[141]
0.92BNT–0.06BT–0.02KNN	BiT	75%	—	—	138	298	[142]
0.91BNT–0.07BT–0.02KNN	BNT	—	29	0.47% (50 kV/cm)	119	250	[143]
0.91BNT–0.06BT–0.03AN	BNT	71%	—	0.38% (50 kV/cm)	95	311	[144]
0.79BNT–0.20BKT–0.01KNN	BIT	70%	30	0.45% (50 kV/cm)	100	300	[145]
$0.97Bi_{1/2}(Na_{0.78}K_{0.22})_{1/2}TiO_3$–$0.03BiAlO_3$	NBiT	95%	—	0.40% (40 kV/cm)	—	—	[146]

Table 9.7 Piezoelectric properties and Curie temperatures of BT-based ceramics textured along different orientations.

Ceramic system	Orientation	F/f	d_{33} (pC/N)	d_{33}^{*} a) (pm/V)	T_c (°C)	References
$BaTiO_3$	001	97%	274	—	132	[147]
	110	85%	788	—	—	[148]
	110	83%	—	445d)	125	[149]
	111	80%	—	—	—	[150]
	111	26%	—	627d)	—	[151]
$(Ba_{0.92}Ca_{0.08})TiO_3$	001	97%	207	621c)	—	[152]
	111	67%	113	234c)	—	[152]
$(Ba_{0.85}Ca_{0.15})(Ti_{0.95}Zr_{0.05})O_3$	001	94%	—	760b)	109	[48]
	111	92%	—	310b)	109	[48]

The (a) d_{33}^{*} values were calculated using S_{max}/E_{max} at E_{max} = (b) 10 kV/cm, (c) 20 kV/cm and (d) 30 kV/cm.

Table 9.8 Phase transition temperatures and piezoelectric properties of A&B sites co-substituted with BT-based ceramics textured along $[001]_c$.

Ceramic system	F_{001}	d_{33} (pC/N)	d_{33}^{*} a) (pm/V)	T_{o-t} (°C)	T_c (°C)	References
$(Ba_{0.85}Ca_{0.15})(Ti_{0.9}Zr_{0.1})O_3$	70%	462	621c)	—	85	[154]
$(Ba_{0.85}Ca_{0.15})(Ti_{0.9}Zr_{0.1})O_3$	82%	470	—	—	78	[155]
$(Ba_{0.85}Ca_{0.15})(Ti_{0.9}Zr_{0.1})O_3$	82%	460	—	—	91	[156]
$(Ba_{0.85}Ca_{0.15})(Ti_{0.9}Zr_{0.1})O_3$	95%	—	767b)	—	—	[157]
$(Ba_{0.94}Ca_{0.06})(Ti_{0.95}Zr_{0.05})O_3$	99%	755	1200b)	42	95	[68]
$(Ba_{0.98}Ca_{0.02})(Ti_{0.96}Sn_{0.04})O_3$	98%	546	970b)	26	90	[78]
$(Ba_{0.972}Ca_{0.028})(Ti_{0.96}Sn_{0.005}Hf_{0.035})$	84%	324	575b)	20	100	[158]

The (a) d_{33}^{*} values were calculated using S_{max}/E_{max} at E_{max} = (b) 20 kV/cm and (c) 30 kV/cm.

with F_{001} ~ 70–95% did not show obvious improvements in d_{33} and d_{33}^{*} [154–157], possibly suggesting a "rotator" ferroelectricity nature. Those textured ceramics were dominated by the tetragonal phase at room temperature, so the formation of a "1T" domain configuration after poling could not benefit from the polarization rotation effect. Some other compositions, such as $(Ba_{0.94}Ca_{0.06})(Ti_{0.95}Zr_{0.05})O_3$, $(Ba_{0.98}Ca_{0.02})(Ti_{0.96}Sn_{0.04})O_3$, and $(Ba_{0.972}Ca_{0.028})(Ti_{0.96}Sn_{0.005}Hf_{0.035})$, were also textured along $[001]_c$ [68, 78, 158]. Especially, a high d_{33} of 755 pC/N was achieved in the orthorhombic structure-dominated $(Ba_{0.94}Ca_{0.06})(Ti_{0.95}Zr_{0.05})O_3$ piezoelectric ceramics [68], which was mainly ascribed to piezoelectric anisotropy, the favorable domain configuration, and the formation of smaller domains (Figure 9.20). Overall, textured A&B co-substituted BT-based ceramics show substantially enhanced

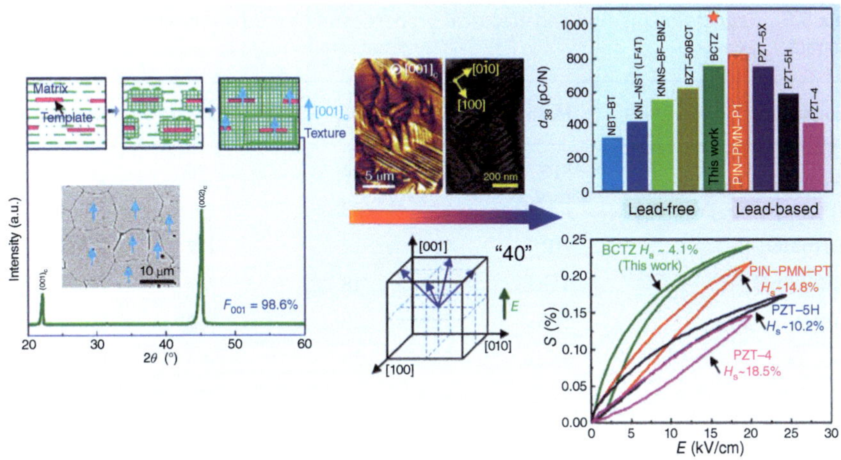

Figure 9.20 Integrating crystallographic texture and domain engineering to enhance piezoelectric properties of $(Ba_{0.94}Ca_{0.06})(Ti_{0.95}Zr_{0.05})O_3$ ceramics. Source: Liu et al. [68]. Reproduced with permission from American Chemical Society.

piezoelectric properties as compared to their randomly oriented counterparts. Moreover, piezoelectric coefficients d_{33} and d_{33}^* are among the highest ones of lead-free ceramics. However, their T_c values are much lower as compared to KNN- and BNT-based ceramics.

9.4.2 Textured Piezoelectrics with Other Structures

Bismuth-layered structured ferroelectrics (BLSF) have merits of high Curie temperature and obvious electromechanical anisotropy, and thus become potential candidates for high-temperature piezoelectric applications. However, the very low piezoelectric response of BLSF ceramics limits their practical applications. The typical bismuth-layered structure (i.e. Aurivillius structure) consists of insulating $(Bi_2O_2)^{2+}$ layers and pseudo-perovskite $(A_{n-1}B_nO_{3n+1})^{2-}$ blocks stacked along *c*-axis, where *n* is the number of BO_6 octahedrons that lie along the [001] direction between the $(Bi_2O_2)^{2+}$ layers. To explore their crystallographic anisotropy, as shown in Table 9.9 [159–171], several methods, including TGG, RTGG, spark plasma sintering (SPS), and hot-pressing (HP), have been used to texture BLSF ceramics, such as $Bi_4Ti_3O_{12}$, $MBi_4Ti_4O_{15}$, and $MBi_2Nb_2O_9$. Those textured ceramics with $F_{001} \sim 70$–98% exhibit d_{33} values ~12–33 pC/N and $T_c \sim 650$–950 °C.

TB family has an oxygen octahedral framework structure that is much more open than the perovskite structure. Generally, this family is an example of a medium-symmetry system in which fiber texture can be achieved. Among this family, tetragonal TB (TTB) crystals with 4 mm symmetry show relatively large dielectric constants, electro-optic constants, and piezoelectric constants. TTB solid solutions such as $KSr_2Nb_5O_{15}$, $Sr_{2-x}Ca_xNaNb_5O_{15}$, $Sr_xBa_{1-x}Nb_2O_6$, and $Ba_{2-x}Sr_xK_{1-y}Na_yNb_5O_{15}$ were fabricated by TGG, RTGG, or RMA, achieving $F_{001} \sim 82$–99% [172–179]. Extrusion, gated doctor blading, and MA are the general

Table 9.9 Fabrication method and electrical properties of bismuth-layered textured ceramics.

Composition	Method	F_{001}	d_{33} (pC/N)	T_c (°C)	References
$Bi_4Ti_3O_{12}$	SPS	95%	14	676	[159]
$Bi_{3\cdot25}La_{0.75}Ti_{2\cdot97}V_{0.03}O_{12}$	RTGG	83%	33	—	[160]
Bi_3NbTiO_9	HP	70%	12	907	[161]
$CaBi_2Nb_2O_9$	TGG	95%	18	930	[162]
$CaBi_2Nb_2O_9$	SPS	70%	20	943	[163]
$Ca_{0.95}Eu_{0.05}Bi_2Nb_2O_9$	TGG	92%	20	945	[164]
$Ca_{0.95}Sm_{0.05}Bi_2Nb_2O_9$	TGG	88%	23	941	[165]
$CaBi_4Ti_4O_{15}$-Mn	TGG	91%	25	790	[166]
$CaBi_4Ti_4O_{15}$	SPS	82%	25	788	[167]
$Ca_{0.85}(Li,Ce)_{0.15}Bi_4Ti_4O_{15}$	TGG	87%	27	775	[168]
$Ca_{0.85}(LiCe)_{0.075}Bi_4Ti_4O_{15}$	TGG	93%	30	790	[169]
$Ca_{0.85}(LiCe)_{0.075}Bi_4Ti_4O_{15}$	RTGG	98%	33	780	[170]
$Sr_{0.9}Nd_{0.1}Bi_2Nb_2O_9$	TGG	95%	27	—	[171]

Table 9.10 Fabrication method and electrical properties of tungsten-bronze and perovskite-layered textured ceramics.

Composition	Method	F_{001}	d_{33} (pC/N)	T_c (°C)	References
Tungsten-bronze structure					
$KSr_2Nb_5O_{15}$	RTGG	91%	65	170	[172]
	RMA	99%	101	—	[173]
	TGG	90%	45	—	[174]
$Sr_{0.53}Ba_{0.47}Nb_2O_6$	TGG	92%	84	128	[175]
	RTGG	82%	78	148	[176]
$Sr_{0.61}Ba_{0.39}Nb_2O_6$	TGG	91%	80	—	[177]
$Sr_2Na_{0.9}K_{0.1}Nb_5O_{15}$	RTGG	86%	82	230	[178]
$Sr_{1.85}Ca_{0.15}NaNb_5O_{15}$	RTGG	88%	162	247	[179]
Perovskite-layered structure					
$La_2Ti_2O_7$	SPS	79%	2.6	1461	[180]
$Sr_2Nb_2O_7$	SPS	86%	2.8	1327	[181]
	SPS	—	2.8	1327	[163]
$Pr_2Ti_2O_7$	SPS	87%	0.5	1555	[182]
$La_{1.85}Ce_{0.15}Ti_2O_7$	SPS	71%	3.9	1440	[183]
$La_{1.9}Bi_{0.1}Ti_2O_7$	SPS	—	3.6	1395	[184]

techniques to align needle-like templates. Those textured samples exhibited d_{33} values around 45–162 pC/N and T_c about 128–247 °C.

Perovskite-layered structured (PLS) compounds with general $A_2B_2O_7$ formula have been reported to have the known highest Curie temperatures, e.g., above 1300 °C. Because of the low crystal symmetry, those compounds usually have very large coercive fields (e.g. above 40 kV/cm) at room temperature. Besides, exaggerated grain growth makes those ceramics difficult to achieve high density. Fabricating dense and textured PLS ceramics will make them easier to pole. As shown in Table 9.10, PLS textured ceramics [163, 181–184] such as $La_2Ti_2O_7$ $Sr_2Nb_2O_7$, $Pr_2Ti_2O_7$, and modified $La_2Ti_2O_7$ have been textured through the SPS method, which shows a texture degree of around 71–87% and a relative density of >95%. In these textured systems, d_{33} values varied from 0.5 to 3.9 pC/N, with ultra-high T_c values above 1300 °C.

9.5 Device Applications of Textured Piezoceramics

Textured ceramics have potential applications in many electronic devices because of their improved piezoelectric properties. In this section, we will review the development of some electronic devices (i.e. transducers, actuators, and energy harvesters) fabricated using the textured piezoceramics.

9.5.1 Ultrasonic Transducers

Relaxor-PT textured ceramics are very promising materials for high-performance transducer applications. Levassort et al. [185, 186] fabricated single-element transducers based on two textured PMN–PT ceramics with $F_{001} \sim 60\%$ and 70% and found that at equivalent relative bandwidth of 40%, sensitivity using the textured ceramics was increased by 3 dB in comparison with that using a PMN–PT ceramic. Brosnan et al. [187] fabricated tonpilz transducers (Figure 9.21b) based on textured PMN-28PT ceramics according to an existing transducer design (Figure 9.21a), and compared their performance with those fabricated based on non-textured ceramics and single crystals of the same composition. In-water characterization of the transducers shows higher source levels, higher in-water coupling, and more usable bandwidth for the textured device than for the non-textured element. The textured PMN-28PT tonpilz element measured under large signals shows linearity in sound pressure levels up to 0.23 MV/m drive field, and its maximum electromechanical coupling under high drive conditions is $k_{eff} = 0.69$. Sun et al. fabricated single-element ultrasonic transducers (Figure 9.21d) with an effective aperture of 2.5 mm in diameter (Figure 9.21c) based on PMN-PZT textured ceramics [188]. The fabricated transducers show a central frequency of 15 MHz, a fractional bandwidth of 67% (at −6 dB), a high k_{eff} of 0.55, and a low insertion loss (IL) of 21 dB. *Ex vivo* ultrasonic imaging of a porcine eyeball result showed that the textured ceramic has great potential for developing ultrasonic devices for biomedical imaging purposes.

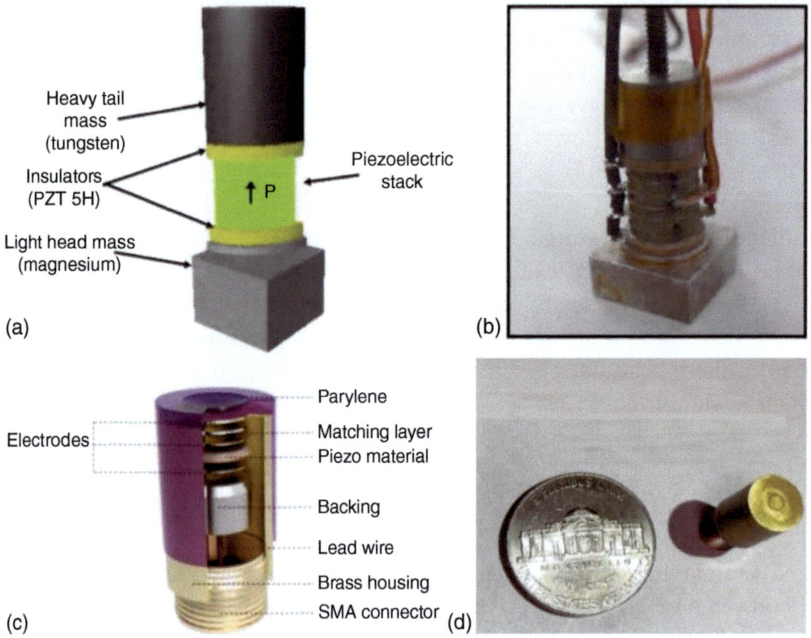

Figure 9.21 (a) The created tonpilz model; and (b) tonpilz element transducer fabricated from six rings of textured PMN-28PT ceramics; (c) schematic illustration of inner structure; and (d) photo of the textured PMN-PZT ceramic single-element transducer. Source: (b) From Brosnan et al. [187]/Reproduced with permission from AIP Publishing LLC. (c, d) From Sun et al. [188]/Reproduced with permission from IEEE.

Lead-free textured ceramics with substantially enhanced piezoelectric properties are also needed for ultrasonic transducer applications. High-frequency needle-type ultrasonic transducers were designed and fabricated based on $0.915(K_{0.45}Na_{0.5}Li_{0.05})NbO_3$–$0.075BaZrO_3$–$0.01(Bi_{0.5}Na_{0.5})TiO_3$ (KNLN–BZ–BNT) textured ceramics (Figure 9.22a,b) [189]. The tightly focused transducers have a center frequency higher than 80 MHz and a 6 dB fractional bandwidth of 52%. The focused process can highly improve their lateral resolution because of the higher bandwidth and low f-number. Its lateral resolution was better than 90 μm by scanning a 15-μm tungsten wire target. Ma et al. [190] fabricated an 80 elements lead-free phased array transducer (Figure 9.22d) based on textured BCTZ ceramics via using a stacking double-layer design (Figure 9.22c). The transducer has the center frequency of 2.9 MHz with over 80% bandwidth. *In vitro* phantom imaging shows excellent performance, including 145 mm penetration depth, 0.86 mm axial resolution, and 7.08 dB CNR, which are comparable to those of a commercial PZT-5H based phased array transducer (4PA, SonoScape Corporation, Shenzhen, China). Satisfactory performance evaluation was also achieved in the live human heart ultrasound imaging. As shown in Figure 9.23, the four-chamber structure and valves were distinctively seen. The color Doppler imaging clearly shows the blood flowing from the left atrium to the left ventricular through the mitral valve [190].

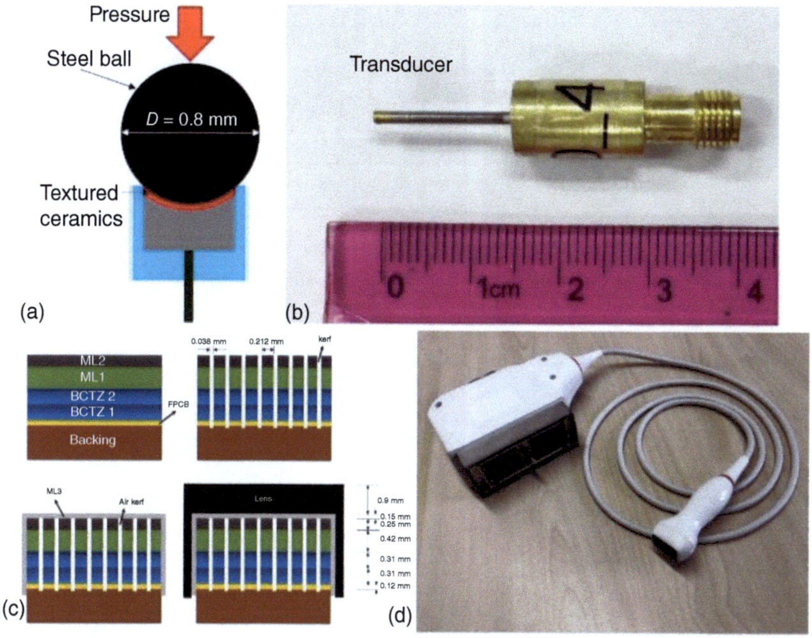

Figure 9.22 (a) Schematic fabrication flow and (b) photograph of press-focused textured KNLN–BZ–BNT ceramic transducers; (c) main fabrication process; and (d) photograph of the 80-element lead-free BCTZ-based phased array transducer. Source: (b) From Quan et al. [189]/Reproduced with permission from IEEE. (d) From Ma et al. [190]/Reproduced with permission from IEEE.

All of these demonstrated the practicality of using lead-free textured ceramics in ultrasonic phased array applications.

9.5.2 Piezoelectric Actuators

Applications of textured piezoelectrics to actuators have been investigated. Kim et al. [191] synthesized textured $0.69Pb(Zr_{0.47}Ti_{0.53})–0.31Pb[(Zn_{0.4}Ni_{0.6})_{1/3}Nb_{2/3}]O_3$ (PZT–PZNN) multilayer ceramics (MLC) with five layers of ceramic thick films and chemically continuous interfaces between the ceramic layers and metal electrodes. Cantilever- and bridge-type actuators were fabricated using the textured MLC (Figure 9.24a,b). The textured cantilever-type actuator exhibited a large acceleration and displacement of 46.9 G and 10.5 mm under a sinusoidal unipolar input voltage of 50 V, which are considerably larger than the acceleration (32.6 G) and displacement (7 mm) of its non-textured counterpart. The actuating properties obtained using the simulations were consistent with the measured ones. Similar results were also obtained from the bridge-type actuators. Tsai et al. [158] fabricated a cylindrical rotary linear piezoelectric motor using a textured $(Ba_{0.968}Ca_{0.032})(Ti_{0.96}Sn_{0.005}\ Hf_{0.035})O_3$ based lead-free plate as the motor vibrator in the first bending vibration mode. The d_{31} and Q_m values of the proposed ceramics were determined to be the key factors affecting ultrasonic motors. The output

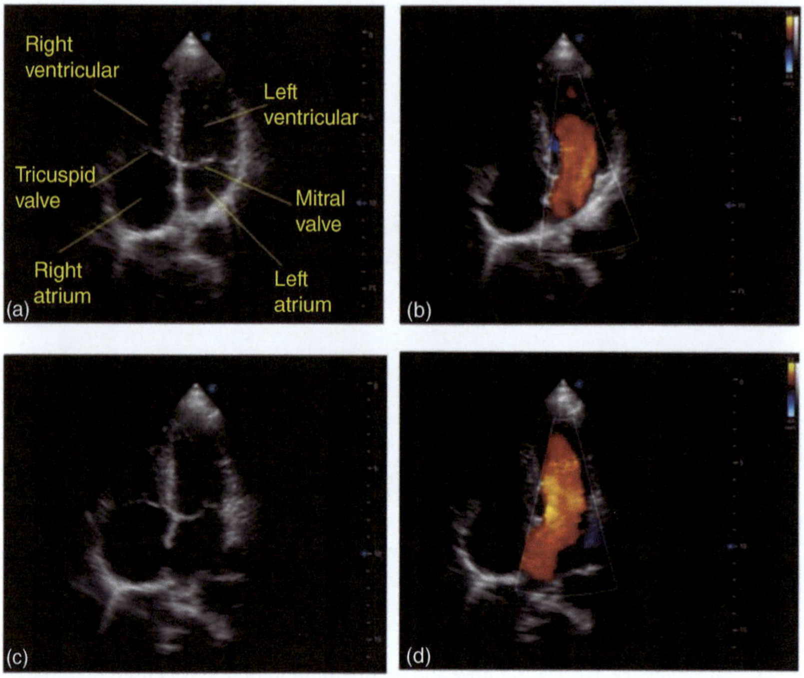

Figure 9.23 Human heart imaging (a) without and (b) with Doppler blood flow, respectively, using the lead-free BCTZ-based phased array; Human heart imaging (c) without and (d) with Doppler blood flow, respectively, using the commercial PZT-5H-based phased array. Source: Ma et al. [190]. Reproduced with permission from IEEE.

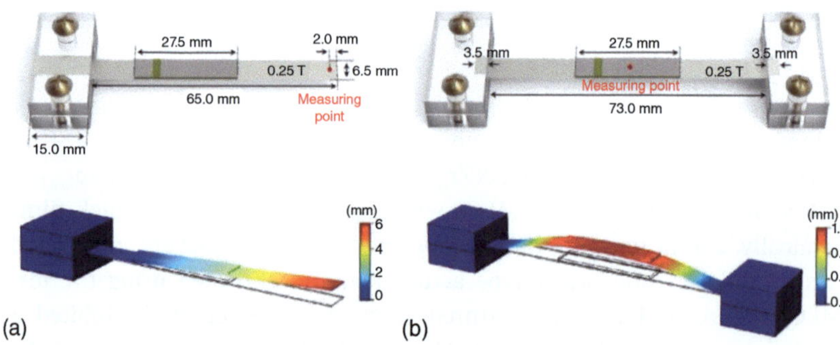

Figure 9.24 Schematic diagrams and simulated displacement distributions of the (a) cantilever-type and (b) bridge-type PZT-PZNN actuators. Source: Kim et al. [191]. Reproduced with permission from Elsevier.

characteristics of the fabricated devices had a left-pull velocity ~ 2.38 mm/s, a right-pull velocity ~ 2.03 mm/s, and a thrust ~2.37 N at 63.6 kHz for an applied voltage of 200 $V_{p\text{-}p}$. The piezoelectric motor can launch a small amount of insulin aqueous solution in a limited time with less linearity error.

9.5.3 Energy Harvesters

Energy-harvesting utilizing textured piezoelectric ceramics has recently attracted extensive attention due to the strong demand for self-powered electronics. The effects of texture engineering on the power generation characteristics of energy harvesters have been investigated in both lead-based and lead-free ceramics. Yan et al. [192] fabricated multilayer textured PMN-PZT piezoelectric ceramic benders, which possessed 700–800% higher $d \times g$ value than that of their random counterpart. The corresponding energy harvesters utilizing trilayer textured ceramics showed output power and power density of 903 μW and 15.5 mW/cm^3 at the acceleration of 0.43 g, respectively. The output power of co-fired multilayer textured samples was 500–600% higher than that of random samples. Berksoy-Yavuz et al. [193] fabricated unimorph energy harvesting devices based on textured and Mn-doped PMN–0.325PT ceramics, which show an over twofold increase in the output power as compared to their undoped counterparts.

For lead-free textured energy harvesters, Nahm et al. fabricated respectively energy harvesters based on textured $0.96(Na_{0.5}K_{0.5})(Nb_{0.93}Sb_{0.07})O_3$–$(0.04–x)BaZrO_3$–$x(Bi_{0.5}Ag_{0.5})ZrO_3$ [NKNS–(0.04–x)BZ–xBAZ], $0.96(Na, K)(Nb_{1-x}Sb_x)$–$0.04SrZrO_3$ and $(K, Na)(Nb, Sb)O_3$-$CaZrO_3$ ceramics [124, 194, 195], which all possessed

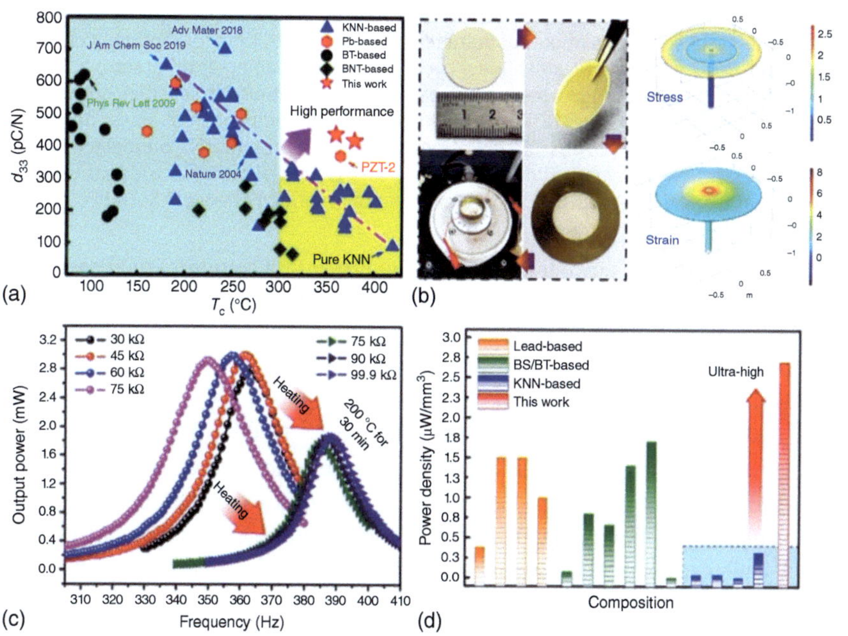

Figure 9.25 (a) The comprehensive performance comparison of T-Ta-9 with other piezoelectric ceramics, including the d_{33} value and Curie temperature; Energy harvesting performance of the T-Ta-9 PCD vibration energy harvester: (b) preparation process and simulation results of the stress and strain distribution. (c) The output power before and after heating at 200 °C for 30 minutes. (d) The output power density comparison of T-Ta-9 with other piezoelectric ceramics. Source: From Lin et al. [196]. Reproduced with permission from Royal Society of Chemistry.

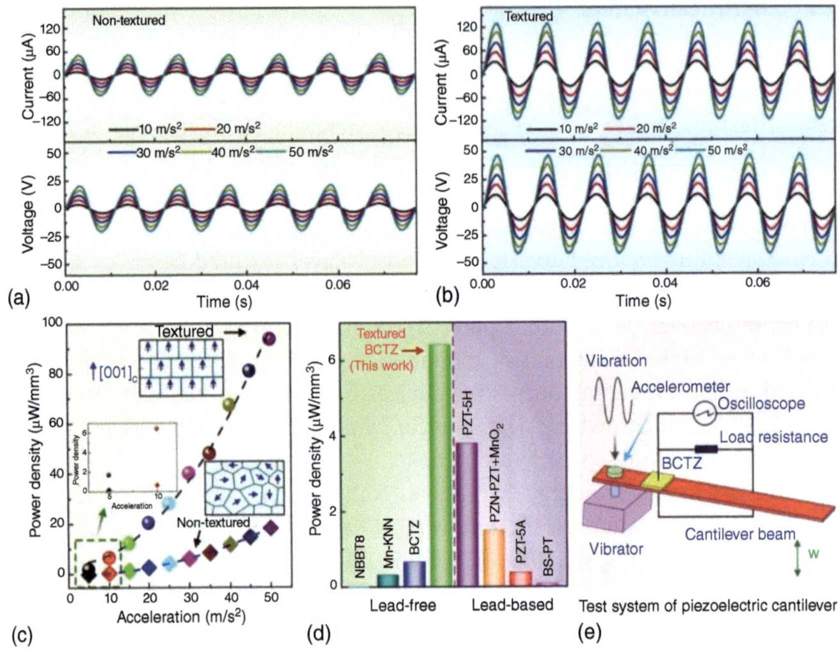

Figure 9.26 (a, b) Variations of output voltage and current with time for randomly oriented and textured BCTZ energy harvesters under cyclic sine-wave loading at different accelerations. (c) Comparison of acceleration-dependent output power densities between randomly oriented and textured BCTZ energy harvesters. (d) Comparison of power densities among our textured BCTZ (10 m/s^2 acceleration) and other representative piezoelectric energy harvesters. (e) Schematic illustration of the test system of piezoelectric cantilever. Source: Liu et al. [197]. Reproduced with permission of American Chemical Society.

substantially enhanced powder density than non-textured counterparts. Especially, the textured NKNS–0.02BZ–0.02BAZ thick film exhibited the largest values for d_{33} (805 pC/N) and $d_{33} \times g_{33}$ (29.5×10^{-12} m^2/N), which are among the largest values of KNN-based piezoceramics to date [124]. The corresponding energy harvester shows a large power density of 4.3 mW/cm^3 at the resonance frequency. Lin et al. [196] fabricated a series of $[001]_c$-textured $0.99K_{0.5}Na_{0.5}Nb_{(1-x)}Ta_xO_3$–$0.01Bi(Ni_{2/3}Nb_{1/3})O_3$ ceramics with both superior piezoelectric properties and high Curie temperature (e.g. T-Ta-9, $d_{33} \sim 435$ pC/N, $T_c \sim 360$ pC/N, as shown in Figure 9.25a). The piezoelectric circular diaphragm (PCD) vibration energy harvester was fabricated based on the T-Ta-9 textured ceramic, with simulation results of the stress and strain distribution shown in Figure 9.25b. The energy harvesters possessed high output voltage ($U \sim 13$ V) and output power (W ~ 3 mW), and could still maintain above 60% after being heated at 200 °C for 30 minutes, as shown in Figure 9.25c,d. Lead-free cantilever energy harvesters have also been fabricated based on highly textured $(Ba_{0.94}Ca_{0.06})(Ti_{0.95}Zr_{0.05})O_3$ (BCTZ) and $(Ba_{0.98}Ca_{0.02})(Ti_{0.96}Sn_{0.04})O_3$ ceramics with $F_{001} \geq 98\%$, which respectively show about 588% and 405% improvements in $d_{33} \times g_{33}$ values compared to non-textured counterparts [78, 197]. As shown in Figure 9.26, much higher output voltage and electric current were achieved in the textured BCTZ energy harvesters as compared to their non-textured counterparts under the same acceleration. Especially, the textured energy harvesters possessed ~9.8 times enhancement in output power density compared to the nontextured counterpart while maintaining stable output features up to 10^6 vibration cycles. The power densities, which increased from 6.4 to 93.6 μW/mm^3 with increasing acceleration excitation from 10 to 50 m/s^2, are even higher than those of some PZT-based energy harvesters reported previously [197].

9.6 Summary and Future Directions

Crystallographic texturing of polycrystalline ceramics allows the exploitation of piezoelectric anisotropy and application of domain engineering, potentially offering an effective means of achieving single-crystal like piezoelectric properties in ceramics. Many efforts have been made to advance the development of texture-engineered piezoelectric ceramics, attempting to solve many existing challenges. Several techniques associated with TGG-type processes yielded high-quality texture, enabled testing of texture orientation–property relationships, and thus led to significantly enhance piezoelectric properties as compared to their non-textured counterparts. All of these greatly accelerated their applications in high-performance electronic devices, especially where the possibility of using single crystals is limited. Therefore, some of the piezoelectric ceramics discussed in this chapter are very promising candidates for future commercialization.

However, challenges still exist for the further development of textured piezoelectric ceramics: (i) currently, ceramics with high texture quality are generally made of oriented grains with ≥10 μm in size. It is still a challenge to fabricate highly textured ceramics with fine grains (especially in nm size) to better explore the microstructure–property relationship; (ii) several piezoelectric compositions with special ions, for example, PZT compositions with higher Zr contents, are still challenging to get textured, owing to the reactions between common templates and matrix before TGG occurs; (iii) more explorations on template-related issues are needed, including synthesis of templates with novel composition and better/controlled morphology, new/improved casting technology for better template alignment and distribution (especially for fine templates), feasibility predicts for TGG between templates and matrix, and more fundamental understanding about texture evolution; (iv) it is still challenging to determine how to accurately quantify texture fraction and grain orientation distribution so that properties can be better correlated with texture; (v) more studies are needed toward shortening the performance gap between textured ceramics and single crystals. For example, even for textured ceramics with $F \sim 99\%$, their d_{33} is still ≤80% of the single-crystal value, and there is always significant hysteresis in the strain-field behavior of textured piezoelectrics; (vi) more factors, such as grain boundaries, templates, template-matrix interface, residual stress, porosity, second phases, and domain characteristics, should also be considered to establish a microstructure–property relationship for textured ceramics; (vii) Although great progress has been made in improving piezoelectric properties of textured ceramics, more research on emerging applications is still needed to promote the development of both materials and device applications.

References

1 Messing, G.L., Trolier-McKinstry, S., Brahmaroutu, E.M. et al. (2004). Templated grain growth of textured piezoelectric ceramics. *Crit. Rev. Solid State Mater. Sci.* 29 (2): 45–96.

2 Kimura, T. (2006). Application of texture engineering to piezoelectric ceramics – a review. *J. Ceram. Soc. Jpn.* 114 (1325): 15–25.

3 Okazaki, K. and Narushima, S. (1968). Electrical properties of the hot-pressed SbSI polycrystals. *J. Ceram. Soc. Jpn.* 76: 19–25.

4 Takenaka, T., Shoji, K., Takai, H. et al. (1976). Ferroelectric and dielectric properties of press forged $Bi_4Ti_3O_{12}$ ceramics. *Proc. Jpn. Cong. Mater. Res.* 19: 230–233.

5 Igarashi, H., Matsunaga, K., Taniai, T. et al. (1978). Dielectric and piezoelectric properties of grain-oriented $PbBi_2Nb_2O_9$ ceramics. *Am. Ceram. Soc. Bull.* 57: 815–817.

6 Swartz, S., Schulze, W., and Biggers, J. (1981). Fabrication and electrical properties of grain oriented $Bi_4Ti_3O_{12}$ ceramics. *Ferroelectrics* 38: 765–769.

7 Nagata, K., Yamamoto, Y., Igarashi, H. et al. (1981). Properties of the hot-pressed strontium barium niobate ceramics. *Ferroelectrics* 38: 853–856.

8 Granahan, M., Holmes, M., Shulze, W.A. et al. (1981). Fabrication of grain-oriented Bi_2WO_6 ceramics. *J. Am. Ceram. Soc.* 64: 68–69.

9 Seabaugh, M.M., Kerscht, I.H., and Messing, G.L. (1997). Texture development by templated grain growth in liquid phase sintered α-alumina. *J. Am. Ceram. Soc.* 80: 1181.

10 Messing, G.L., Poterala, S., Chang, Y.F. et al. (2017). Texture-engineered ceramics – property enhancements through crystallographic tailoring. *J. Mater. Res.* 32 (17): 3219–3241.

11 Wu, J., Zhang, S., and Li, F. (2022). Prospect of texture engineered ferroelectric ceramics. *Appl. Phys. Lett.* 121 (12): 120501.

12 Kang, S. (2005). *Sintering: Densification, Grain Growth, and Microstructure.* London: Elsevier Butterworth-Heinemann.

13 Hillert, M. (1965). On the theory of normal and abnormal grain growth. *Acta Metall.* 13 (3): 227–238.

14 Saetre, T. (2002). On the theory of normal grain growth in two dimensions. *Acta Mater.* 50 (6): 1539–1546.

15 Li, J., Zhang, Y., Fan, H. et al. (2021). Simulation for microstructure regulation of textured ferroelectric ceramics by grain boundary energy anisotropy. *Ceram. Int.* 47 (14): 20362–20370.

16 Seabaugh, M., Suvaci, E., Brahmaroutu, B. et al. (2000). Modeling anisotropic single crystal growth kinetics in liquid phase sintered α-Al_2O_3. *Interface Sci.* 8 (2–3): 257–267.

17 Kabakov, P., Dean, C., Kurusingal, V. et al. (2020). Solid-state crystal growth of lead-free ferroelectrics. *J. Mater. Chem. C* 8 (23): 7606–7649.

18 Richter, T., Schuh, C., Moos, R. et al. (2008). Single crystal growth and texturing of lead-based piezoelectric ceramics via templated grain growth process. *Funct. Mater. Lett.* 1 (2): 127–132.

19 Chang, Y.F., Sun, Y., Wu, J. et al. (2016). Formation mechanism of highly $[001]_C$ textured $Pb(In_{1/2}Nb_{1/2})O_3$-$Pb(Mg_{1/3}Nb_{2/3})O_3$-$PbTiO_3$ relaxor ferroelectric ceramics with giant piezoelectricity. *J. Eur. Ceram. Soc.* 36 (8): 1973–1981.

20 Wang, H., Xia, S., Ma, M. et al. (2022). Anisotropic growth kinetics and electric properties of PZT-5H single crystal by solid-state crystal growth method. *J. Am. Ceram. Soc.* 105 (5): 3238–3251.

21 Wenk, H. and Van Houtte, P. (2004). Texture and anisotropy. *Rep. Prog. Phys.* 67: 1367–1428.

22 Deluca, M. (2016). Microscopic texture characterisation in piezoceramics. *Adv. Appl. Ceram.* 115 (2): 112–122.

23 Lotgering, F.K. (1959). Topotactical reactions with ferrimagnetic oxides having hexagonal crystal structures-I. *J. Inorg. Nucl. Chem.* 9 (2): 113–123.

24 Jones, J.L., Iverson, B.J., and Bowman, K.J. (2007). Texture and anisotropy of polycrystalline piezoelectrics. *J. Am. Ceram. Soc.* 90 (8): 2297–2314.

25 Dollase, W.A. (1986). Correction of intensities for preferred orientation in powder diffractometry: application of the march model. *J. Appl. Crystallogr.* 19: 267–272.

26 Chang, Y.F., Wu, J., Liu, Z. et al. (2020). Grain-oriented ferroelectric ceramics with single-crystal-like piezoelectric properties and low texture temperature. *ACS Appl. Mater. Interfaces* 12 (34): 38415–38424.

27 Moriana, A.D. and Zhang, S. (2018). Lead-free textured piezoceramics using tape casting: a review. *J. Mater.* 4 (4): 277–303.

28 Sanjaya Ranmohotti, K.G., Josepha, E., Choi, J. et al. (2011). Topochemical manipulation of perovskites: low-temperature reaction strategies for directing structure and properties. *Adv. Mater.* 23 (4): 442–460.

29 Kimura, T. (2011). *Advances in Ceramics: Synthesis and Characterization, Processing and Specific Applications*. Shanghai: CoorsTek.

30 Li, L., Deng, J., Chen, J. et al. (2016). Topochemical molten salt synthesis for functional perovskite compounds. *Chem. Sci.* 7 (2): 855–865.

31 Schaak, R.E. and Mallouk, T.E. (2000). Topochemical synthesis of three-dimensional perovskites from lamellar precursors. *J. Am. Chem. Soc.* 122 (12): 2798–2803.

32 Ebrahimi, M.E., Allahverdi, M., and Safari, A. (2005). Synthesis of high aspect ratio platelet $SrTiO_3$. *J. Am. Ceram. Soc.* 88 (8): 2129–2132.

33 Saito, Y. and Takao, H. (2007). Synthesis of polycrystalline platelike $KNbO_3$ particles by the topochemical micro-crystal conversion method and fabrication of grain-oriented $(K_{0.5}Na_{0.5})NbO_3$ ceramics. *J. Eur. Ceram. Soc.* 27 (13–15): 4085–4092.

34 Li, L., Zhang, Y., Bai, W. et al. (2015). Synthesis of high aspect ratio (K, Na)NbO_3 plate-like particles and study on the synthesis mechanism. *Dalton Trans.* 44 (25): 11621–11625.

35 Poterala, S.F., Meyer, R.J. Jr., and Messing, G.L. (2011). Synthesis of high aspect ratio $PbBi_4Ti_4O_{15}$ and topochemical conversion to $PbTiO_3$-based microplatelets. *J. Am. Ceram. Soc.* 94 (8): 2323–2329.

36 Madaro, F., Tolchard, J.R., Yu, Y. et al. (2011). Synthesis of anisometric $KNbO_3$ and $(K_{0.5}Na_{0.5})NbO_3$ single crystals by chemical conversion of non-perovskite templates. *CrystEngComm* 13 (5): 1350–1359.

37 Hu, D., Kong, X., Mori, K. et al. (2013). Ferroelectric mesocrystals of bismuth sodium titanate: formation mechanism, nanostructure, and application to piezoelectric materials. *Inorg. Chem.* 52 (18): 10542–10551.

38 Hu, D., Luo, X., Kong, X. et al. (2015). Topochemical conversion of protonated titanate single crystals into platelike $Ba_{0.5}Sr_{0.5}TiO_3$ mesocrystals with controllable microstructures. *CrystEngComm* 17 (8): 1758–1764.

39 Poterala, S.F., Chang, Y.F., Clark, T. et al. (2010). Mechanistic interpretation of the Aurivillius to perovskite topochemical microcrystal conversion process. *Chem. Mater.* 22 (6): 2061–2068.

40 Chang, Y.F., Ning, H.B., Wu, J. et al. (2014). Formation mechanism of (001) oriented perovskite $SrTiO_3$ microplatelets synthesized by topochemical microcrystal conversion. *Inorg. Chem.* 53 (20): 11060–11067.

41 Wu, J., Chang, Y.F., Lv, W.M. et al. (2018). Topochemical transformation of single crystalline $SrTiO_3$ microplatelets from $Bi_4Ti_3O_{12}$ precursors and their orientation-dependent surface piezoelectricity. *CrystEngComm* 20 (22): 3084–3095.

42 Li, J., Shen, Z., Chen, X. et al. (2020). Grain-orientation-engineered multilayer ceramic capacitors for energy storage applications. *Nat. Mater.* 19 (9): 999–1005.

43 Liu, D., Yan, Y., and Zhou, H. (2007). Synthesis of micron-scale platelet $BaTiO_3$. *J. Am. Ceram. Soc.* 90 (4): 1323–1326.

44 Sato, T., Yoshida, Y., and Kimura, T. (2007). Preparation of <110>-textured $BaTiO_3$ ceramics by the reactive-templated grain growth method using needle-like TiO_2 particles. *J. Am. Ceram. Soc.* 90 (9): 3005–3008.

45 Lv, D., Zuo, R., and Su, S. (2012). Processing and morphology of (111) $BaTiO_3$ crystal platelets by a two-step molten salt method. *J. Am. Ceram. Soc.* 95 (6): 1838–1842.

46 Fu, J., Hou, Y., Zheng, M. et al. (2019). Topochemical conversion of (111) $BaTiO_3$ piezoelectric microplatelets using $Ba_6Ti_{17}O_{40}$ as the precursor. *Cryst. Growth Des.* 19 (2): 1198–1205.

47 Yan, X., Gao, F., and Liu, Z. (2014). Synthesis of high aspect ratio platelike $Ba_{1-x}Ca_xTiO_3$ particles by topochemical microcrystal conversion method. *Ceram. Int.* 40 (3): 4927–4932.

48 Kou, Q.W., Yang, B., Sun, Y. et al. (2022). Tetragonal (Ba, Ca) (Zr, Ti)O_3 textured ceramics with enhanced piezoelectric response and superior temperature stability. *J. Mater.* 8 (2): 366–374.

49 Saito, Y. and Takao, H. (2006). Synthesis of platelike {100} $SrTiO_3$ particles by topochemical microcrystal conversion and fabrication of grain-oriented ceramics. *Jpn. J. Appl. Phys.* 45 (9B): 7377–7381.

50 Li, L., Liu, J., Cao, S. et al. (2021). Microstructure evolution and reaction mechanism of $Pb(Zr_{1/2}Ti_{1/2})O_3$-$Pb(Zn_{1/3}Nb_{2/3})O_3$-$Pb(Ni_{1/3}Nb_{2/3})O_3$ piezoelectric ceramics with plate-like $PbTiO_3$ template. *Ceram. Int.* 47 (1): 470–478.

51 Setasuwon, P. and Kijamnajsuk, S. (2009). Synthesis of anisotropic micro-crystals of cubic $K_{0.5}Na_{0.5}NbO_3$. *CrystEngComm* 11 (9): 1947.

52 Dursun, S., Mensur-Alkoy, E., Sabuncu, A. et al. (2017). Growth of NBT template particles through topochemical microcrystal conversion and their structural characterization. *J. Am. Ceram. Soc.* 100 (3): 937–944.

53 Cha, H.A., Kim, Y., and Jeon, J. (2017). Mechanism of $Bi_{0.5}Na_{0.5}TiO_3$ and $Bi_{4.5}Na_{0.5}Ti_4O_{15}$ template synthesis during topochemical microcrystal conversion and texturing of $Bi_{0.5}(Na_{0.8}K_{0.2})_{0.5}TiO_3$ piezoelectric ceramics. *J. Eur. Ceram. Soc.* 37 (3): 967–974.

54 Jiang, C., Zhang, D., Zhou, K. et al. (2017). Topochemical transformation of two-dimensional single crystalline $Na_{0.5}Bi_{0.5}TiO_3$-$BaTiO_3$ platelets from $Na_{0.5}Bi_{4.5}Ti_4O_{15}$ precursors and their piezoelectricity. *J. Mater. Chem. A* 5 (30): 15780–15788.

55 Saito, Y., Takao, H., and Tani, T. (2004). Lead-free piezoceramics. *Nature* 432: 84–87.

56 Yan, Y., Liu, D., Zhao, W. et al. (2007). Topochemical synthesis of a high-aspect-ratio platelet $NaNbO_3$ template. *J. Am. Ceram. Soc.* 90 (8): 2399–2403.

57 Chang, Y.F., Yang, Z.P., Chao, X.L. et al. (2008). Synthesis and morphology of anisotropic $NaNbO_3$ seed crystals. *Mater. Chem. Phys.* 111 (2–3): 195–200.

58 Horn, J.A., Zhang, S.C., Selvaraj, U. et al. (1999). Templated grain growth of textured bismuth titanate. *J. Am. Ceram. Soc.* 82 (4): 921–926.

59 Tang, Q., Kan, Y., Wang, P. et al. (2007). Nd/V co-doped $Bi_4Ti_3O_{12}$ powder prepared by molten salt synthesis. *J. Am. Ceram. Soc.* 90 (10): 3353–3356.

60 Duran, C., Messing, G.L., and Trolier-McKinstry, S. (2004). Molten salt synthesis of anisometric particles in the SrO-Nb_2O_5-BaO system. *Mater. Res. Bull.* 39 (11): 1679–1689.

61 Lv, R., Liu, L., Wang, Y. et al. (2020). A-site cation and morphology control of $KSr_2Nb_5O_{15}$ microcrystalline by a modified molten salt method. *Adv. Powder Technol.* 31 (8): 3256–3266.

62 Yang, Z.P., Wei, L.L., and Chang, Y.F. (2007). Preparation and morphology of anisometric $KSr_2Nb_5O_{15}$ particles. *J. Eur. Ceram. Soc.* 27 (1): 267–272.

63 Chang, Y.F., Lee, S., Poterala, S. et al. (2011). A critical evaluation of reactive templated grain growth (RTGG) mechanisms in highly [001] textured $Sr_{0.61}Ba_{0.39}Nb_2O_6$ ferroelectric-thermoelectrics. *J. Mater. Res.* 26 (24): 3044–3050.

64 Le Ferrand, H. (2021). Magnetic slip casting for dense and textured ceramics: a review of current achievements and issues. *J. Eur. Ceram. Soc.* 41 (1): 24–37.

65 Andreeta, E.R.M., Dos Santos, H.F.L., Andreeta, M.R.B. et al. (2007). Anisotropy on $SrTiO_3$ templated textured PMN-PT monolithic ceramics. *J. Eur. Ceram. Soc.* 27 (6): 2463–2469.

66 Bouville, F., Portuguez, E., Chang, Y.F. et al. (2014). Templated grain growth in macroporous materials. *J. Am. Ceram. Soc.* 97 (6): 1736–1742.

67 Walton, R.L., Kupp, E.R., and Messing, G.L. (2021). Additive manufacturing of textured ceramics: a review. *J. Mater. Res.* 36 (18): 3591–3606.

68 Liu, Y.C., Chang, Y.F., Li, F. et al. (2017). Exceptionally high piezoelectric coefficient and low strain hysteresis in grain-oriented (Ba, Ca)(Ti, Zr)O_3 through integrating crystallographic texture and domain engineering. *ACS Appl. Mater. Interfaces* 9 (35): 29863–29871.

69 Chen, W., Hotta, Y., Tamura, T. et al. (2006). Effect of suction force and starting powders on microstructure of $Bi_4Ti_3O_{12}$ ceramics prepared by magnetic alignment via slip casting. *Scr. Mater.* 54 (12): 2063–2068.

70 Poterala, S.F., Meyer, R.J., and Messing, G.L. (2013). Low-field dynamic magnetic alignment and templated grain growth of diamagnetic PMN-PT ceramics. *J. Mater. Res.* 28 (21): 2960–2969.

71 Chartier, T., Badev, A., Abouliatim, Y. et al. (2012). Stereolithography process: influence of the rheology of silica suspensions and of the medium on polymerization kinetics – cured depth and width. *J. Eur. Ceram. Soc.* 32 (8): 1625–1634.

72 Lüchtenborg, J., Mühler, T., Leonard, F. et al. (2017). Laser-induced slip casting (LIS) – a new additive manufacturing process for dense ceramics demonstrated with Si_3N_4. *J. Ceram. Sci. Technol.* 8 (4): 531–540.

73 Zocca, A., Lima, P., and Günster, J. (2017). LSD-based 3D printing of alumina ceramics. *J. Ceram. Sci. Technol.* 8 (1): 141–148.

74 Walton, R.L., Fanton, M.A., Meyer, R.J. et al. (2020). Dispersion and rheology for direct writing lead-based piezoelectric ceramic pastes with anisotropic template particles. *J. Am. Ceram. Soc.* 103 (11): 6157–6168.

75 Walton, R.L., Brova, M.J., Watson, B.H. et al. (2021). Direct writing of textured ceramics using anisotropic nozzles. *J. Eur. Ceram. Soc.* 41 (3): 1945–1953.

76 Zhou, J.E., Yan, Y., Priya, S. et al. (2017). Computational study of textured ferroelectric polycrystals: texture development during templated grain growth. *J. Appl. Phys.* 121 (6): 064108.

77 Yan, Y., Cho, K.H., Maurya, D. et al. (2013). Giant energy density in [001]-textured $Pb(Mg_{1/3}Nb_{2/3})O_3$-$PbZrO_3$-$PbTiO_3$ piezoelectric ceramics. *Appl. Phys. Lett.* 102: 042903.

78 Sun, Y., Chang, Y.F., Wu, J. et al. (2019). Ultrahigh energy harvesting properties in textured lead-free piezoelectric composites. *J. Mater. Chem. A* 7 (8): 3603–3611.

79 Tutuncu, G., Chang, Y., Poterala, S. et al. (2012). In situ observations of templated grain growth in $(Na_{0.5}K_{0.5})_{0.98}Li_{0.02}NbO_3$ piezoceramics: texture development and template-matrix interactions. *J. Am. Ceram. Soc.* 95 (8): 2653–2659.

80 Chang, Y.F., Poterala, S., Yener, D. et al. (2013). Fabrication of highly textured fine-grained α-alumina by templated grain growth of nanoscale precursors. *J. Am. Ceram. Soc.* 96 (5): 1390–1397.

81 Yan, Y., Wang, Y.U., and Priya, S. (2012). Electromechanical behavior of [001]-textured $Pb(Mg_{1/3}Nb_{2/3})O_3$-$PbTiO_3$ ceramics. *Appl. Phys. Lett.* 100: 192905.

82 Brosnan, K.H. (2011). *Processing, Properties, and Applications of Textured 0.72Pb($Mg_{1/3}Nb_{2/3}$)O_3–0.28PbTiO_3 Ceramics.* Doctor thesis. The Pennsylvania State University.

83 Poterala, S.F., Trolier-McKinstry, S., Meyer, R.J. et al. (2011). Processing, texture quality, and piezoelectric properties of <001>$_C$ textured (1–*x*)$Pb(Mg_{1/3}Nb_{2/3})$ TiO_3-*x*$PbTiO_3$ ceramics. *J. Appl. Phys.* 110: 014105.

84 Kwon, S., Sabolsky, E.M., Messing, G.L. et al. (2005). High strain, <001> textured 0.675$Pb(Mg_{1/3}Nb_{2/3})O_3$–0.325$PbTiO_3$ ceramics: templated grain growth and piezoelectric properties. *J. Am. Ceram. Soc.* 88 (2): 312–317.

85 Amorín, H., Uršič, H., Ramos, P. et al. (2014). $Pb(Mg_{1/3}Nb_{2/3})O_3$-$PbTiO_3$ textured ceramics with high piezoelectric response by a novel templated grain growth approach. *J. Am. Ceram. Soc.* 97 (2): 420–426.

86 Miwa, Y., Kawada, S., Kimura, M. et al. (2015). Processing and enhanced piezoelectric properties of highly oriented compositionally modified $Pb(Zr,Ti)O_3$ ceramics fabricated by magnetic alignment. *Appl. Phys. Express* 8 (4): 041501.

87 Kang, W.S., Lee, T.G., Kang, J.H. et al. (2021). Bi-templated grain growth maximizing the effects of texture on piezoelectricity. *J. Eur. Ceram. Soc.* 41 (4): 2482–2487.

88 Chang, Y.F., Wu, J., Sun, Y. et al. (2015). Enhanced electromechanical properties and phase transition temperatures in [001] textured $Pb(In_{1/2}Nb1/2)O_3$-$Pb(Mg_{1/3}Nb_{2/3})O_3$-$PbTiO_3$ ternary ceramics. *Appl. Phys. Lett.* 107: 082902.

89 Wei, D., Yuan, Q., Zhang, G. et al. (2015). Templated grain growth and piezoelectric properties of <001>-textured PIN-PMN-PT ceramics. *J. Mater. Res.* 30 (14): 2144–2150.

90 Leng, H., Yan, Y., Wang, B. et al. (2022). High performance high-power textured Mn/Cu-doped PIN-PMN-PT ceramics. *Acta Mater.* 234: 118015.

91 Chang, Y., Watson, B., Fanton, M. et al. (2017). Enhanced texture evolution and piezoelectric properties in CuO-doped $Pb(In_{1/2}Nb_{1/2})O_3$-$Pb(Mg_{1/3}Nb_{2/3})O_3$-$PbTiO_3$ grain-oriented ceramics. *Appl. Phys. Lett.* 111: 232901.

92 Duran, C., Dursun, S., and Akça, E. (2016). High strain, <001>-textured $Pb(Mg_{1/3}Nb_{2/3})O_3$-$Pb(Yb_{1/2}Nb_{1/2})O_3$-$PbTiO_3$ piezoelectric ceramics. *Scr. Mater.* 113: 14–17.

93 Chang, Y.F., Wu, J., Yang, B. et al. (2021). Large, thermally stabilized and fatigue-resistant piezoelectric strain response in textured relaxor-$PbTiO_3$ ferroelectric ceramics. *J. Mater. Chem. C* 9 (6): 2008–2015.

94 Liu, L., Yang, B., Yang, S. et al. (2022). Cu-modified $Pb(Mg_{1/3}Nb_{2/3})O_3$-$PbZrO_3$-$PbTiO_3$ textured ceramics with enhanced electromechanical properties and improved thermal stability. *J. Eur. Ceram. Soc.* 42 (6): 2743–2751.

95 Zate, T.T., Kim, M., and Jeon, J.H. (2022). Outstanding unipolar strain of textured $Pb(Mg_{1/3}Nb_{2/3})O_3$-$PbZrO_3$-$PbTiO_3$ piezoelectric ceramics manufactured by particle size distribution control of the plate-like $BaTiO_3$ template. *Sensors Actuators A Phys.* 335: 113373.

96 Dursun, S., Mensur-Alkoy, E., Unver, M.U. et al. (2019). Enhancement of electrical properties in the ternary PMN-PT-PZ through compositional variation, crystallographic texture, and quenching. *J. Am. Ceram. Soc.* 103 (4): 2499–2508.

97 Yang, S., Qiao, L., Wang, J. et al. (2022). Full matrix electromechanical properties of textured $Pb(In_{1/2}Nb_{1/2})O_3$-$Pb(Sc_{1/2}Nb_{1/2})O_3$-$PbTiO_3$ ceramic. *J. Appl. Phys.* 131: 124104.

98 Yang, S., Li, J., Liu, Y. et al. (2021). Textured ferroelectric ceramics with high electromechanical coupling factors over a broad temperature range. *Nat. Commun.* 12: 1414.

99 Yan, Y., Geng, L.D., Liu, H. et al. (2022). Near-ideal electromechanical coupling in textured piezoelectric ceramics. *Nat. Commun.* 13: 3565.

100 Takao, H., Saito, Y., Aoki, Y. et al. (2006). Microstructural evolution of crystalline-oriented $(K_{0.5}Na_{0.5})NbO_3$ piezoelectric ceramics with a sintering aid of CuO. *J. Am. Ceram. Soc.* 89 (6): 1951–1956.

101 Li, Y., Hui, C., Wu, M. et al. (2012). Textured $(K_{0.5}Na_{0.5})NbO_3$ ceramics prepared by screen-printing multilayer grain growth technique. *Ceram. Int.* 38S: S283–S286.

102 Gupta, S., Belianinov, A., Baris Okatan, M. et al. (2014). Fundamental limitation to the magnitude of piezoelectric response of $(001)_{pc}$ textured $K_{0.5}Na_{0.5}NbO_3$ ceramic. *Appl. Phys. Lett.* 104: 172902.

103 Lv, D., Zuo, R., and Su, S. (2011). Reactive templated grain growth and anisotropic electrical properties of $(K_{0.5}Na_{0.5})NbO_3$ ceramics without sintering aids. *J. Mater. Sci. Mater. Electron.* 23 (7): 1367–1372.

104 Lv, D. and Zuo, R. (2013). Evolution of crystallographic grain orientation and anisotropic properties of $(K_{0.5}Na_{0.5})NbO_3$ ceramics using $BaTiO_3$ templates by reactive templated grain growth. *J. Alloys Compd.* 560: 62–66.

105 Ishii, K. and Tashiro, S. (2013). Orientation control of $(K,Na)NbO_3$ ceramics using $NaNbO_3$ particles prepared by single-step molten salt synthesis. *Jpn. J. Appl. Phys.* 52: 09KD04.

106 Li, Y.L., Hui, C., Li, Y.X. et al. (2012). Enhanced ferroelectric and piezoelectric properties of textured $K_{0.45}Na_{0.55}NbO_3$ ceramics prepared by screen-printing technique. *J. Inorg. Mater.* 27 (2): 214–218.

107 Chang, Y.F., Poterala, S.F., Yang, Z. et al. (2011). Microstructure development and piezoelectric properties of highly textured CuO-doped KNN by templated grain growth. *J. Mater. Res.* 25 (4): 687–694.

108 Chang, Y.F., Poterala, S.F., Yang, Z. et al. (2009). (001) textured $(K_{0.5}Na_{0.5})(Nb_{0.97}Sb_{0.03})O_3$ piezoelectric ceramics with high electromechanical coupling over a broad temperature range. *Appl. Phys. Lett.* 95: 232905.

109 Rafiq, M.A., Costa, M.E., and Vilarinho, P.M. (2016). Pairing high piezoelectric coefficients, d_{33}, with high Curie temperature (T_c) in lead-free $(K,Na)NbO_3$. *ACS Appl. Mater. Interfaces* 8 (49): 33755–33764.

110 Chang, Y.F., Poterala, S., Yang, Z. et al. (2011). Enhanced electromechanical properties and temperature stability of textured $(K_{0.5}Na_{0.5})NbO_3$-based piezoelectric ceramics. *J. Am. Ceram. Soc.* 94 (8): 2494–2498.

111 Hussain, A., Kim, J.S., Song, T.K. et al. (2013). Fabrication of textured KNNT ceramics by reactive template grain growth using NN templates. *Curr. Appl. Phys.* 13 (6): 1055–1059.

112 Li, P., Liu, B., Shen, B. et al. (2018). Mechanism of significantly enhanced piezoelectric performance and stability in textured potassium-sodium niobate piezoelectric ceramics. *J. Eur. Ceram. Soc.* 38 (1): 75–83.

113 Li, P., Fu, Z., Wang, F. et al. (2020). High piezoelectricity and stable output in $BaHfO_3$ and $(Bi_{0.5}Na_{0.5})ZrO_3$ modified $(K_{0.5}Na_{0.5})(Nb_{0.96}Sb_{0.04})O_3$ textured ceramics. *Acta Mater.* 199: 542–550.

114 Li, L., Bai, W., Zhang, Y. et al. (2015). The preparation and piezoelectric property of textured KNN-based ceramics with plate-like $NaNbO_3$ powders as template. *J. Alloys Compd.* 622: 137–142.

115 Liu, B., Li, P., Zhang, Y. et al. (2017). Enhanced piezoelectricity in (K, Na) NbO_3-based ceramics by optimizing composition and texture process. *J. Alloys Compd.* 695: 2207–2214.

116 Zhang, H., Zhu, Y., Fan, P. et al. (2018). Temperature-insensitive electric field induced strain and enhanced piezoelectric properties of <001> textured (K, Na)NbO_3-based lead-free piezoceramics. *Acta Mater.* 156: 389–398.

117 Kim, D.S., Eum, J.M., Go, S.H. et al. (2021). Remarkable piezoelectric performance and good thermal stability of <001>-textured $0.96(K_{0.5}Na_{0.5})(Nb_{1-y}Sb_y)O_3$–$0.04SrZrO_3$ lead-free piezoelectric ceramics. *J. Alloys Compd.* 882: 160662.

118 Quan, Y., Wang, L., Ren, W. et al. (2022). Effect of template amounts on the orientation degree and electrical properties of lead-free piezoelectric textured KNN-based ceramics. *J. Appl. Phys.* 131: 094101.

119 Liu, B., Li, P., Shen, B. et al. (2017). Enhanced piezoelectric properties and temperature-insensitive strain behavior of <001>-textured KNN-based ceramics. *Ceram. Int.* 43 (11): 8004–8009.

120 Liu, B., Li, P., Shen, B. et al. (2018). Simultaneously enhanced piezoelectric response and piezoelectric voltage coefficient in textured KNN-based ceramics. *J. Am. Ceram. Soc.* 101 (1): 265–273.

121 Li, P., Zhai, J., Shen, B. et al. (2018). Ultrahigh piezoelectric properties in textured (K,Na)NbO_3-based lead-free ceramics. *Adv. Mater.* 30 (8): 1705171.

122 Yang, W., Wang, Y., Li, P. et al. (2020). Improving electromechanical properties in KNANS-BNZ ceramics by the synergy between phase structure modification and grain orientation. *J. Mater. Chem. C* 8 (18): 6149–6158.

123 Chae, S.J., Kim, D.S., Kim, H. et al. (2021). Structural and piezoelectric properties of textured NLKNS-CZ thick films and their application in planar piezoactuator. *J. Am. Ceram. Soc.* 105 (2): 1185–1197.

124 Go, S.H., Kim, H., Kim, D.S. et al. (2022). Improvement of piezoelectricity of (Na, K)Nb-based lead-free piezoceramics using [001]-texturing for piezoelectric energy harvesters and actuators. *J. Eur. Ceram. Soc.* 42 (14): 6478–6492.

125 Chen, P.Y., Chen, C.S., Tu, C.S. et al. (2016). Effects of texture on microstructure, Raman vibration, and ferroelectric properties in 92.5%$(Bi_{0.5}Na_{0.5})TiO_3$-7.5%$BaTiO_3$ ceramics. *J. Eur. Ceram. Soc.* 36 (7): 1613–1622.

126 Su, S., Zuo, R., and Lv, D. (2012). Densification and texture evolution of $Bi_4Ti_3O_{12}$ templated $Na_{0.5}Bi_{0.5}TiO_3$-$BaTiO_3$ ceramics: effects of excess Bi_2O_3. *J. Alloys Compd.* 519: 25–28.

127 Yan, Y.K., Zhou, H.P., Zhao, W. et al. (2007). Fabrication and electrical properties of textured $Na_{1/2}Bi_{1/2}TiO_3$-$BaTiO_3$ ceramics by reactive-templated grain growth. *J. Electroceram.* 21 (1): 246–250.

128 Bai, W., Wang, L., Zheng, P. et al. (2018). Pairing high piezoelectric properties and enhanced thermal stability in grain-oriented BNT-based lead-free piezoceramics. *Ceram. Int.* 44 (10): 11402–11409.

129 Bai, W., Zhao, X., Ding, Y. et al. (2020). Giant field-induced strain with low hysteresis and boosted energy storage performance under low electric field in $(Bi_{0.5}Na_{0.5})TiO_3$-based grain orientation-controlled ceramics. *Adv. Electron. Mater.* 6 (9): 2000332.

130 Bai, W., Zhao, X., Huang, Y. et al. (2020). Integrating chemical engineering and crystallographic texturing design strategy for the realization of practically viable lead-free sodium bismuth titanate-based incipient piezoceramics. *Dalton Trans.* 49 (25): 8661–8671.

131 Huseyin, Y., Trolier-McKinstry, S., and Messing, G.L. (2003). (Reactive) templated grain growth of textured sodium bismuth titanate ($Na_{1/2}Bi_{1/2}TiO_3$-$BaTiO_3$) ceramics-II dielectric and piezoelectric properties. *J. Electroceram.* 11: 217–226.

132 Su, S. and Zuo, R. (2012). Fabrication and electrical properties of $0.94Na_{0.5}Bi_{0.5}TiO_3$-$0.06BaTiO_3$ textured ceramics by RTGG method using micrometer sized $BaTiO_3$ plate-like templates. *J. Alloys Compd.* 525: 133–136.

133 Maurya, D., Zhou, Y., Yan, Y. et al. (2013). Synthesis mechanism of grain-oriented lead-free piezoelectric $Na_{0.5}Bi_{0.5}TiO_3$-$BaTiO_3$ ceramics with giant piezoelectric response. *J. Mater. Chem.* C1 (11): 2102–2111.

134 Zhao, W., Ya, J., Xin, Y. et al. (2009). Fabrication of $Na_{0.5}Bi_{0.5}TiO_3$-$BaTiO_3$-textured ceramics templated by plate-like $Na_{0.5}Bi_{0.5}TiO_3$ particles. *J. Am. Ceram. Soc.* 92 (7): 1607–1609.

135 Jiang, C., Zhou, X., Zhou, K. et al. (2016). Grain oriented $Na_{0.5}Bi_{0.5}TiO_3$-$BaTiO_3$ ceramics with giant strain response derived from single-crystalline $Na_{0.5}Bi_{0.5}TiO_3$-$BaTiO_3$ templates. *J. Eur. Ceram. Soc.* 36 (6): 1377–1383.

136 Bai, W., Xi, J., Zhang, J. et al. (2015). Effect of different templates on structure evolution and large strain response under a low electric field in <00l>-textured lead-free BNT-based piezoelectric ceramics. *J. Eur. Ceram. Soc.* 35 (9): 2489–2499.

137 Maurya, D., Zhou, Y., Wang, Y. et al. (2015). Giant strain with ultra-low hysteresis and high temperature stability in grain oriented lead-free $K_{0.5}Bi_{0.5}TiO_3$-$BaTiO_3$-$Na_{0.5}Bi_{0.5}TiO_3$ piezoelectric materials. *Sci. Rep.* 5: 8595.

138 Dong, N., Gao, X., Xia, F. et al. (2019). Dielectric and piezoelectric properties of textured lead-free $Na_{0.5}Bi_{0.5}TiO_3$-based ceramics. *Crystals* 9 (4): 206.

139 Si, Y., Li, Y., Li, L. et al. (2019). Giant electro-strain in textured Li^+-doped 0.852BNT-0.11BKT-0.038BT ternary lead-free piezoelectric ceramics. *J. Am. Ceram. Soc.* 103 (3): 1765–1772.

140 Zhao, Z.H., Ye, M.Y., Ji, H.M. et al. (2018). Enhanced piezoelectric properties and strain response in <001> textured BNT-BKT-BT ceramics. *Mater. Des.* 137: 184–191.

141 Ma, S., Zhang, Y., Liu, Z. et al. (2015). Preparation and enhanced electric-field-induced strain of textured 91BNT-6BT-3KNN lead-free piezoceramics by TGG method. *J. Mater. Sci. Mater. Electron.* 27 (3): 3076–3081.

142 Fu, F., Zhai, J., Xu, Z. et al. (2011). Electric properties of high strain textured $Na_{0.5}Bi_{0.5}TiO_3$-$BaTiO_3$-$K_{0.5}Na_{0.5}NbO_3$ thick films. *Solid State Sci.* 13 (5): 934–937.

143 Fancher, C.M., Jo, W., Rödel, J. et al. (2014). Effect of texture on temperature-dependent properties of $K_{0.5}Na_{0.5}NbO_3$ modified $Bi_{1/2}Na_{1/2}TiO_3{-}x BaTiO_3$. *J. Am. Ceram. Soc.* 97 (8): 2557–2563.

144 Zhang, H., Xu, P., Patterson, E. et al. (2015). Preparation and enhanced electrical properties of grain-oriented $(Bi_{1/2}Na_{1/2})TiO_3$-based lead-free incipient piezoceramics. *J. Eur. Ceram. Soc.* 35 (9): 2501–2512.

145 Ye, C., Hao, J., Shen, B. et al. (2012). Large strain response in <001> textured 0.79BNT-0.20BKT-0.01NKN lead-free piezoelectric ceramics. *J. Am. Ceram. Soc.* 95 (11): 3577–3581.

146 Ahn, C.W., Choi, G., Kim, I.W. et al. (2017). Forced electrostriction by constraining polarization switching enhances the electromechanical strain properties of incipient piezoceramics. *NPG Asia Mater.* 9 (1): e346.

147 Vriami, D., Damjanovic, D., Vleugels, J. et al. (2015). Textured $BaTiO_3$ by templated grain growth and electrophoretic deposition. *J. Mater. Sci.* 50 (24): 7896–7907.

148 Wada, S., Takeda, K., Muraishi, T. et al. (2007). Preparation of [110] grain oriented barium titanate ceramics by templated grain growth method and their piezoelectric properties. *Jpn. J. Appl. Phys.* 46 (10B): 7039–7043.

149 Khanal, G.P., Fujii, I., Ueno, S. et al. (2021). Phase evolution and <110>-orientation mechanism in RTGG-processed $BaTiO_3$ ceramics with electrical properties. *J. Am. Ceram. Soc.* 104 (9): 4649–4658.

150 Sugawara, T., Nomura, Y., Kimura, T. et al. (2001). Fabrication of <111> oriented $BaTiO_3$ bulk ceramic by reactive templated grain growth method. *J. Ceram. Soc. Jpn.* 109 (10): 897–900.

151 Mallik, H.S., Fujii, I., Khanal, G.P. et al. (2020). Fabrication of <111>-oriented $BaTiO_3$ ceramics by high magnetic field electrophoretic deposition using hexagonal-tetragonal co-existing $BaTiO_3$ powder. *J. Ceram. Soc. Jpn.* 128 (8): 469–474.

152 Haugen, A.B., Morozov, M.I., Johnsson, M. et al. (2014). Effect of crystallographic orientation in textured $Ba_{0.92}Ca_{0.08}TiO_3$ piezoelectric ceramics. *J. Appl. Phys.* 116: 134102.

153 Liu, W. and Ren, X. (2009). Large piezoelectric effect in Pb-free ceramics. *Phys. Rev. Lett.* 103 (25): 257602.

154 Bai, W., Chen, D., Li, P. et al. (2016). Enhanced electromechanical properties in <00l>-textured $(Ba_{0.85}Ca_{0.15})(Zr_{0.1}Ti_{0.9})O_3$ lead-free piezoceramics. *Ceram. Int.* 42 (2): 3429–3436.

155 Ye, S.K., Fuh, J.Y.H., and Lu, L. (2012). Structure and electrical properties of <001>-textured $(Ba_{0.85}Ca_{0.15})(Ti_{0.9}Zr_{0.1})O_3$ lead-free piezoelectric ceramics. *Appl. Phys. Lett.* 100 (25): 252906.

156 Çakmak, Ö., Mensur-Alkoy, E., Toprak, G. et al. (2020). Investigation of the electrical properties of textured $0.5[Ba(Zr_{0.2}Ti_{0.8})]O_3$-$0.5[(Ba_{0.7}Ca_{0.3})TiO_3]$ piezoceramics. *J. Mater. Sci. Mater. Electron.* 31 (5): 4184–4192.

157 Zhao, Z.H., Li, X.L., Dai, Y.J. et al. (2016). Texture development in $Ba_{0.85}Ca_{0.15}Ti_{0.90}Zr_{0.10}O_3$ lead-free ceramics prepared by reactive template grain growth with different Ba and ca sources. *Ceram. Int.* 42 (16): 18756–18763.

158 Tsai, C.C., Liao, W.H., Chu, S.Y. et al. (2021). Investigation of the piezoelectric and anti-reduction properties of (Ba, Ca) (Ti, Sn, Hf) textured ceramics prepared under low oxygen partial pressure conditions at low sintering temperatures. *J. Eur. Ceram. Soc.* 41 (4): 2472–2481.

159 He, X., Chen, C., Boda, M.A. et al. (2022). Textured $Bi_4Ti_3O_{12}$ ceramics: one-step spark plasma sintering and their single-crystal-like polar anisotropy. *Adv. Eng. Mater.* 24 (9): 2200058.

160 Ahn, C.W., Jeong, E.D., Kim, Y.H. et al. (2008). Piezoelectric properties of textured $Bi_{3.25}La_{0.75}Ti_{2.97}V_{0.03}O_{12}$ ceramics fabricated by reactive templated grain growth method. *J. Electroceram.* 23 (2–4): 392–396.

161 Spitsin, A.I., Bush, A.A., and Kamentsev, K.E. (2020). Piezoelectric and dielectric properties of Bi_3TiNbO_9 prepared by hot pressing from powders activated using the serial dilution method. *Sci. Rep.* 10 (1): 22198.

162 Chen, H., Fu, F., and Zhai, J. (2011). Fabrication and piezoelectric property of highly textured $CaBi_2Nb_2O_9$ ceramics by tape casting. *Jpn. J. Appl. Phys.* 50 (5): 050207.

163 Yan, H.X., Ning, H.P., Zhang, H.T. et al. (2013). Textured high Curie point piezoelectric ceramics prepared by spark plasma sintering. *Advances in Applied Ceramics* 109 (3): 139–142.

164 Chen, H.B. and Zhai, J.W. (2012). Enhanced piezoelectric properties of $CaBi_2Nb_2O_9$ with Eu modification and templated grain growth. *Key Eng. Mater.* 512–515: 1367–1371.

165 Chen, H., Guo, X., Cui, Z. et al. (2013). Effects of A-site Sm substitution and textured structure on electric properties of $CaBi_2Nb_2O_9$-based high-curie-temperature ceramics. *Int. J. Appl. Ceram. Technol.* 10 (S1): E151–E158.

166 Tamura, K., Kimura, T., and Damjanovic, D. (2012). Tailoring of grain size in textured $CaBi_4Ti_4O_{15}$ ceramics prepared by templated grain growth process. *J. Am. Ceram. Soc.* 95 (11): 3421–3427.

167 Li, G., Tian, G., Zhai, Y. et al. (2022). Textured $CaBi_4Ti_4O_{15}$ ceramics with large piezoelectricity, excellent thermal stability and high resistivity. *J. Eur. Ceram. Soc.* 42 (13): 5584–5590.

168 Chen, H., Shen, B., Xu, J. et al. (2012). Textured $Ca_{0.85}(Li,Ce)_{0.15}Bi_4Ti_4O_{15}$ ceramics for high temperature piezoelectric applications. *Mater. Res. Bull.* 47 (9): 2530–2534.

169 Li, T., Li, X., Zhao, Z. et al. (2015). Evolution of textured $Ca_{0.85}(LiCe)_{0.075}Bi_4Ti_4O_{15}$ ceramics via templated grain growth using a rolling-extended method. *J. Mater. Sci. Mater. Electron.* 26 (4): 2082–2089.

170 Li, T., Li, X., Zhao, Z. et al. (2015). Structures and electrical properties of textured $Ca_{0.85}(LiCe)_{0.075}Bi_4Ti_4O_{15}$ ceramics prepared by the reactive templated grain growth. *Integr. Ferroelectr.* 162 (1): 1–7.

171 Kawada, S., Ogawa, H., Kimura, M. et al. (2007). Texture fraction dependences of high-power characteristics for textured $SrBi_2Nb_2O_9$ ceramics. *Ferroelectrics* 356 (1): 175–179.

172 Alkoy, S., Duran, C., and Hall, D.A. (2008). Electrical properties of textured potassium strontium niobate ($KSr_2Nb_5O_{15}$) ceramics fabricated by reactive templated grain growth. *J. Am. Ceram. Soc.* 91 (5): 1597–1602.

173 Tanaka, S., Takahashi, T., Furushima, R. et al. (2010). Fabrication of c-axis-oriented potassium strontium niobate ($KSr_2Nb_5O_{15}$) ceramics by a rotating magnetic field and electrical property. *J. Ceram. Soc. Jpn.* 118 (8): 722–725.

174 Dursun, S. and Alkoy, S. (2012). Processing, structural and electrical properties of textured potassium strontium niobate ($KSr_2Nb_5O_{15}$) piezoceramic fibers. *Adv. Mater. Res.* 445: 481–486.

175 Duran, C., Trolier-McKinstry, S., and Messing, G.L. (2002). Dielectric and piezoelectric properties of textured $Sr_{0.53}Ba_{0.47}Nb_2O_6$ ceramics prepared by templated grain growth. *J. Mater. Res.* 17 (9): 2399–2409.

176 Duran, C., Trolier-McKinstry, S., and Messing, G.L. (2000). Fabrication and electrical properties of textured $Sr_{0.53}Ba_{0.47}Nb_2O_6$ ceramics by templated grain growth. *J. Am. Ceram. Soc.* 83 (9): 2203–2213.

177 Dursun, S., Mensur-Alkoy, E., and Alkoy, S. (2016). Fabrication of textured lead-free strontium barium niobate (SBN61) bulk ceramics and their electrical properties. *J. Eur. Ceram. Soc.* 36 (10): 2479–2487.

178 Wei, L., Han, X., Li, Z. et al. (2012). Fabrication of textured $Sr_2Na_{0.9}K_{0.1}Nb_5O_{15}$ ceramics: anisotropy in structures and electrical properties. *J. Eur. Ceram. Soc.* 32 (14): 3773–3780.

179 Wei, L., Chao, X., Han, X. et al. (2014). Structure and electrical properties of textured $Sr_{1.85}Ca_{0.15}NaNb_5O_{15}$ ceramics prepared by the reactive templated grain growth. *Mater. Res. Bull.* 52: 65–69.

180 Yan, H., Ning, H., Kan, Y. et al. (2009). Piezoelectric ceramics with super-high Curie points. *J. Am. Ceram. Soc.* 92 (10): 2270–2275.

181 Ning, H., Yan, H., and Reece, M.J. (2010). Piezoelectric strontium niobate and calcium niobate ceramics with super-high Curie points. *J. Am. Ceram. Soc.* 93 (5): 1409–1413.

182 Gao, Z.P., Yan, H.X., Ning, H.P. et al. (2013). Ferroelectricity of $Pr_2Ti_2O_7$ ceramics with super high Curie point. *Adv. Appl. Ceram.* 112 (2): 69–74.

183 Gao, Z.P., Yan, H.X., Ning, H.P. et al. (2013). Piezoelectric and dielectric properties of Ce substituted $La_2Ti_2O_7$ ceramics. *J. Eur. Ceram. Soc.* 33 (5): 1001–1008.

184 Li, C., Xiang, H., Chen, J. et al. (2016). Phase transition, dielectric relaxation and piezoelectric properties of bismuth doped $La_2Ti_2O_7$ ceramics. *Ceram. Int.* 42 (9): 11453–11458.

185 Levassort, F., Pham Thi, M., Hemery, H. et al. (2006). Piezoelectric textured ceramics: effective properties and application to ultrasonic transducers. *Ultrasonics* 44: e621–e626.

186 Levassort, F., Thi, M.P., Marechal, P. et al. (2007). Ultrasonic transducer based on highly textured PMN-PT piezoelectric ceramic. *J. Electroceram.* 19 (4): 375–381.

187 Brosnan, K.H., Messing, G.L., Markley, D.C. et al. (2009). Comparison of the properties of tonpilz transducers fabricated with <001> fiber-textured lead magnesium niobate-lead titanate ceramic and single crystals. *J. Acoust. Soc. Am.* 126 (5): 2257–2265.

188 Sun, Y., Jiang, L., Chen, R. et al. (2022). Design and fabrication of 15-MHz ultrasonic transducers based on a textured $Pb(Mg_{1/3}Nb_{2/3})O_3$-$Pb(Zr, Ti)O_3$ ceramic. *IEEE Trans. Ultrason. Ferroelectr. Freq. Control* 69 (11): 3095–3101.

189 Quan, Y., Fei, C., Ren, W. et al. (2021). Lead-free KNN-based textured ceramics for high-frequency ultrasonic transducer application. *IEEE Trans. Ultrason. Ferroelectr. Freq. Control* 68 (5): 1979–1987.

190 Ma, X., Liu, Y., Ruan, J. et al. (2022). Lead-free ultrasonic phased array transducer for human heart imaging. *IEEE Trans. Ultrason. Ferroelectr. Freq. Control* 69 (2): 751–760.

191 Kim, E.J., Lee, T.G., Kim, D.S. et al. (2020). Textured $Pb(Zr,Ti)O_3$-$Pb[(Zn,Ni)_{1/3}Nb_{2/3}]O_3$ multilayer ceramics and their application to piezoelectric actuators. *Appl. Mater. Today* 20: 100695.

192 Yan, Y., Marin, A., Zhou, Y. et al. (2014). Enhanced vibration energy harvesting through multilayer textured $Pb(Mg_{1/3}Nb_{2/3})O_3$-$PbZrO_3$-$PbTiO_3$ piezoelectric ceramics. *Energy Harvest. Syst.* 1 (3–4): 189–195.

193 Berksoy-Yavuz, A., Savacı, U., Turan, S. et al. (2020). Structural features and energy harvester device applications of textured 0.675PMN-0.325PT piezoceramics. *J. Mater. Sci. Mater. Electron.* 31 (12): 9650–9659.

194 Chae, S.J., Na, J.W., Kim, D.S. et al. (2022). Application of [001]-textured (K, Na)(Nb, Sb)O_3-$CaZrO_3$ thick films to piezoelectric energy harvesters. *Ceram. Int.* 48 (13): 18401–18414.

195 Kim, S.W., Chae, S.J., Kim, D.S. et al. (2022). High-power piezoelectric energy harvester produced using [0 0 1]-textured (Na, K)NbO_3-based lead-free piezoceramics. *J. Am. Ceram. Soc.* 106 (1): 488–500.

196 Lin, J., Cao, Y., Zhu, K. et al. (2022). Ultrahigh energy harvesting properties in temperature-insensitive eco-friendly high-performance KNN-based textured ceramics. *J. Mater. Chem. A* 10 (14): 7978–7988.

197 Liu, Y., Chang, Y., Sun, E. et al. (2018). Significantly enhanced energy-harvesting performance and superior fatigue-resistant behavior in $[001]_c$-textured $BaTiO_3$-based lead-free piezoceramics. *ACS Appl. Mater. Interfaces* 10 (37): 31488–31497.

10

Piezoelectric Thin Films

Huajun Liu

*Institute of Materials Research and Engineering (IMRE), Agency for Science, Technology and Research (A*STAR), 2 Fusionopolis Way, Innovis #08-03, Singapore 138634, Republic of Singapore*

10.1 Introduction

In previous chapters, bulk piezoelectric materials, including single crystals and ceramics, have been introduced and reviewed. In this chapter, we will focus on the piezoelectric thin films. The research interests in piezoelectric thin films mainly arise from device applications in modern electronics and the novel science of new material phases in thin films that are not available in bulk materials.

10.1.1 Applications of Piezoelectric Thin Film Devices

Thin films with piezoelectric properties intrinsically convert electrical energy to mechanical energy and vice versa in a reversible manner. This unique property adds mechanical functionality when integrated to complementary metal–oxide–semiconductor (CMOS) platform in microelectronics, contributing to the route of "more-than-Moore" in the post-Moore's law era [1]. The continuous drive toward miniaturization and lower power consumption in electronic devices will lead to a strong demand for high-performance piezoelectric thin films.

The largest market for piezoelectric thin film device is the acoustic filters for radio frequency (RF) wireless communications, such as in fifth-generation (5G) networks [2]. The RF front-end modules in the mobile devices serve the function of receiving and transmitting wireless signals. They typically consist of antenna, diplexer, power amplifier (PA), low noise amplifier (LNA), and acoustic filters, as shown in Figure 10.1a [3]. Among these components, the acoustic filters play the key roles of filtering out unwanted frequency bands and only allowing carrier frequency bands to pass through. The acoustic filters, therefore, define the bandwidth of RF communication (Figure 10.1b), which is proportional to the speed of wireless data transfer according to Shannon's theory [6]. The bandwidths of acoustic filters are directly related to the electromechanical coupling coefficient k_t^2 of the piezoelectric thin films. Therefore, piezoelectric thin films with large k_t^2 are highly desired to increase the speed of wireless networks.

Piezoelectric Materials: From Fundamentals to Emerging Applications, First Edition. Edited by Jiagang Wu.

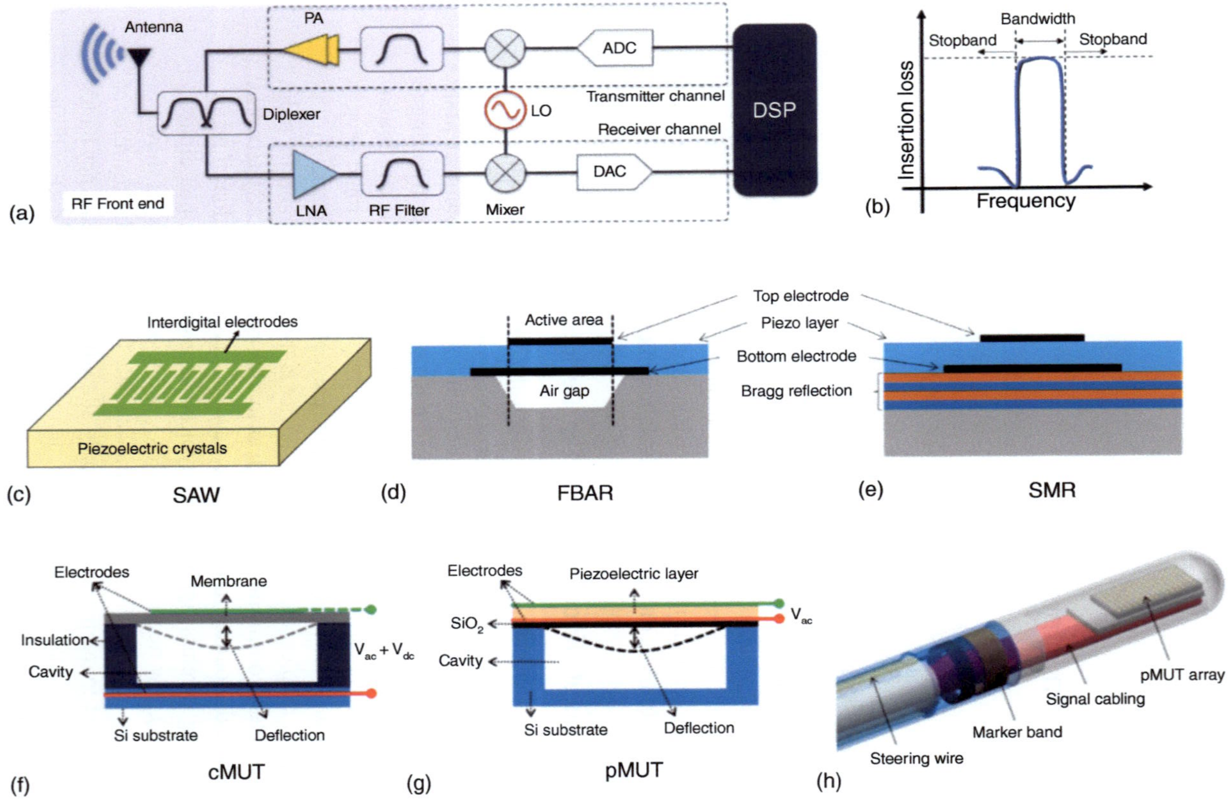

Figure 10.1 Device applications of piezoelectric thin films: (a) schematic diagram of RF front-end modules in mobile devices. PA, power amplifier; LNA, low noise amplifier; ADC, analog to digital converter; DAC, digital to analog converter; LO, local oscillator; DSP, digital signal processor. Source: Will-Cole et al. [3]. Reproduced with permission of AIP Publishing, (b) the filtering function of acoustic filters in the frequency bands. Source: Huajun Liu. Schematic diagrams of (c) SAW, (d) FBAR, and (e) SMR resonators for RF wireless communications. Source: (c) Huajun Liu. (d) and (e) Liu et al. [4] Reproduced under the open access Creative Commons CC BY license from MDPI. Cross-sectional structures of (f) cMUT and (g) pMUT, (h) mechanical model of the distal end of the steerable catheter containing pMUT matrix arrays for intracardiac imaging. Source: (f–h) Qiu et al. [5] Reproduced under the open access Creative Commons CC BY license from MDPI.

There are two types of acoustic filters in the market [4, 7]. The first type is the surface acoustic wave (SAW) resonators (Figure 10.1c) [8]. The RF electrical field is converted by piezoelectric layers to mechanical waves, which propagate mainly in the in-plane directions along the surface of the device. The working frequencies of SAW resonators are defined by the widths of interdigital electrodes, which are currently limited up to ~2.5 GHz. As 5G network is working toward higher frequency ranges above 3 GHz, SAW resonators cannot meet the requirements. The second type is the bulk acoustic wave (BAW) resonators, with two subtypes: thin-film bulk acoustic resonators (FBAR) (Figure 10.1d) and solidly mounted resonators (SMR) (Figure 10.1e) [4]. The acoustic wave propagates along the vertical direction in BAW resonators, where the piezoelectric thin films are sandwiched by top and bottom electrodes. Therefore, the resonance frequency is determined by the film thickness, thus showing the potential to achieve high frequencies with thinner films. For FBAR resonators, devices are separated by air gaps to confine the acoustic waves inside the films. For SMR devices, the multilayer Bragg reflectors below the piezoelectric layers are used to prevent leakage of acoustic waves to the substrates. AlN thin films are the most widely used materials in commercial acoustic filters due to their high mechanical quality factor, mature deposition process, and compatibility with the CMOS platform [7]. However, the low electromechanical coupling coefficients of AlN limit the bandwidth of commercial acoustic filters. Two strategies are explored to enhance the coupling coefficients. The first one is by chemical doping of AlN thin films, among which Sc-doped AlN shows the best performance [9]. The second one is to use oxide piezoelectric thin films with high coupling coefficients. The leading candidate is $LiNbO_3$ thin films [10, 11].

The second major application of piezoelectric thin films is the piezoelectric micromachined ultrasonic transducers (pMUT) [5, 12]. The main idea is to utilize micromachining techniques to fabricate ultrasonic transducers in thin films with higher resolution and smaller pixel size than bulk ceramics yet in an easier manner [13]. Compared with capacitive micromachined ultrasonic transducers (cMUT) in Figure 10.1f, pMUT (Figure 10.1g) can operate at lower voltage with lower loss and higher output signals. This leads to increased sensitivity of pMUT, which has shown promising performance in devices for medical ultrasonic imaging (Figure 10.1h) [14], fingerprint sensors [15], and rangefinders [16].

The other common applications of piezoelectric thin films are microphones and speakers [17], inkjet printheads [18], and energy harvesters [19, 20]. Overall, piezoelectric thin films provide a unique linkage between electrical and mechanical information and energy in microelectronic devices. With the growing demands of microelectromechanical devices from 5G communication, the Internet of Things, and biomedical imaging, research on piezoelectric thin film devices is expected to be substantially increased in the coming years.

10.1.2 Thin Films as Unique Playgrounds for New Science

In addition to their device applications, piezoelectric thin films also provide unique playgrounds for studying the novel science of new material phases in thin films that are not available in bulk materials. There are two major types of piezoelectric thin films. The first type has wurtzite structure, such as AlN and ZnO, which is only piezoelectric but not ferroelectric. The second type has a perovskite structure, such as $PbTiO_3$ and $BaTiO_3$, which is both piezoelectric and ferroelectric. We focus on the second type of piezoelectric thin films for the discussions below in this chapter.

As piezoelectric thin films are typically deposited on substrates with different lattice parameters, the lattice mismatch between the film and substrate could introduce significant distortions to the crystal structure and, thus, the material phases that are not stable in bulk materials. This is known as strain engineering, which is especially powerful in ferroelectric and piezoelectric thin films due to the strong coupling between polarization and strain [21, 22]. For example, highly strained $SrTiO_3$ thin films show ferroelectric behavior, whereas their bulk is not ferroelectric [23]. $BaTiO_3$ epitaxial thin films show greatly enhanced ferroelectric polarization, approximately 250% higher than bulk single crystals due to a compressive strain of 1.7% from $DyScO_3$ substrate [24]. Tensile strain increases the in-plane lattice dimension while reducing the out-of-plane lattice dimension of the films so that the axial ratio c/a is reduced as compared to bulk. On the other hand, compressive strain reduces the in-plane lattice dimension while increasing the out-of-plane lattice dimension so that the axial ratio c/a is increased compared to bulk. Under large compressive strain above 4.5%, $BiFeO_3$ thin films show a strain-induced morphotropic phase boundary and much enhanced electric field-induced strain [25]. Figure 10.2a lists available single crystal substrates with different in-plane lattice dimensions for strain engineering of piezoelectric thin films with perovskite structures.

The thicknesses of piezoelectric thin films typically range from a few nanometers to a few micrometers. This dimension is substantially smaller than the lateral dimensions on the order of a few to hundreds of millimeters. This quasi-two-dimensional geometry leads to interesting physics in piezoelectric thin films. Due to insufficient compensation of the depolarization field by electrodes, the loss of polarization was predicted below a critical thickness [28]. Without contact with electrodes, the depolarization field can be effectively compensated by the formation of 180° up–down strip domains with regular periodicity (Figure 10.2b). These periodic stripe domains give rise to characteristic satellite peaks in diffuse X-ray scattering around the Bragg peaks, as shown in Figure 10.2c. Using these satellite peaks as an indication of the existence of polarization, ferroelectricity is shown to be stable down to a thickness of three unit cells (UC) $PbTiO_3$ thin films grown on $SrTiO_3$ substrates [29]. The 3 UC thickness is the minimum requirement to contain TiO_2 and PbO layers with the same nearest neighbor environment as in the bulk $PbTiO_3$. In this ultrathin thickness range, the substrate termination, interface atomic structure, and surface reconstruction are all critical factors to determine the physical properties of the films. With progress in the control of these extrinsic factors and improvements in film quality, recent works have shown ferroelectric polarization can be stable in the 2 UC $BaTiO_3$ [30], 1.5 UC $PbZr_{0.2}Ti_{0.8}O_3$ [31], and 1 UC $BiFeO_3$ thin films [32, 33]. It is, therefore, very promising to integrate these ultrathin ferroelectric and piezoelectric thin films into electronic devices without losing their inherent functionalities.

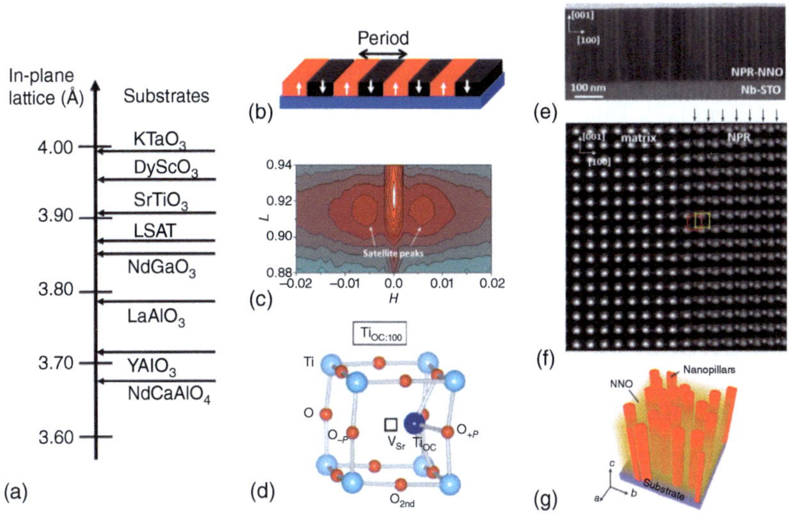

Figure 10.2 Engineering piezoelectric thin films by tuning misfit strain, size effect, and functional defects: (a) single crystal substrates for epitaxial growth of piezoelectric thin films with perovskite structures. LSAT: $(LaAlO_3)_{0.3}-(SrAl_{0.5}Ta_{0.5}O_3)_{0.7}$. Source: Huajun Liu. (b) 180° up–down strip domains formed in ferroelectric thin films grown on insulating substrates. Source: Huajun Liu. (c) Satellite peaks in diffuse X-ray scattering around the Bragg peaks in the reciprocal space. Source: Huajun Liu. (d) Atomic configuration of off-centered Ti_{Sr} antisite defects, leading to local polarization in Sr-deficient $SrTiO_3$ thin films. Source: Choi et al. [26]. Reproduced with permission of American Physical Society. (e) Cross-sectional transmission electron microscopy image of vertical nanopillars in Na-deficient $NaNbO_3$ thin films. (f) Atomically resolved image of matrix and nanopillar regions (NPR). (g) 3D model of vertical nanopillar structures embedded in perovskite matrix. Source: (e–g) Liu et al. [27]. Reproduced with permission from The American Association for the Advancement of Science - AAAS.

Defects are known to play critical roles in determining the properties of perovskite oxides [34]. By rational design and controlled synthesis, functional defects can be beneficial instead of being detrimental to the performance of materials. In thin oxide films, the most common defects are oxygen vacancies [35] due to loss of oxygen during the film deposition process. Recently, antisite defects, where B-site cations occupy A-site positions (B_A) or vice versa (A_B) in perovskites (ABO_3), have been attracting attention due to enhanced ferroelectric and piezoelectric properties [26, 27, 36]. As a result of the differences in ionic radii between A-site and B-site cations, antisite defects lead to significant distortions of atomic structures, especially the ionic displacements. In the prototype perovskite material $SrTiO_3$, Sr deficiency in the thin films leads to Ti atoms occupying Sr vacancies, as shown in Figure 10.2d. As the size of Ti is too small for the A-site space to bond with 12 oxygens, off-center displacements (Ti_{OC}) are energetically more favorable. This leads to the ferroelectric behavior of otherwise non-ferroelectric $SrTiO_3$ at room temperature. Another recent work on Y-rich $YFeO_3$ thin films shows the case of A_B antisite defects (A-site occupying B-site positions), where Y_{Fe} defects cause ferroelectric polarization lead to multiferroic behavior in these magnetic films [37]. More interestingly, our recent work shows that antisite defects can grow into extended defects to create new nanoscale heterogeneous structures, such as nanopillars through direct self-assembly in Na-deficient $NaNbO_3$ thin films (Figure 10.2e–g) [27]. These nanopillar structures give rise to more than 50 times the enhancement of piezoelectric coefficients while maintaining a high Curie temperature. This defect engineering strategy in piezoelectric thin films could be extended to other material systems with volatile A-site elements, such as Pb, Bi, K, and Li, for improved piezoelectric properties.

10.2 Growth of Piezoelectric Thin Films

10.2.1 $Pb(Zr,Ti)O_3$ Thin Films

$Pb(Zr,Ti)O_3$ (PZT) thin films are the most widely studied piezoelectric thin films [38, 39]. The major methods to grow PZT thin films are sputtering, pulsed laser deposition (PLD), metal-organic chemical vapor deposition (MOCVD), and chemical solution deposition (CSD). Due to the volatility of Pb, Pb excesses of 10–30% are typically used to compensate for the loss during the film growth process. The deposition conditions, such as temperature and gas pressure, are known to critically affect the microstructure and piezoelectric properties of PZT thin films. For example, the effects of oxygen pressure and deposition temperature on the transverse piezoelectric coefficient e_{31} were systematically studied to find the optimal deposition conditions [40]. As PZT is a solid solution between $PbTiO_3$ and $PbZrO_3$, the Zr/Ti ratio is the key factor to tune the composition of the thin films near the morphotropic phase boundary to obtain high piezoelectric coefficients [18]. The highest field-induced strain is obtained near 50% Zr in PZT thin films. This work also shows that the crystallographic orientations strongly affect the piezoelectric coefficients of PZT thin films. In addition to the PZT films themselves, the top and bottom electrode layers were found to significantly change the piezoelectric properties [41]. The PZT thin films grown on Pt bottom electrodes have a (110) preferred orientation, while those on $LaNiO_3$ bottom electrodes have a (001) preferred orientation. This orientation difference leads to a change in the piezoelectric properties of PZT. In addition, the top electrode layers also change the piezoelectric coefficient d_{33} of PZT thin films by ~50%.

The effect of substrate misfit strain has been studied in PZT thin films by theoretical calculations [42]. The strain phase diagram is shown in Figure 10.3a. Under compressive strain, the ferroelectric polarization is purely along the out-of-plane (001) direction, which is called *c* phase with a tetragonal symmetry. Under tensile strain, the ferroelectric polarization lies down in plane, which is called *aa* phase with orthorhombic symmetry. In a small strain range, the PZT film shows a monoclinic *r* phase, with both in-plane and out-of-plane components of polarization. Experimental results confirm these phases exist when PZT films are grown on substrates with different mismatch strains [44]. Ferroelectric polarization shows a strong dependence on the lattice dimensions of the PZT films, while piezoelectric

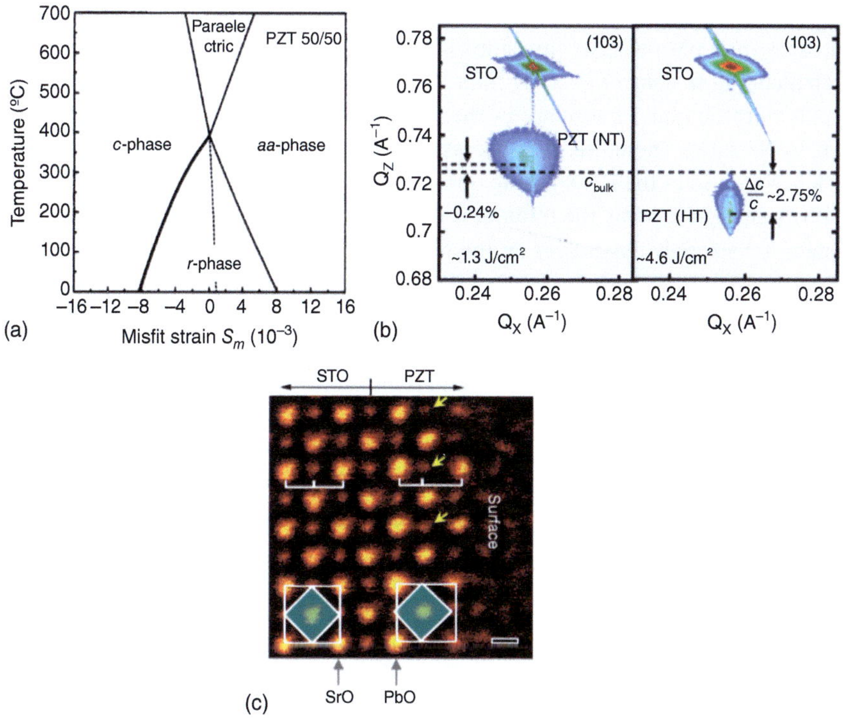

Figure 10.3 Growth of Pb(Zr,Ti)O_3 thin films: (a) Misfit strain phase diagram of PZT thin films as calculated by a nonlinear thermodynamic theory. Source: Pertsev et al. [42]. Reproduced with permission of American Physical Society. (b) Reciprocal space mappings of PZT thin films grown on (001) STO substrates with different laser power by PLD. Source: Huang et al. [43]. Reproduced under the open access Creative Commons CC BY license from Wiley. (c) High-resolution TEM image of 1.5 UC PZT grown on (001) STO substrate. Source: Gao et al. [31]. Reproduced from Springer Nature / CC BY 4.0.

coefficients d_{33} hardly change with misfit strain. In addition to substrate misfit strain, the lattice dimensions of PZT thin films can be further tuned by laser power during film growth by PLD [43]. The out-of-plane lattice parameter c increased from 4.13 to 4.27 Å as the laser power increased (Figure 10.3b), leading to the enhancement of polarization from 70 to 100 μC/cm^2. It will be interesting to investigate the piezoelectric properties of these PZT thin films with much enhanced out-of-plane lattice dimensions. For the ultrathin PZT, high-resolution transmission electron microscopy (TEM) images show that the ferroelectric polarization is retained even at a thickness of 1.5 UC (Figure 10.3c) [31], partially due to the relatively large compressive strain of ~1.2% from SrTiO_3 (STO) substrates.

10.2.2 (K,Na)NbO_3 Thin Films

(K,Na)NbO_3 (KNN) is the most studied lead-free piezoelectric material system due to its high Curie temperature and large piezoelectric coefficients [45]. KNN thin films have also been heavily investigated to achieve better performance than lead-containing piezoelectric thin films [46–48]. The volatile elements (Na and K) are challenging to control in KNN films during the growth process. Furthermore, the polymorphic phase boundary in the KNN system also requires fine-tuning of the K to Na ratio. Therefore, it is necessary to control both the (K + Na)/Nb and Na/(K + Na) ratio in the KNN film to optimize the piezoelectric properties [49, 50]. As shown in Figure 10.4a, the transverse piezoelectric coefficient d_{31} and dielectric constant are strongly dependent on the Na/(K + Na) ratio. The optimal properties are obtained at a Na/(K+Na) ratio of ~0.55, which is close to the bulk KNN ceramics [45]. The orientations of the KNN films also affect the piezoelectric properties. KNN films grown on (100) Nb-doped STO substrates show a higher piezoelectric coefficient d_{33} than the other two orientations along (110) and (111) [53]. This difference could come from the contributions of non-180° domain motion in (100) films. Similar to PZT thin films, the top electrodes are also found to significantly change the piezoelectric coefficients of KNN thin films [51]. The $LaNiO_3$ top electrodes increase d_{33} to 58 pm/V as compared to 26 pm/V when Pt top electrodes are used (Figure 10.4b).

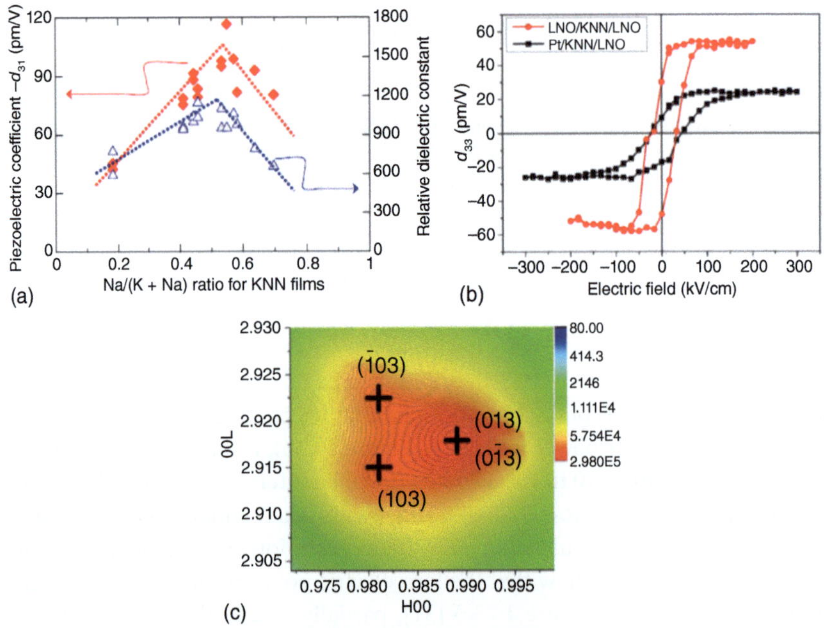

Figure 10.4 Growth of $(K,Na)NbO_3$ thin films. (a) Dependence of transverse piezoelectric coefficients d_{31} and dielectric constants of KNN thin films on the chemical composition of Na/(K+Na) ratio. Source: Shibata et al. [49]. Reproduced with permission of Japanese Society of Applied Physics. (b) The effect of top electrodes on the piezoelectric coefficients d_{33} of KNN thin films. LNO, $LaNiO_3$. Source: Li et al. [51]. Reproduced with permission of American Ceramic Society. (c) Reciprocal space mapping around (103) diffraction peak for KNN thin films grown on Nb-doped $SrTiO_3$ (001) substrate. Source: Luo et al. [52]. Reproduced with permission of Wiley.

The effect of substrate misfit strain on the crystalline phases of KNN thin films has been investigated by single-domain thermodynamics [54]. The temperature–strain phase diagram for KNN thin films grown on substrates with equal in-plane lattice dimensions has been calculated. There are four stable phases in the strain range from −2% to 2%: the paraelectric tetragonal T^P phase at high temperature, the tetragonal T_3 phase with only out-of-plane polarization in a high compressive strain range, the monoclinic M_{12} phase with only in-plane polarizations in high tensile strain range, the rhombohedral R^D phase with polarizations in both in-plane and out-of-plane directions. The KNN thin films grown on a highly strained Nb-doped $SrTiO_3$ (001) substrate exhibit a monoclinic phase that is not available in bulk KNN materials [52]. Figure 10.4c shows the splitting of (103) diffraction peaks into three peaks, which is a characteristic feature of the monoclinic *Mc* phase. More recently, epitaxial KNN thin films have been grown on rare-earth scandate substrates with systematically varied in-plane misfit strains [55]. By using a combination of characterization methods, such as piezoelectric force microscopy (PFM) and high-resolution X-ray diffraction, the phase transition temperatures from monoclinic ferroelectric *Mc* phase to orthorhombic ferroelectric *c*-phase have been determined. This transition temperature is greatly shifted by ~400 °C as the substrate misfit strain changes from ~0% to a compressive strain of ~2%. This huge change in the transition temperature demonstrates the effectiveness of strain engineering in KNN thin films, which is promising for further improvements in the piezoelectric coefficients by tuning crystalline phases.

10.3 Measurements of Piezoelectric Coefficients in Thin Films

One major difficulty in the research of piezoelectric thin films is the accurate measurement of piezoelectric coefficients, owing to the very small changes in displacements (as low as ~1 pm) or electric charges (as low as ~1 pC) from the thin film samples under applied electric fields or forces [39, 56–58]. This is one of the reasons that more data on ferroelectric polarizations than piezoelectric coefficients are reported for piezoelectric thin films. Here, we review methods to characterize piezoelectric coefficients in thin films. Based on the mechanisms of measurement, it is classified into two sessions: direct piezoelectric effect and reverse piezoelectric effect.

10.3.1 Direct Piezoelectric Effect

As the piezoelectric thin films are clamped by the substrate, most methods do not measure the real piezoelectric coefficient d_{33} of the free thin films but an effective piezoelectric coefficient $d_{33,f}$ [59]. Using the direct piezoelectric effect, $d_{33,f}$ can be derived by measuring the induced electric charge Q from applied force F. In the normal load method (Figure 10.5a), a force F is applied to the metallic tip on the top of the piezoelectric thin film. This metallic tip, piezoelectric thin film, and bottom electrode form a capacitor, which is connected in parallel to a standard capacitor

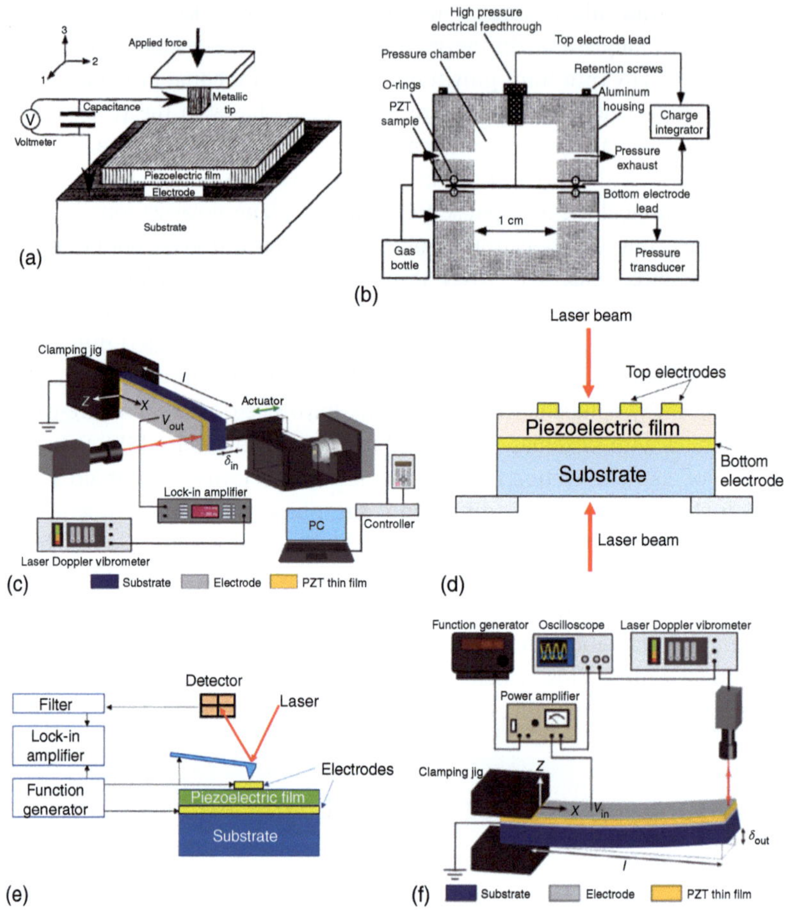

Figure 10.5 Measurements of piezoelectric coefficients for thin films by direct piezoelectric effect (a–c) and reverse piezoelectric effect (d–f): (a) Schematic drawing of the normal load method to measure piezoelectric coefficient $d_{33,f}$. Source: Lefki and Dormans [59]. Reproduced with permission of AIP Publishing. (b) Setup of pneumatic pressure rig method for $d_{33,f}$ measurement. Source: Xu et al. [60]. Reproduced with permission of AIP Publishing. (c) Schematic setup of the cantilever method to measure transverse piezoelectric coefficient $e_{31,f}$ by direct piezoelectric effect. Source: Tsujiura et al. [61]. Reproduced with permission of Japanese Society of Applied Physics. (d) Schematic illustration of double-beam laser interferometer to eliminate the contributions from the bending motion of the substrate. Source: Huajun Liu. (e) Schematic setup of atomic force microscopy to measure longitudinal piezoelectric coefficients. Source: Huajun Liu. (f) Schematic drawing of the setup to measure transverse piezoelectric coefficient $e_{31,f}$ by reverse piezoelectric effect. Source: Tsujiura et al. [61]. Reproduced with permission of Japanese Society of Applied Physics.

with a much larger capacitance. The induced electric charge Q can thus be derived by measuring the voltage across the standard capacitor. This method is direct and simple. However, the stress distribution over the sample is nonuniform. This leads to the bending of substrates, thus additional contributions of charges from transverse piezoelectric effect and errors in the measurements. To improve the uniformity of stress applied to piezoelectric thin films, a pneumatic pressure rig was developed to apply stress from gas pressure [60]. As shown in Figure 10.5b, uniform forces were imposed on the thin films by introducing high-pressure nitrogen gas into the cavities. The potential problem with this method is the inflated $d_{33,f}$ value due to remnant in-plane stress produced between O-rings and the test samples. In addition to the longitudinal piezoelectric coefficient $d_{33,f}$, measurements of transverse piezoelectric coefficients $d_{31,f}$ or $e_{31,f}$ are often necessary for thin film-based sensors and actuators [62]. In particular, $e_{31,f}$ coefficients are more relevant parameters to evaluate for devices that utilize the deflections of cantilevers or membranes as functioning units. Figure 10.5c shows the setup for measuring $e_{31,f}$ coefficients by the direct piezoelectric effect [61]. An actuator was used to apply periodic displacement at the end of the cantilever. The output voltage from the piezoelectric capacitor was measured by a lock-in amplifier to calculate the $e_{31,f}$ coefficients.

10.3.2 Reverse Piezoelectric Effect

The piezoelectric coefficients can also be measured by the reverse piezoelectric effect through the measurement of surface displacements of piezoelectric thin films under applied electric fields. Due to high vertical resolution (better than 1 pm), a laser interferometer is a widely used method to obtain longitudinal piezoelectric coefficients $d_{33,f}$ of thin films [63]. Due to the large contribution from the bending motion of the substrate, a single-beam laser interferometer is difficult to measure the accurate value of piezoelectric coefficients. By using a double-beam laser interferometer (Figure 10.5d), substrate bending contributions can be eliminated by detecting displacements from both sides of the testing samples. Another method to measure surface displacement is by laser Doppler vibrometer, which has been extended to scan the two-dimensional profile of the whole electrode area to improve reliability [64]. Atomic force microscopy (AFM) has also been used to measure surface displacements of piezoelectric thin films under applied electric fields. Figure 10.5e shows the schematic setup to measure surface displacements by an AFM tip in contact with a top electrode [65]. The electric field is well defined by the top electrode, instead of the conductive tip in piezoelectric force microscopy, to improve reliability. Similar to the direct piezoelectric effect, a cantilever structure is necessary to measure transverse piezoelectric coefficients $e_{31,f}$ by the reverse piezoelectric effect. The electric field was applied to the piezoelectric layers by a function generator, and the displacement of the cantilever was measured by a laser Doppler vibrometer (Figure 10.5f) [61]. It was found that $e_{31,f}$ measured by the reverse piezoelectric effect is larger than that measured by the direct piezoelectric effect. This could be due to the possible extrinsic piezoelectric responses from domain motion under applied electric fields, while no such domain motion is expected under applied force in the measurement by the direct piezoelectric effect.

10.4 Conclusions

In this chapter, we introduced research works on piezoelectric thin films, motivated by both piezoelectric thin film device applications and interesting physics involved when growing piezoelectric materials into thin films. We have also briefly reviewed two important piezoelectric material systems: PZT and KNN thin films. Finally, we summarized the challenges and methods to characterize piezoelectric coefficients in thin films, which are very different from bulk ceramics or single crystals due to the presence of substrates.

With the fast-growing developments of 5G wireless networks, Internet of Things, virtual reality, autonomous vehicles, and portable biomedical imaging, the demand for high-performance piezoelectric thin films is expected to increase exponentially in the near future. Furthermore, recent works continue to report novel types of piezoelectric behavior in thin films from interface symmetry breaking [66] and the rearrangement of defects under electric field [67]. We, therefore, anticipate an accelerated expansion of global research efforts in piezoelectric thin films.

References

1 Khan, H.N., Hounshell, D.A., and Fuchs, E.R.H. (2018). Science and research policy at the end of Moore's law. *Nat. Electron.* 1 (1): 14–21. https://doi.org/10.1038/s41928-017-0005-9.

2 Resonant. 5G: technology innovation under the hood. Accessed 2 Jan 2022. www.resonant.com/resources/technical-presentations.

3 Will-Cole, A.R., Hassanien, A.E., Calisgan, S.D. et al. (2022). Tutorial: Piezoelectric and magnetoelectric N/MEMS—Materials, devices, and applications. *J. Appl. Phys.* 131 (24): 241101.

4 Liu, Y., Cai, Y., Zhang, Y. et al. (2020). Materials, design, and characteristics of bulk acoustic wave resonator: a review. *Micromachines* 11 (7): 630.

5 Qiu, Y., Gigliotti, J.V., Wallace, M. et al. (2015). Piezoelectric micromachined ultrasound transducer (PMUT) arrays for integrated sensing, actuation and imaging. *Sensors* 15: 8020–8041.

6 Shannon, C.E. (1998). Communication in the presence of noise. *Proc. IEEE* 86 (2): 447–457.

7 Bhugra, H. and Piazza, G. (2017). *Piezoelectric MEMS Resonators*. Springer.

8 Wikipedia. Surface acoustic wave. https://en.wikipedia.org/wiki/Surface_acoustic_wave Accessed 9 Jan 2024.

9 Akiyama, M., Kamohara, T., Kano, K. et al. (2009). Enhancement of piezoelectric response in scandium aluminum nitride alloy thin films prepared by dual reactive cosputtering. *Adv. Mater.* 21 (5): 593–596.

10 Gong, S. and Piazza, G.10.1.2. Large frequency tuning of lithium niobate laterally vibrating MEMS resonators via electric boundary reconfiguration. *2013 Transducers Eurosensors XXVII 17th International Conference on Solid-State Sensors,*

Actuators Microsystems, TRANSDUCERS EUROSENSORS 2013, Barcelona, Spain (16–20 June 2013). IEEE. pp. 2465–2468.

11 Plessky, V., Yandrapalli, S., Turner, P.J. et al. (2019). 5 GHz laterally-excited bulk-wave resonators (XBARs) based on thin platelets of lithium niobate. *Electron. Lett.* 55 (2): 98–100.

12 Jung, J., Lee, W., Kang, W. et al. (2017). Review of piezoelectric micromachined ultrasonic transducers and their applications. *J. Micromech. Microeng.* 27 (11): aa851b. https://doi.org/10.1088/1361-6439/aa851b.

13 Muralt, P., Ledermann, N., Paborowski, J. et al. (2005). Piezoelectric micromachined ultrasonic transducers based on PZT thin films. *IEEE Trans. Ultrason. Ferroelectr. Freq. Control* 52 (12): 2276–2288.

14 Dausch, D.E., Gilchrist, K.H., Carlson, J.B. et al. (2014). In vivo real-time 3-D intracardiac echo using PMUT arrays. *IEEE Trans. Ultrason. Ferroelectr. Freq. Control* 61 (10): 1754–1764.

15 Lu, Y., Tang, H., Fung, S. et al. (2015). Ultrasonic fingerprint sensor using a piezoelectric micromachined ultrasonic transducer array integrated with complementary metal oxide semiconductor electronics. *Appl. Phys. Lett.* 106 (26): 263503.

16 Przybyla, R.J., Tang, H.Y., Guedes, A. et al. (2015). 3D ultrasonic rangefinder on a chip. *IEEE J. Solid State Circuits* 50 (1): 320 334.

17 Lang, Y., Liu, C., Fawzy, A. et al. (2022). Piezoelectric bimorph MEMS speakers. *Nanotechnol. Precis. Eng.* 5 (3): https://doi.org/10.1063/10.0013406.

18 Funakubo, H., Dekkers, M., Sambri, A. et al. (2012). Epitaxial PZT films for MEMS printing applications. *MRS Bull.* 37 (11): 1030–1038.

19 Muralt, P., Polcawich, R.G., and Trolier-McKinstry, S. (2009). Piezoelectric thin films for sensors, actuators, and energy harvesting. *MRS Bull.* 34 (9): 658–664.

20 Khan, A., Abas, Z., Kim, H.S., and Oh, I.-K. (2016). Piezoelectric thin films: an integrated review of transducers and energy harvesting. *Smart Mater. Struct.* 25 (5): 53002.

21 Schlom, D.G., Chen, L.Q., Eom, C.B. et al. (2007). Strain tuning of ferroelectric thin films. *Annu. Rev. Mater. Res.* 37: 589–626.

22 Schlom, D.G., Chen, L.Q., Fennie, C.J. et al. (2014). Elastic strain engineering of ferroic oxides. *MRS Bull.* 39 (2): 118–130.

23 Haeni, J.H., Irvin, P., Chang, W. et al. (2004). Room-temperature ferroelectricity in strained $SrTiO_3$. *Nature* 430 (7001): 758–761.

24 Choi, K.J., Biegalski, M., Li, Y.L. et al. (2004). Enhancement of ferroelectricity in strained $BaTiO_3$ thin films. *Science* 306 (5698): 1005–1009.

25 Zeches, R.J., Rossell, M.D., Zhang, J.X. et al. (2009). A strain-driven morphotropic phase boundary in $BiFeO_3$. *Science* 324 (November): 63–67.

26 Choi, M., Oba, F., and Tanaka, I. (2009). Role of Ti antisitelike defects in $SrTiO_3$. *Phys. Rev. Lett.* 103: 185502.

27 Liu, H., Wu, H., Ong, K.P. et al. (2020). Giant Piezoelectricity in oxide thin films with nanopillar structure. *Science* 369: 292.

28 Junquera, J. and Ghosez, P. (2003). Critical thickness for ferroelectricity in perovskite ultrathin films. *Nature* 422 (6931): 506–509.

29 Fong, D.D., Stephenson, G.B., Streiffer, S.K. et al. (2004). Ferroelectricity in ultrathin perovskite films. *Science* 304 (5677): 1650–1653.

30 Lee, S.R., Baasandorj, L., Chang, J.W. et al. (2019). First observation of ferroelectricity in ~1 nm ultrathin semiconducting $BaTiO_3$ films. *Nano Lett.* 19 (4): 2243–2250.

31 Gao, P., Zhang, Z., Li, M. et al. (2017). Possible absence of critical thickness and size effect in ultrathin perovskite ferroelectric films. *Nat. Commun.* 8 (June): 1–8. https://doi.org/10.1038/ncomms15549.

32 Wang, H., Liu, Z.R., Yoong, H.Y. et al. (2018). Direct observation of room-temperature out-of-plane ferroelectricity and tunneling electroresistance at the two-dimensional limit. *Nat. Commun.* 9 (1): 3319.

33 Ji, D., Cai, S., Paudel, T.R. et al. (2019). Freestanding crystalline oxide perovskites down to the monolayer limit. *Nature* 570 (7759): 87–90.

34 Kalinin, S.V. and Spaldin, N.A. (2013). Functional ion defects in transition metal oxides. *Science* 341: 858–859.

35 Hwang, J., Rao, R.R., Giordano, L. et al. (2017). Perovskites in catalysis and electrocatalysis. *Science* 358: 751.

36 Lee, D., Lu, H., Gu, Y. et al. (2015). Emergence of room-temperature ferroelectricity at reduced dimensions. *Science* 349: 1314.

37 Ning, S., Kumar, A., Klyukin, K. et al. (2021). An antisite defect mechanism for room temperature ferroelectricity in orthoferrites. *Nat. Commun.* 12: 4298. https://doi.org/10.1038/s41467-021-24592-w.

38 Kanno, I. (2019). Piezoelectric PZT thin films: deposition, evaluation and their applications. *2019 20th International Conference on Solid-State Sensors, Actuators Microsystems Eurosensors XXXIII, TRANSDUCERS 2019 EUROSENSORS XXXIII*, Berlin, Germany (23–27 June 2019). IEEE. pp. 785–788.

39 Song, L., Glinsek, S., and Defay, E. (2021). Toward low-temperature processing of lead zirconate titanate thin films: advances, strategies, and applications. *Appl. Phys. Rev.* 8 (4): 041315.

40 Schatz, A., Pantel, D., and Hanemann, T. (2017). Towards low-temperature deposition of piezoelectric $Pb(Zr,Ti)O_3$: Influence of pressure and temperature on the properties of pulsed laser deposited $Pb(Zr,Ti)O_3$. *Thin Solid Films* 636: 680–687.

41 Sama, N., Herdier, R., Jenkins, D. et al. (2008). On the influence of the top and bottom electrodes – a comparative study between Pt and LNO electrodes for PZT thin films. *J. Cryst. Growth* 310 (14): 3299–3302.

42 Pertsev, N.A., Kukhar, V.G., Kohlstedt, H., and Waser, R. (2003). Phase diagrams and physical properties of single-domain epitaxial (formula presented) thin films. *Phys. Rev. B: Condens. Matter Mater. Phys.* 67 (5): 1–10.

43 Huang, C., Liao, Z., Li, M. et al. (2021). A highly strained phase in $PbZr_{0.2}Ti_{0.8}O_3$ films with enhanced ferroelectric properties. *Adv. Sci.* 8 (8): 1–8.

44 Nguyen, M.D., Dekkers, M., Houwman, E. et al. (2011). Misfit strain dependence of ferroelectric and piezoelectric properties of clamped (001) epitaxial $Pb(Zr_{0.52},Ti_{0.48})O_3$ thin films. *Appl. Phys. Lett.* 99 (25): 0–4.

45 Wu, J., Xiao, D., and Zhu, J. (2015). Potassium–sodium niobate lead-free piezoelectric materials: past, present, and future of phase boundaries. *Chem. Rev.* 115 (7): 2559–2595.

46 Seog, H.J., Ullah, A., Ahn, C.W. et al. (2018). Recent progress in potassium sodium niobate lead-free thin films. *J. Korean Phys. Soc* 72 (12): 1467–1483.

47 Zhang, S.W., Zhou, Z., Luo, J., and Li, J.F. (2019). Potassium-sodium-niobate-based thin films: lead free for micro-piezoelectrics. *Ann. Phys.* 531 (7): 1800525.

48 Burns, S.R. and Dolgos, M.R. (2021). Sizing up $(K_{1-x}Na_x)NbO_3$ films: a review of synthesis routes, properties & applications. *New J. Chem.* 45 (17): 7408–7436.

49 Shibata, K., Suenaga, K., Watanabe, K. et al. (2011). Improvement of piezoelectric properties of $(K,Na)NbO_3$ films deposited by sputtering. *Jpn. J. Appl. Phys.* 50 (4 Part 1): 041503.

50 Fujii, I., Tagata, S., Nakao, T. et al. (2015). Fabrication of $(K,Na)NbO_3$ films on $SrRuO_3$/(001) $SrTiO_3$ substrates by pulsed laser deposition. *Jpn. J. Appl. Phys.* 54: 10NA13.

51 Li, T., Wang, G., Li, K. et al. (2013). Influence of LNO top electrodes on electrical properties of KNN/LNO thin films prepared by RF magnetron sputtering. *J. Am. Ceram. Soc.* 96 (3): 787–790.

52 Luo, J., Sun, W., Zhou, Z. et al. (2017). Monoclinic $(K,Na)NbO_3$ ferroelectric phase in epitaxial films. *Adv. Electron. Mater.* 3 (10): 1–8.

53 Yu, Q., Li, J.F., Sun, W. et al. (2013). Electrical properties of $K_{0.5}Na_{0.5}NbO_3$ thin films grown on Nb:$SrTiO_3$ single-crystalline substrates with different crystallographic orientations. *J. Appl. Phys.* 113 (2): 024101.

54 Zhou, M.J., Wang, J.J., Chen, L.Q., and Nan, C.W. (2018). Strain, temperature, and electric-field effects on the phase transition and piezoelectric responses of $K_{0.5}Na_{0.5}NbO_3$ thin films. *J. Appl. Phys.* 123 (15): 0–6. https://doi.org/10.1063/1.5027505.

55 Von Helden, L., Bogula, L., Janolin, P.E. et al. (2019). Huge impact of compressive strain on phase transition temperatures in epitaxial ferroelectric $K_xNa_{1-x}NbO_3$ thin films. *Appl. Phys. Lett.* 114 (23): 232905.

56 Fernandez, A., Acharya, M., Lee, H.G. et al. (2022). Thin-film ferroelectrics. *Adv. Mater.* 34 (30): 1–48.

57 Liu, J.M., Chan, H.L.W., Zhu, S.N. et al. (2002). Piezoelectric coefficient measurement of piezoelectric thin films: an overview. *Mater. Chem. Phys.* 75: 12–18.

58 Shetty, S., Yang, J.I., Stitt, J., and Trolier-Mckinstry, S. (2015). Quantitative and high spatial resolution d_{33} measurement of piezoelectric bulk and thin films. *J. Appl. Phys.* 118: 174104. https://doi.org/10.1063/1.4935140.

59 Lefki, K. and Dormans, G.J.M. (1994). Measurement of piezoelectric coefficients of ferroelectric thin films. *J. Appl. Phys.* 76: 1764.

60 Xu, F., Chu, F., and Trolier-McKinstry, S. (1999). Longitudinal piezoelectric coefficient measurement for bulk ceramics and thin films using pneumatic pressure rig. *J. Appl. Phys.* 86: 588.

61 Tsujiura, Y., Kawabe, S., Kurokawa, F. et al. (2015). Comparison of effective transverse piezoelectric coefficients $e_{31,f}$ of $Pb(Zr,Ti)O_3$ thin films between direct and converse piezoelectric effects. *Jpn. J. Appl. Phys.* 54 (10): 10NA04.

62 Dubois, M.A. and Muralt, P. (1999). Measurement of the effective transverse piezoelectric coefficient $e_{31,f}$ of AlN and $Pb(Zr_x,Ti_{1-x})O_3$ thin films. *Sens. Actuators, A* 77 (2): 106–112.

63 Kholkin, A.L., Wütchrich, C., Taylor, D.V., and Setter, N. (1996). Interferometric measurements of electric field-induced displacements in piezoelectric thin films. *Rev. Sci. Instrum.* 67 (5): 1935–1941.

64 Yao, K. and Tay, F.E.H. (2003). Measurement of longitudinal piezoelectric coefficient of thin films by a laser-scanning vibrometer. *IEEE Trans. Ultrason. Ferroelectr. Freq. Control* 50 (2): 113–116.

65 Christman, J.A., Maiwa, H., Kim, S.H. et al. (1999). Piezoelectric measurements with atomic force microscopy. *Appl. Phys. Lett.* 73 (26): 3851–3853.

66 Yang, M.-M., Luo, Z.-D., Mi, Z. et al. (2020). Piezoelectric and pyroelectric effects induced by interface polar symmetry. *Nature* 584 (7821): 377–381.

67 Park, D.-S., Hadad, M., Riemer, L.M. et al. (2022). Induced giant piezoelectricity in centrosymmetric oxides. *Science* 375 (6581): 653–657.

11

Lead-free KNN and KTN Single Crystals: Piezoelectric and Electro-Optic Properties

Chengpeng Hu, Yu Wang, and Hao Tian

Harbin Institute of Technology, Department of Physics, 92 West Dazhi Street, Nangang District, Harbin 150001, PR China

11.1 Introduction

The element-doped potassium sodium niobate ($K_{1-x}Na_xNbO_3$, that is, KNN) piezoelectric system material is the piezoelectric material with relatively good piezoelectric performance in the lead-free piezoelectric material system at present. So far, most of the research on KNN-based materials has focused on ceramic materials. At present, the origin of piezoelectricity has not been solved, and the analysis of the mechanism needs more suitable conditions to improve the piezoelectric properties. The improvement of piezoelectric properties of ceramics involves far more problems than the improvement of piezoelectric properties with single crystals, mainly due to the existence of grain boundaries, doping uniformity, density, and other factors affecting piezoelectric properties in ceramics. Therefore, single crystals are the best experimental materials for studying the origin of piezoelectric. KNN crystal has multiple ferroelectric phase structures. With the decrease in temperature, the crystal will undergo a phase transition from paraelectric phase to tetragonal phase to orthorhombic phase to trigonal phase. Its Curie temperature is 400–450 °C (related to K/Na), so it has excellent temperature stability, but its piezoelectric properties are not high.

Under the action of an external electric field, the dielectric crystal induces polarized charges at both ends of the electrode, making the gravity centers of the positive and negative bound charges not coincide, thus resulting in the phenomenon of polarization. When the material is in a non-centrosymmetric structure, spontaneous polarization occurs, and the direction of spontaneous polarization changes with the change of the external electric field. One of its main features is the hysteresis loop. A complete and self-consistent measurement of the dielectric constant, piezoelectric coefficient, elastic constant, and electromechanical coupling coefficient is very necessary for the basic research of materials and device design and is also a very practical work.

In order to study the physical mechanism of high-piezoelectric properties of KNN materials and further improve the piezoelectric properties of materials, the

Piezoelectric Materials: From Fundamentals to Emerging Applications, First Edition. Edited by Jiagang Wu.

KNN single crystal system was studied. This chapter takes KNN crystals of various components grown by the TSSG method as the research object and studies the change of dielectric constant and loss of KNN crystals of different components with temperature, ferroelectric properties, and the ultralarge strain of KNN crystals.

The current mainstream electro-optic (EO) materials can be divided into four categories, including inorganic crystalline electro-optic materials (such as $LiNbO_3$, $BaTiO_3$, and PMN–PT), semiconductor electro-optic materials (such as silicon-based materials, InP, and graphene), organic electro-optic materials (such as organic-doped PMMA and Im-ClO_4 molecular crystals), and organic–inorganic mixed electro-optic materials (such as $CH_3NH_3PbI_3$ and PMA_2PbCl_4).

Inorganic crystalline electro-optic materials are the main force in electro-optic materials, especially perovskite crystalline materials, such as $LiNbO_3$, $BaTiO_3$, $KTa_{1-x}Nb_xO_3$, and PMN–PT, due to their excellent electro-optic properties (linear electro-optic coefficient up to 10^2–10^3 pm/V), (quadratic electro-optic coefficient up to 10^{-15}–10^{-14} m^2/V^2), excellent performance and material stability, loss less than 0.2 dB/cm, and modulation bandwidth greater than 30 GHz (to several hundred GHz), they make them the preferred material for current electro-optic applications.

Perovskite single crystal material is an important part of inorganic crystal electro-optic materials, which have excellent electro-optic properties and are a good medium for current electro-optic devices. The potassium tantalum niobate ($KTa_{1-x}Nb_xO_3$, that is KTN) single crystal, as a lead-free perovskite single crystal, has excellent electro-optic properties, especially the quadratic electro-optic coefficient of 10^{-14} m^2/V^2 [1], and its holographic grating diffraction efficiency is over 80% [2], much higher than other electro-optic materials. Due to its intrinsic solid solution property, its properties can be regulated by components, so it is considered to be one of the most excellent electro-optic materials with the most potential applications.

11.2 KNN and KTN-Based Single Crystals

11.2.1 Top-seeded Solution Crystal Growth

The crystal growth methods can be divided into solution growth method, melt growth method, gas phase growth method, solid phase growth method, and thin film growth method according to the phase state. Crystal growth from melt is one of the earliest and most widely used growth methods. The pull-up method belongs to the melt growth method. This growth method has high growth speed, high purity, and integrity of crystal. When the temperature of the crystalline solid is higher than the melting point, it melts. The crystal growth process involves the solid–liquid phase change process. In this phase transition process, the random state of atoms or molecules is transformed into an ordered arrangement, which is the growth process of crystals.

In the process of crystal growth, the most important thing is to choose a suitable and stable temperature gradient, which is the primary and necessary condition for growing high-quality crystals. Without a proper temperature field, it is impossible to grow crystals or obtain high-quality crystals. A proper temperature gradient can reduce the thermal stress during crystal growth and suppress the occurrence of component supercooling, which can reduce the occurrence of various defects.

For a cylindrical crystal, assuming its diameter is d, then $A = \frac{\pi d^2}{4}$. Therefore

$$d = 2\left(\frac{\Delta Q}{\pi \Delta H \rho R}\right)^{\frac{1}{2}} \tag{11.1}$$

where A is the cross-sectional area of the crystal interface. That is, the diameter of the crystal is proportional to the square root of the pulling rate. Although formula (11.1) is only an approximate expression, it can qualitatively reflect the relationship between the pulling rate and the crystal diameter. Decreasing the growth rate will lead to an increase in the crystal diameter and the realization of the shouldering process of the crystal. When the crystal diameter reaches a predetermined value, a stable crystal diameter can be maintained by appropriately increasing the pulling rate. In addition, during the crystal growth process, we can also adjust the diameter of the growing crystal by controlling the heating power. When we increase the power, the latent heat of crystallization increases, making the diameter of the crystal smaller. Conversely, reducing the power makes the diameter of the crystal larger.

Crystal cannot be widely used because of the problem of crystal quality. Therefore, it is very important to choose a method suitable for crystal growth. Up to now, according to the literature, the methods suitable for single crystal growth generally include Czochralski method, flux method, and Bridgeman method.

Top seeded-solution growth method, also known as Czochralski method, can be widely used in the growth of optical crystals, mainly because of the following advantages: First, the crystal growth can be clearly observed, so that the parameters can be adjusted according to the actual situation; Second, this method is applicable to most cases. For some materials, as long as a suitable cosolvent can be found, the single crystals of these materials can be grown; Third, the crystal grows at the center away from the crucible wall, and the growth interface is only at the contact position between the solution and air, which can effectively reduce the crystal stress; Fourth, many compounds may not be able to directly grow complete single crystals due to some problems, such as their extreme volatility at the melting point. However, the top seed cosolvent method (TSSG) can effectively reduce the growth temperature with cosolvent, so that the above problems can be well solved. Therefore, this method has a unique charm for the single crystal growth of these materials. Fifth, directional seed crystal and "necking" processes can be conveniently used.

11.2.2 Crystal Growth of KNN and KTN Single Crystals

Which crystal growth method is appropriate depends on the physical and chemical properties and phase diagrams of the system to be grown. The crystal phase of KNN and KTN single crystal system is complex and nonuniform. Solid solution refers to a substance with the same structure, position, and ion replacement. In the process of crystal growth, KNN and KTN solid melt enter into the unit cell at different speeds due to the difference of basic properties such as ion radius, which makes the grown crystal different from the original component, which is called the segregation phenomenon of single crystal components. It can be seen from the phase diagram of the KNN and KTN systems that the segregation coefficients of potassium and sodium in KNN single crystals or the segregation coefficients of tantalum and niobium in KTN single crystals are quite different, so it is difficult to grow high-quality and large-size KNN and KTN single crystals. The top seeded-solution growth method is selected to grow the required KNN and KTN single crystals. This method has good repeatability, fast growth speed, and large single crystal size, which can meet the experimental requirements.

The use of high-purity raw materials is the primary condition for ensuring crystal quality. The raw materials of KNN and KTN crystals are K_2CO_3, Na_2CO_3, Nb_2O_5, and Nb_2O_5. See Table 11.1 for the raw materials and raw material purity used in the growth.

Weigh various raw materials with an electronic balance, fully mix the weighed raw materials, then put them into a ball mill for grinding, and finally add the milled powder into a crucible and fully squeeze it. Polycrystalline synthesis was carried out by burning at 950–1050 °C for 10–20 hours. The reaction equation of the synthesis process is as follows:

$$(1-x)K_2CO_3 + xNa_2CO_3 + Nb_2O_5 = 2K_{1-x}Na_xNbO_3 + CO_2 \uparrow \tag{11.2}$$

$$K_2CO_3 + xNb_2O_5 + (1-x)Ta_2O_5 = 2KTa_{1-x}Nb_xO_3 + CO_2 \uparrow \tag{11.3}$$

When the crystal is put into the single crystal growth furnace and the temperature is raised, the crystal will start to grow. Crystal growth includes feeding, diameter reduction, shoulder setting, isometric growth, and cooling. The processing of each process has a great impact on the quality of crystals. Finding the operating parameters in each process is the fundamental guarantee for obtaining high-quality crystals and is also the basic task of crystal growth. During crystal growth:

First, heat the polycrystalline powder put into the platinum crucible after grinding in the single crystal furnace and slowly raise the temperature. After observing that all the polycrystalline materials melt, continue to raise the temperature to overheat the temperature by about 50 °C, so as to eliminate floating crystals and stabilize the surface tension, density, viscosity, and other physical properties of the melt. However, the overheating time should not be too long to prevent the solution from volatilizing

Table 11.1 Raw materials and purity required to grow KNN crystals.

Raw material	Purity (%)
K_2CO_3	99.99
Na_2CO_3	99.99
Ta_2O_5	99.99
Nb_2O_5	99.99

at high temperatures. Then, keep the temperature at 30 °C higher than the melting point for a period of time, fully homogenize the raw materials, and then lower the temperature to near the melting point to prepare for planting. In this process, it is required to ensure that the polycrystalline powder put into the crucible is sufficiently fine and evenly mixed. When raising the temperature to melt the polycrystal, the melting point of the crystal should be accurately identified, and the temperature change during growth should be controlled as described above.

However, after the temperature drops to near the melting point, the seed crystal is slowly lowered to near the liquid surface of the melt and kept for a period of time, so as to avoid the generation of hybrid crystals due to local quenching when the seed crystal enters the melt. When the seed crystal contacts the melt, it should be kept for a period of time, so that the end of the seed crystal will melt slightly, and ensure that the cross-section where the seed crystal contacts the melt has a fresh atomic surface. This process often needs to be repeated many times to find a suitable growth temperature. Pulling starts when bright tetragonal edges appear around the seed crystal. Subsequently, the new seed crystal should be narrowed to obtain a more complete new seed crystal, which is called the necking process. This process can prevent the mechanical defects and dislocations carried by the seed crystal from continuing into the underlying crystal. The growth parameters to be controlled in the necking stage require that the pulling speed is 0.2 mm/h, the rotation speed is 20 rpm, and the necking length should not be too long, generally 1–2 mm.

Shoulder placement is a key process for growing high-quality crystals. Generally, the lifting speed and the system temperature should be reduced in the process of shoulder laying. These two parameters are used to control the shoulder laying speed so that the process should not be too fast; too fast is easy to generate defects such as hybrid crystals, and too slow will cause time waste. Therefore, proper cooling rate, drawing speed, and rotating speed should be controlled for shoulder laying.

When the crystal is shouldered to a predetermined size, it should naturally turn to equal-diameter growth by adjusting the temperature and increasing the pulling speed. Whether the crystal has an equal diameter can be determined by observing the aperture at the interface between the crystal and the melt. Constant temperature growth is generally controlled in the equal-diameter growth stage. However, due to the condensation of components in the growth process, the equiaxed process needs to be cooled slightly, and the temperature change will lead to the change of crystal components, the formation of stripe defects, and the application of crystals. Therefore, at this stage, we try to adjust the pulling speed to ensure the stable growth of the crystal and try not to control it by temperature adjustment.

When the crystal size reaches our expected size, appropriately increase the system temperature, quickly pull the crystal out of the liquid surface, and then start cooling. The cooling rate of crystal also plays an important role in ensuring the quality of crystal. A fast cooling rate will generate large thermoelastic stress in the crystal, leading to dislocation, nucleation, and proliferation, and even crystal cracking in serious cases. Therefore, slow cooling (cooling rate is 30 °C/hour) can ensure the quality of the crystal.

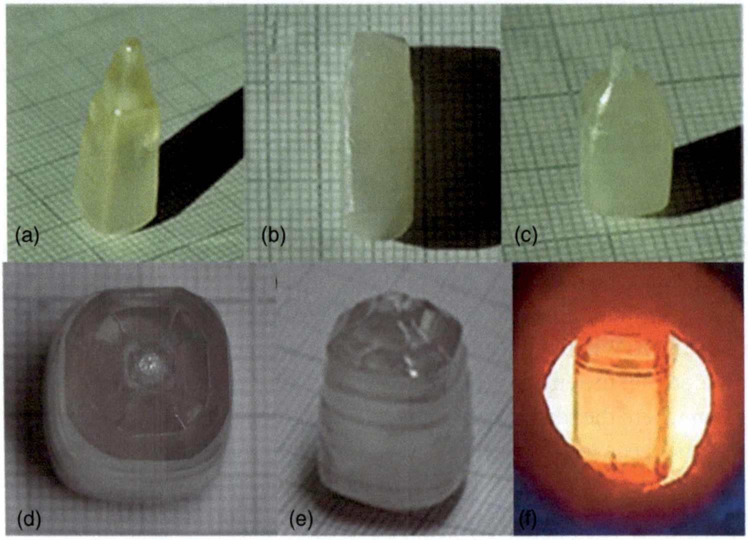

Figure 11.1 KNN crystals grown by the TSSG method. (a–e) $K_{1-x}Na_xNbO_3$ crystals with $x = 0.11–0.70$, (f) $K_{0.882}Na_{0.118}NbO_3$ crystal at 800 °C. Source: Tian et al. [3]/Reproduced with permission of American Chemical Society.

Through the above process procedures, KNN and KTN crystals with large size, complete appearance, no macro defects, and good quality can be grown in a short time (five to six days), as shown in Figures 11.1 and 11.2. It can be seen that the single crystal grown by top seeded-solution growth method has regular shape, no cracks, and milky color. This is because the crystal is in the ferroelectric phase, and there are a large number of ferroelectric domain walls inside the crystal. At the same time, Figure 11.1f reflects that under high temperature, the $K_{0.882}Na_{0.118}NbO_3$ crystal is transparent in the paraelectric phase, indicating that there is no inclusion in the crystal. To sum up, it can be considered that the KNN and KTN crystals with different components grown in this paper have large blocks and high quality.

11.2.3 Segregation Effect of KNN and KTN Crystals

11.2.3.1 Segregation Effect of KNN Crystals

Electron microprobe analysis (EMPA) is a means to test and analyze the kinds and contents of elements in the microregion of materials. Its test principle is to use each element to have its own unique X-ray wavelength and intensity. Study the component condensation phenomenon during crystal growth by testing the components of the growing single crystal and determining whether the crystal components are uniform.

The segregation coefficient is a very important parameter in the crystal growth system, and the definition of segregation coefficient (α) refers to the ratio of the concentration of a certain ion in the crystal to the concentration in the melt, that is

$$\alpha = \frac{x}{X} \tag{11.4}$$

Figure 11.2 KTN-based crystals grown by the TSSG method: (a) KLTN: Fe, (b,c) KTN, Source: Tian et al. [4]/Reproduced with permission of Royal Society of Chemistry. (d,e) KTN: Mn. Source: (d) Wang et al. [2]/Reproduced with permission of IOP Publishing. (e) From Wu (Book Author). (f) KNTN. Source: Meng et al. [5]/Reproduced with permission of AIP Publishing LLC.

where x refers to the concentration of an ion in the crystal; X refers to the concentration of the ion in the melt. Therefore, the K and Na segregation coefficients of KNN crystals are listed in Table 11.2.

From Table 11.2, it can be seen that the segregation coefficient of Na is greater than 1, while that of K is less than 1, which means that as KNN crystal is continuously pulled out of the melt, the proportion of K element content in KNN crystal keeps increasing. On the contrary, the content of Na element keeps decreasing. Therefore, in a KNN crystal, the content of K element near the seed crystal is less than that far away from the seed crystal. At the same time, we also know that when the K:Na ratio in the melt is about 73 : 27, the difference between the segregation coefficients of K and Na elements is the largest, which is 0.458 and 2.47, that is, at

Table 11.2 Segregation coefficient of elements K and Na and growth temperature in KNN crystal.

Ratio of potassium to sodium in the melt	Ratio of potassium to sodium in the crystal	Segregation coefficient of K	Segregation coefficient of Na	Growth temperature (°C)
73 : 27	33.4 : 66.6	0.458	2.47	1170
76 : 24	45.5 : 54.5	0.600	2.27	1148.5
79 : 21	53.8 : 46.2	0.681	2.20	1127.5
82 : 18	62.2 : 37.8	0.759	2.10	1116
92 : 8	88.2 : 11.8	0.959	1.47	1065

this ratio of raw materials. In the process of crystal growth, the degree of nonuniformity of components is the largest, which requires us to grow KNN crystals with relatively small size as much as possible to ensure the high quality of crystals. When the K:Na ratio in the melt is about 92 : 8, the difference between the segregation coefficients of K and Na elements is the smallest, which is 0.950 and 1.47. That is to say, under the raw material ratio of this ratio, the degree of nonuniformity of components in the crystal growth process is the smallest, and relatively large KNN crystals can be grown.

Table 11.3 shows the growth temperatures of crystal shoulder growth under different raw material ratios when the cosolvents K_2CO_3 and Na_2CO_3 are present at 10 mol%. It can be observed from the table that a higher content of K in the raw material leads to a lower growth temperature. Every 10 mol% increase in the K content reduces the growth temperature by 40–50 °C. These differences may be due to slight variations caused by the relocation of the crucible during the re-growth of the crystal. Additionally, to verify the influence of growth temperature on crystal composition, an additional 20 mol% of cosolvent was added to the raw material ratio. The resulting growth temperature and crystal composition were then recorded, as

Table 11.3 Shows the dependence of KNN crystal growth temperature on solvents.

Ratio of potassium to sodium in the melt	Ratio of excess $K_2CO_3 + Na_2CO_3$ mixed flux (mol %)	Growth temperature (°C)	Crystal
73 : 27	10	1170	$K_{0.334}Na_{0.666}NbO_3$
	30	1132	$K_{0.350}Na_{0.650}NbO_3$
82 : 18	10	1127.5	$K_{0.622}Na_{0.378}NbO_3$
	30	1085	$K_{0.643}Na_{0.357}NbO_3$
92 : 8	10	1065	$K_{0.882}Na_{0.118}NbO_3$
	4	1075	$K_{0.854}Na_{0.146}NbO_3$

shown in Table 11.3. It can be seen that adding 10 mol% mol of cosolvent decreases the growth temperature by about 20 °C, and the crystal composition grown under the same K and Na element ratio changes only slightly. However, significant changes in the crystal composition occur without changing the raw material ratio. This suggests that the crystal composition may be slightly affected by temperature changes, with a lower temperature resulting in a higher K content in the crystal composition. It is also noted that excessive use of cosolvent can negatively impact the quality of the crystal, causing it to yellow.

11.2.3.2 Equal-diameter Growth of KTN Crystal

In Figure 11.3, the "Top" end corresponds to the zero point (starting point of growth), and the "Bottom" corresponds to the end point of growth. As shown in Figure 11.3, KTN-1 crystal grows at a constant cross-section size from the origin, and a growth step is generated only at the position of 20.3 mm (0.0 mm from the origin of growth). However, the cross-section size of KTN-2 crystal increases at the initial growth stage, and the crystal begins to enter the equal-diameter growth stage when it reaches 6.5 mm.

Due to the component segregation of crystals, KTN-1 and KTN-2 crystals have different components along the pulling direction, corresponding to different actual growth temperatures T_g, as shown in Figure 11.4a,b. The average value of component weighting coefficient could be calculated to obtain the corresponding component KTN parameters. In addition, the growth equilibrium temperature T_e of a specific component is known according to the phase diagram. The $(T_e - T_g)$ values corresponding to the growth states of adjacent "diameter reduction" and

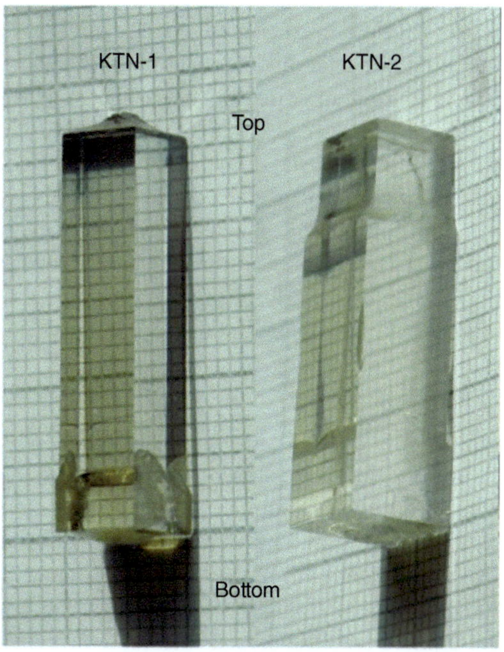

Figure 11.3 Photos of KTN single crystals and corresponding growth stages. Source: Tian et al. [6]/Reproduced with permission of American Chemical Society.

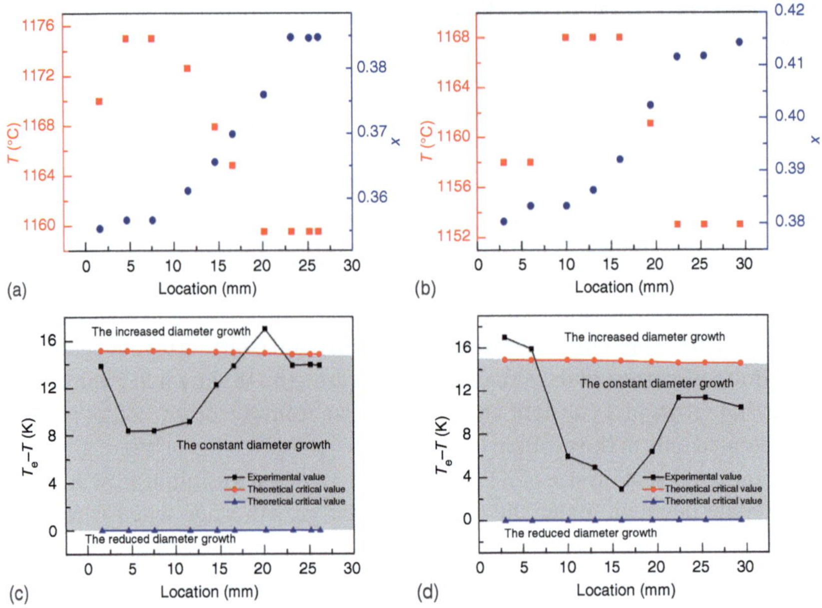

Figure 11.4 (a,b) Growth temperature T and Nb composition x and (c,d) relationships between size fluctuations and T along the pulling direction of different KTN, respectively. Source: Tian et al. [6]. Reproduced with permission of American Chemical Society.

"diameter expansion" crystals can be calculated. The experimental values and the critical values calculated by the model along the pulling direction of KTN crystals ($T_e - T_g$) are shown in Figure 11.3c,d.

Comparing the experimental value of ($T_e - T_g$) with the critical value based on the equal-diameter model in Figure 11.4c,d, it can be seen that the crystal shows "diameter expansion" when the black line value is greater than the red line value (the position of 20 mm of KTN-1 and the part of 0–6.5 mm of KTN-2), which is consistent with the actual crystal growth results. In addition, it can be found that KTN-2 crystal has a weak local growth edge at ~22 mm, corresponding to the peak of ($T_e - T_g$) black line at the critical external temperature, which is related to the deviation of the crystal interface from the center of the temperature field. The large local area ($T_e - T_g$) leads to the local external growth of the crystal, reflecting the high temperature sensitivity of the crystal isocratic growth.

11.2.4 Structure of KNN Single Crystals

In order to verify the structure of KNN crystal, a small sample was cut from each component crystal, ground into uniform micron-sized powder, and then analyzed by X-ray powder diffraction (XRD) experiments. Because the crystal has an ordered lattice arrangement, when the lattice is arranged in parallel to the crystal plane, it can be equivalent to an X-ray diffraction grating. When X-rays with a short wavelength are emitted onto the crystal surface, X-rays interfere with each other and diffract.

The diffraction of different crystal axis orientations and lattice constants is superimposed to form a specific diffraction pattern. The instrument records the diffraction intensity of different deflection angles so that the diffraction pattern of the test sample can be read. Therefore, XRD is an important research method to obtain information about material composition, internal atomic structure, or morphology. Bragg equation based on crystal diffraction is:

$$2d \sin \theta = n\lambda \tag{11.5}$$

where:

d	crystal plane spacing;
θ	diffraction angle of X-ray;
n	diffraction order (0, 1, 2, 3, 4, …)
λ	the wavelength of the X-ray.

The powder XRD pattern of the grown KNN crystal is shown in Figure 11.5. It can be seen from the figure that the grown KNN ferroelectric single crystals are all of pure perovskite structure, and there are no other heterophases.

For the orthogonal crystal phase structure in ferroelectric materials, it can be judged by comparing the strength of the $[001]_C$ and $[110]_C$ peaks in the XRD spectrum [113]. For the orthogonal phase structure, the peak intensity ratio (I_{001}/I_{110}) is around 2 : 1, whereas for the tetragonal phase structure, the peak intensity ratio (I_{001}/I_{100}) is around 1 : 2. It can be seen from the diffraction peak intensity in the figure that all KNN single crystals are in the orthogonal phase structure, which is mutually verified with the results of dielectric temperature spectrum.

By comparing with the standard card, the lattice constants of KNN crystals with different components are calculated, as shown in Table 11.4. It can be found that with the increase of K element content, the lattice constants a, b, and c of KNN crystal are all increasing, which can be understood as the reason that the radius of K ion is

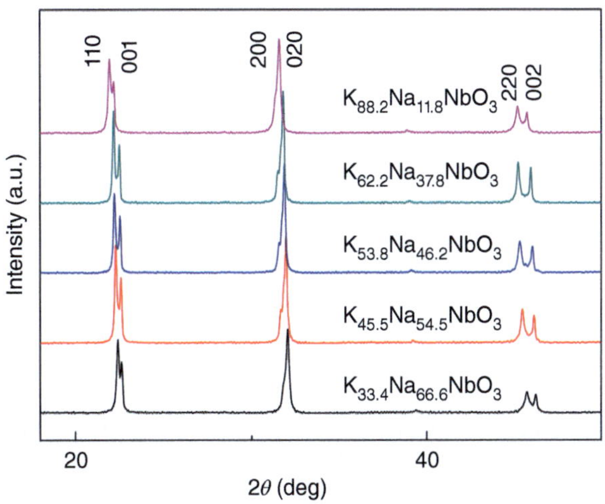

Figure 11.5 X-ray powder diffraction patterns of the $K_{1-x}Na_xNbO_3$ crystals with different x values. Source: Tian et al. [3]. Reproduced with permission of American Chemical Society.

Table 11.4 Lattice parameters of $K_{1-x}Na_xNbO_3$ crystals with different x values.

Crystal	*a* (Å)	*b* (Å)	*c* (Å)
$K_{0.334}Na_{0.666}NbO_3$	5.5988	3.9380	5.6587
$K_{0.455}Na_{0.545}NbO_3$	5.6265	3.9386	5.6697
$K_{0.538}Na_{0.462}NbO_3$	5.6258	3.9437	5.7157
$K_{0.622}Na_{0.378}NbO_3$	5.6429	3.9547	5.7211
$K_{0.882}Na_{0.118}NbO_3$	5.6851	3.9571	5.7227
0.95KNN–0.05LN [7]	5.5706[c)] (3.939)	3.919	5.5706[c)] (3.941)
$K_{0.25}Na_{0.75}NbO_3$ [8]	5.598 45[e)] (3.9748)	3.9391	5.6366[e)] (3.9697)
$K_{0.3}Na_{0.7}NbO_3$ [9]	5.612 60	3.931 87	5.643 04

c), e) Calculated using the equations $a = 2a' \sin\left(\frac{\beta}{2}\right)$, $c = 2c' \cos\left(\frac{\beta}{2}\right)$, where a and c are the crystal lattice parameters of the orthorhombic system and a' and c' are the crystal lattice parameters of the pseudo-monoclinic system.

larger than that of Na ion. Compared with the lattice constants of KNN materials in other literature, the lattice constants of KNN crystals calculated by us are reasonable and accurate. At the same time, we also measured the XRD and XRD rocking curve of $K_{0.882}Na_{0.118}NbO_3$ single crystal along the $[001]_C$ direction. From the single crystal XRD, we can see that KNN crystal has only a single peak, and from the rocking curve,

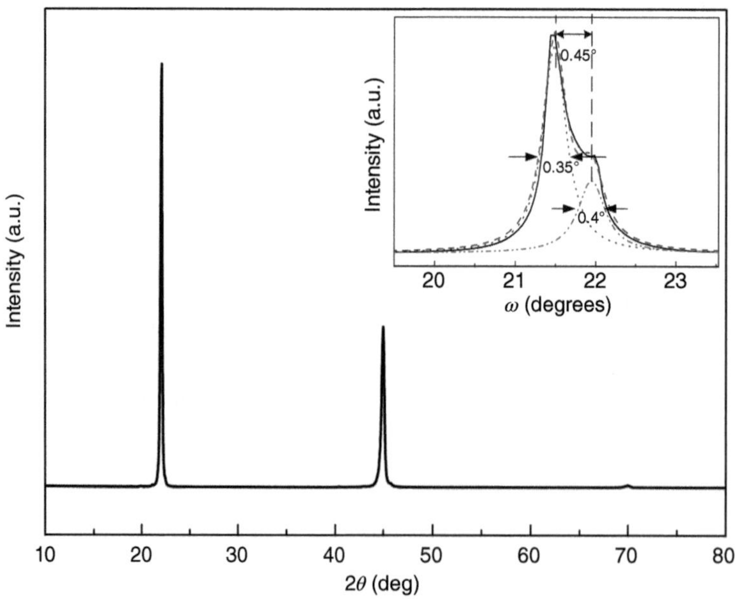

Figure 11.6 X-ray single-crystal pattern of a $K_{0.882}Na_{0.118}NbO_3$ single crystal for the (001) diffraction plane at room temperature. The inset shows the rocking curve of the (001) peak of the $K_{0.882}Na_{0.118}NbO_3$ single crystal. Source: Tian et al. [3]. Reproduced with permission of American Chemical Society.

we can get that the half maximum width of $[001]_C$ and $[110]_C$ peaks of KNN crystal is 0.35° and 0.4°, which indicates that $K_{0.882}Na_{0.118}NbO_3$ single crystal has good quality (Figure 11.6).

11.3 Piezoelectric Properties of KNN Single Crystal

11.3.1 Dielectric Properties of KNN Single Crystal

There is generally a critical temperature (Curie temperature) T_c in ferroelectric crystals. When the temperature is higher than T_c, the crystal will undergo a structural phase transition from ferroelectric phase to paraelectric phase, and the ferroelectricity of the crystal will also disappear. When an uncharged dielectric crystal is placed in an electric field, a certain amount of charge will be induced in its volume and surface, which is called electrowinning. The dielectric coefficient suddenly becomes large near the Curie point, which is often used to determine the Curie point of crystals in experiments. In practical applications, materials with high dielectric constant are mainly used in the fields of ultrahigh-power capacitors and microwave devices. Its use can make the device miniaturized, but at the same time, the loss is relatively large. Therefore, it is of great significance to find materials with high dielectric con stant and low loss for various devices.

Inductance–resistance–capacitance (LCR) digital bridge impedance meter (Agilent E4980A) was used to measure the dielectric constant of samples with different components ε Curve varying with temperature T. As shown in Figure 11.7, the dielectric temperature spectra of KNN with different components along the $[001]_C$ direction are given.

It can be seen from Figure 11.7 that the dielectric loss is obviously different. Especially when the temperature is higher than 300 °C, the dielectric constant loss measured at 10 kHz is significantly greater than the loss measured at 100 kHz. However, the dielectric constant changes with temperature at two test frequencies of 10 and 100 kHz, basically consistent, indicating that KNN crystal is non-relaxation and is a traditional type of ferroelectric material. Table 11.5 records the room temperature dielectric value, loss, temperature rise drop $T_{O\text{-}T}$ (orthorhombic–tetragonal phase transition temperature) and T_C (tetragonal–cubic phase transition temperature) in the dielectric spectrum measured during temperature rise and temperature drop at 10 kHz test frequency. It can be seen from the table that the $T_{O\text{-}T}$ difference between the heating process and the cooling process is 10–20 °C. In addition, the $T_{O\text{-}T}$ of $K_{0.538}Na_{0.462}NbO_3$ crystal is the lowest, and the heating process is 198 °C (the cooling process is 184 °C). It can be seen from the dielectric loss that, compared with KNN, KNN, and piezoelectric materials of the same type, our KNN crystal has very good quality, and the minimum dielectric loss is 0.17% of $K_{0.455}Na_{0.545}NbO_3$ crystal.

In addition, we measured the dielectric temperature spectrum of $K_{0.8}Na_{0.2}NbO_3$ crystal along $[001]_C$ and $[011]_C$ directions, as shown in Figure 11.8. At the same time, the dielectric value, loss, temperature rise–temperature drop $T_{O\text{-}T}$ and T_C data of related room temperature are listed in Table 11.5. It can be seen from the figure

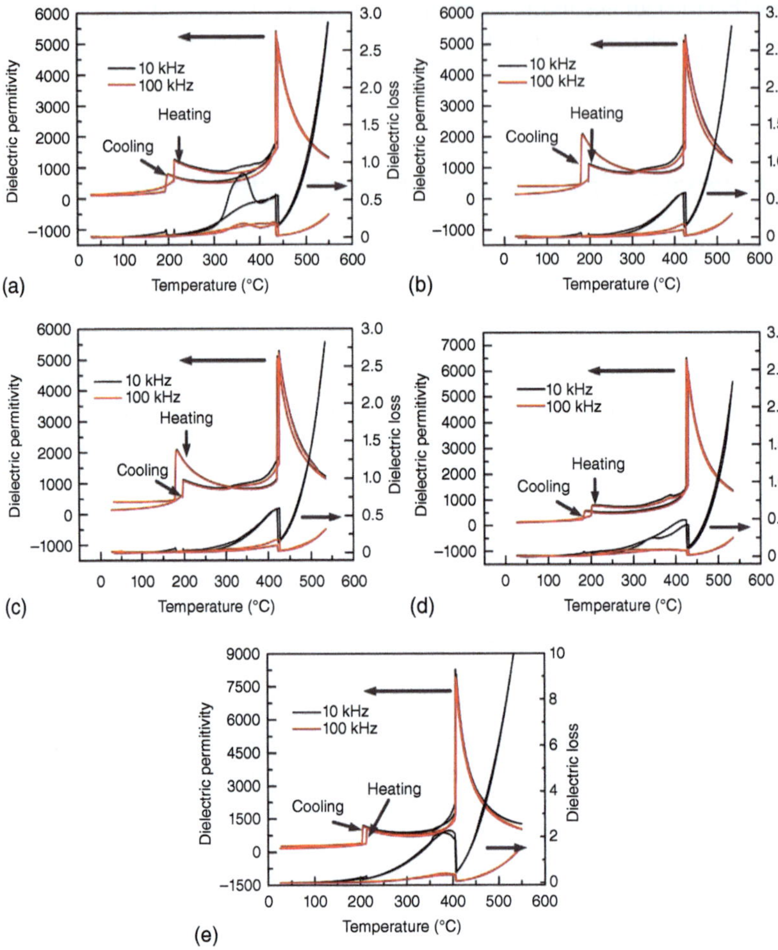

Figure 11.7 Temperature dependence of the dielectric permittivity of KNN single crystals at 10 and 100 kHz. (a) $K_{0.334}Na_{0.666}NbO_3$, (b) $K_{0.455}Na_{0.545}NbO_3$, (c) $K_{0.538}Na_{0.462}NbO_3$, (d) $K_{0.622}Na_{0.378}NbO_3$, (e) $K_{0.882}Na_{0.118}NbO_3$.

and the table that the dielectric behavior along the $[001]_C$ and $[011]_C$ directions is basically the same, but the dielectric constant in the $[011]_C$ direction at room temperature is four times that in the $[001]_C$ direction, reaching 392.

11.3.2 Piezoelectric and Ferroelectric Properties of KNN Crystals

Hysteresis loop, that is, the relationship loop between polarization intensity and applied electric field intensity, is an important basis for judging ferroelectrics. At the same time, by measuring the hysteresis loop, we can obtain the basic ferroelectric parameters of ferroelectrics, such as the remanent polarization P_r and the coercive field E_c, and use them to judge the magnitude of the electric field applied to the crystal polarization. The testing equipment used in this paper is a precision ferroelectric

Table 11.5 Values of orthorhombic–tetragonal T_{O-T} and tetragonal–cubic T_C phase transition temperatures, the dielectric constant ε, and dielectric loss $\tan\delta$ of $K_{1-x}Na_xNbO_3$ crystals at a frequency of 10 kHz.

Crystal	T_{O-T} (°C) of heating	T_{O-T} (°C) of cooling	T_C (°C)	ε	$\tan\delta$ (%)
$K_{0.334}Na_{0.666}NbO_3$	214	199	440	130	1
$K_{0.455}Na_{0.545}NbO_3$	207	190	435	69	0.17
$K_{0.538}Na_{0.462}NbO_3$	198	184	427	143	0.2
$K_{0.622}Na_{0.378}NbO_3$	208	190	431	133	0.4
$K_{0.8}Na_{0.2}NbO_3$	239	219	455	104	0.4
$K_{0.8}Na_{0.2}NbO_3$-[011]	237	217	454	392	0.5
$K_{0.882}Na_{0.118}NbO_3$	216	206	409	169	0.3
KNN–LN [10]	192	—	406	—	1.1
$Li_{0.02}(Na_{0.5}K_{0.5})_{0.98}NbO_3$ [11]	177	—	—	205	33
$K_{0.5}Na_{0.5}NbO_3$ [12]	192	—	410	1015	1
$K_{0.5}Na_{0.5}NbO_3$ [13]	205	—	393	240	2

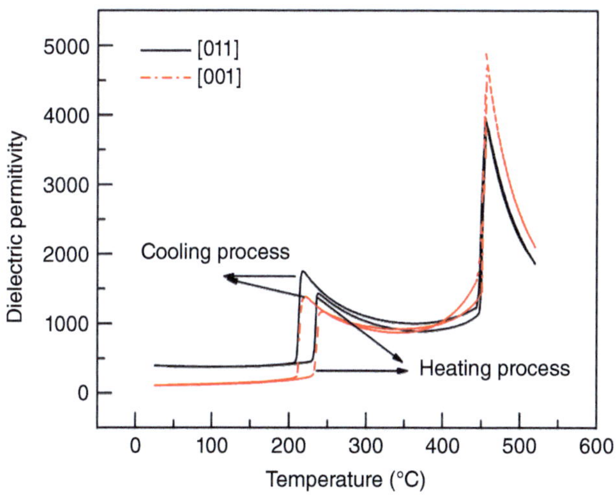

Figure 11.8 Temperature dependence of dielectric constants of $[001]_C$- and $[011]_C$-oriented $K_{0.8}Na_{0.2}NbO_3$ single crystals at 10 kHz frequency for heating and cooling processes. Source: Tian et al. [14]. Reproduced with permission of Royal Society of Chemistry.

comprehensive tester (Radiant Tech., USA). Before testing, high-temperature silver paste electrodes are coated on the upper and lower sides of single crystal samples.

Generally, in KNN crystal, there will inevitably be oxygen vacancies. The main reasons are: (i) When KNN crystal is grown at high temperature, it is found that a large number of 4-valent Nb ions exist in the crystal, and 4d electrons in Nb^{4+} ions will increase the leakage current; (ii) Elements K and Na are easy to volatilize at high temperature, which will form a large number of oxygen vacancies $V_O^{\cdot\cdot}$ during crystal growth, and also lead to the formation of electronic holes, thus leading to the increase of leakage current. The leakage current is caused by the electron hole effect $h^{\cdot}$, which can be explained by Formulas (11.6) and (11.7) [15]

$$\frac{1}{2}O_2 + 2Nb'_{Nb} + V_O^{\cdot\cdot} \rightarrow O_O^* + 2Nb_{Nb}^* \tag{11.6}$$

$$V_O^{\cdot\cdot} + \frac{1}{2}O_2 \rightarrow O_O^* + 2h^{\bullet} \tag{11.7}$$

The dependence curve between the leakage current density J and the electric field E of $K_{0.334}Na_{0.666}NbO_3$ and $K_{0.882}Na_{0.118}NbO_3$ crystals is shown in Figure 11.9. The leakage current is tested with a ferroelectric comprehensive tester, and the maximum applied external electric field is 3 kV/mm. It can be seen from the figure that the leakage current density J of $K_{0.334}Na_{0.666}NbO_3$ and $K_{0.882}Na_{0.118}NbO_3$ crystals is about 10^{-6} A/cm^2. Compared with the work of Kizaki et al. in 2006, it is equivalent to the leakage current measured after annealing KNN crystal in high-pressure oxygen (35 MPa) at 750 °C. However, compared with the leakage current after annealing at 1100 °C, the leakage current is large, indicating that there are few oxygen vacancies in the crystals of $K_{0.334}Na_{0.666}NbO_3$ and $K_{0.882}Na_{0.118}NbO_3$ grown by us, but the oxygen vacancies are not completely eliminated.

Figure 11.10 shows the relationship between polarization intensity and electric field at different frequencies. From Figure 11.10, we can see that the hysteresis loop is round under the low-frequency 10 Hz (100 ms) test, that is, the KNN crystal has a large leakage current during the test of the hysteresis loop. This is due to the existence of oxygen vacancies in KNN crystals. Under the low-frequency applied electric

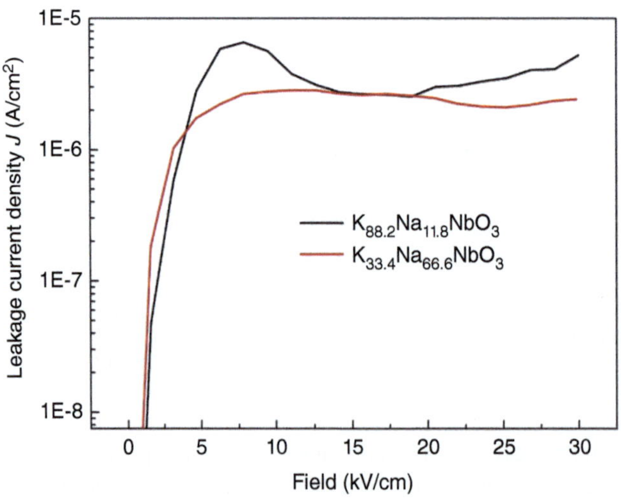

Figure 11.9 Leakage current densities (J) measured at 25 °C along the $[100]_C$ direction of the as-grown $K_{1-x}Na_xNbO_3$ ($x = 0.118$ and 0.666) crystals. Source: Tian et al. [3]. Reproduced with permission of American Chemical Society.

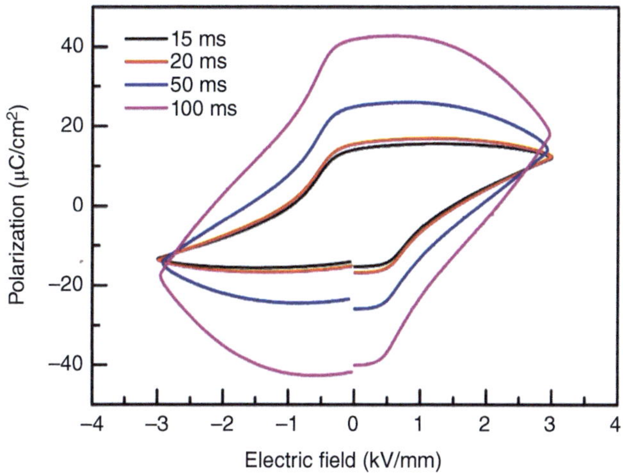

Figure 11.10 The hysteresis loop at different frequencies of $K_{0.334}Na_{0.666}NbO_3$ crystal.

field, the space charge formed by oxygen vacancies will make a great contribution to the dielectric properties. At the same time, we can also observe that the space charge will have different responses to different test frequencies. When the test frequency is greater than 50 Hz (20 ms), the hysteresis loop performs very well, that is, there is basically no leakage current. At this time, the space charge has little effect on the hysteresis loop. Therefore, in the subsequent experiment to measure the hysteresis loop of KNN crystal with different components, the measurement frequency is 100 Hz (10 ms).

Figure 11.11 shows the hysteresis loops measured under different electric fields for KNN crystals with five different components: $K_{0.334}Na_{0.666}NbO_3$, $K_{0.455}Na_{0.545}NbO_3$, $K_{0.538}Na_{0.462}NbO_3$, $K_{0.622}Na_{0.378}NbO_3$, and $K_{0.882}Na_{0.118}NbO_3$. It can be seen from the figure that all the hysteresis loops are similar, showing saturated hysteresis loops, indicating that these crystals have the essence of ferroelectrics. Residual polarization intensity (P_r) and coercive field (E_c) of all crystals are shown in Table 11.6. It is worth noting that the remanent polarization intensity decreases first and then increases with the increase of K content. The remanent polarization intensity of $K_{0.538}Na_{0.462}NbO_3$ crystal is the lowest, only 8.2 μC/cm². The remanent polarization intensity of $K_{0.8}Na_{0.12}NbO_3$ crystal is half of16.1 μC/cm². At the same time, the coercive field of most KNN crystals is about 1.0–1.3 kV/mm, and only $K_{0.882}Na_{0.118}NbO_3$ crystals have a coercive field of 1.54 kV/mm. The coercive field size reflects the ease of domain inversion and domain wall movement in ferroelectrics, indicating that all KNN crystals are not very easy to polarize, but it also indicates that the polarized crystals are not easy to depolarize due to external influence. As shown in Figure 11.12, the hysteresis loop of $K_{0.8}Na_{0.2}NbO_3$ single crystal along $[001]_C$ and $[011]_C$ directions shows that the coercive field E_c in $[001]_C$ direction is 8.6 kV/cm, which is less than the coercive field in $[011]_C$ direction, indicating that the single domain structure state of $K_{0.8}Na_{0.2}NbO_3$ single crystal is harder to polarize than the multi domain state, and the residual polarization

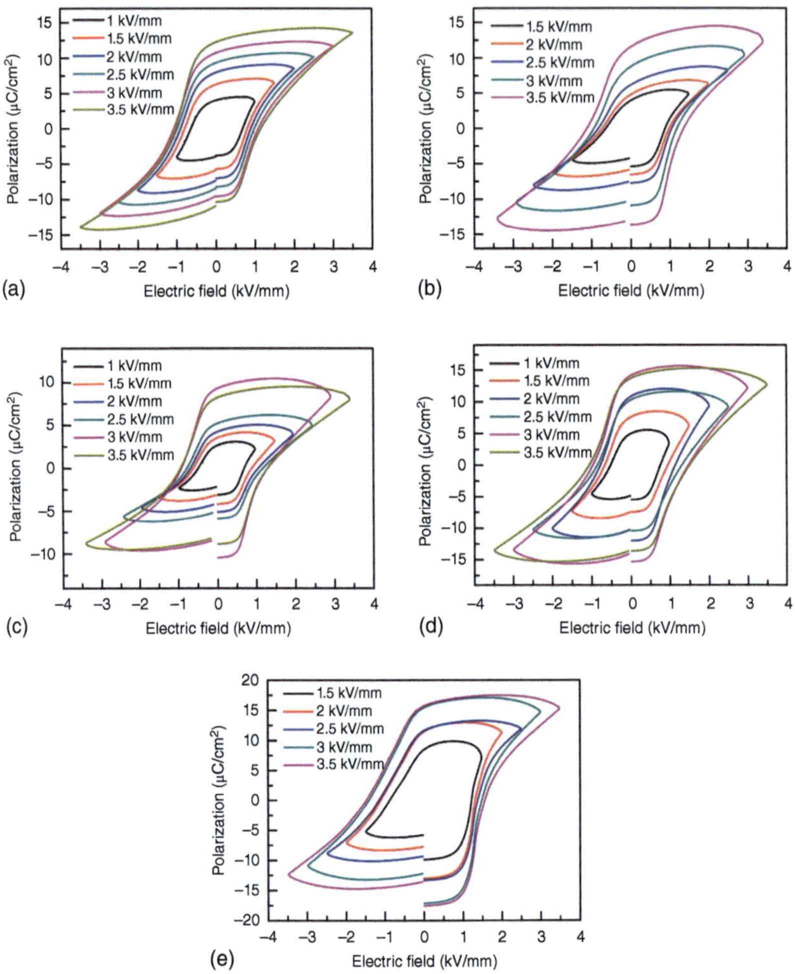

Figure 11.11 Hysteresis loops of $K_{1-x}Na_xNbO_3$ crystals with different values of x; the measurements were performed at room temperature and at a frequency of 100 Hz. (a) $x = 0.666$, (b) $x = 0.545$, (c) $x = 0.462$, (d) $x = 0.378$, (e) $x = 0.118$.

intensity in $[011]_C$ direction is 27 μC/cm^2, 1.68 times of the remanent polarization intensity in $[001]_C$ direction. The establishment of E_c provides an important basis for the selection of polarization electric field.

11.3.3 Giant Electro-strain of KNN Single Crystal

The strain study of ferroelectrics is also an important parameter reflecting ferroelectric properties. We measured the strain of KNN in different components with a ferroelectric tester and found that super strain was found in the crystal with K elements of 0.43 and 0.46. S reached 1%, almost the extreme value of perovskite ferroelectrics. Under an electric field of 1 kV/mm, the strain could reach 0.9%, as shown

Table 11.6 Ferroelectric properties of $K_{1-x}Na_xNbO_3$ crystals with different values of x and KNN-based materials at a frequency of 100 Hz.

Materials	E_c (kV/cm)	P_r (μC/cm²)
$K_{0.334}Na_{0.666}NbO_3$	11.6	12.3
$K_{0.455}Na_{0.545}NbO_3$	12.5	11.5
$K_{0.538}Na_{0.462}NbO_3$	11.3	8.2
$K_{0.622}Na_{0.378}NbO_3$	12.8	13.7
$K_{0.8}Na_{0.12}NbO_3$	10.8	16.1
$K_{0.8}Na_{0.2}NbO_3$-$[011]_C$	8.6	27.
$K_{0.882}Na_{0.118}NbO_3$	15.4	15.7
$K_{0.5}Na_{0.5}NbO_3{}^{b}$-$[1\bar{3}1]$ [12]	24	17
KNN [16]	14.6	18.5
KNN-LN [10]	22	9.05

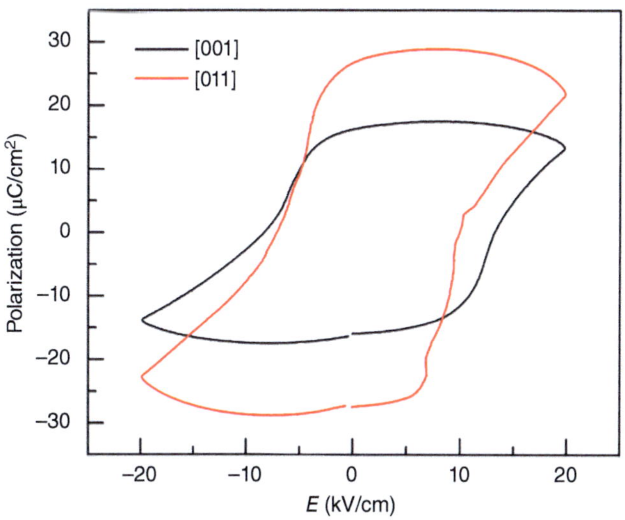

Figure 11.12 Hysteresis loops of $[001]_C$- and $[011]_C$-oriented $K_{0.8}Na_{0.2}NbO_3$ single crystals at 100 Hz frequency. Source: Tian et al. [14]. Reproduced with permission of Royal Society of Chemistry.

in Figure 11.13. Compared with all ferroelectric materials with large strain reported at present [17], the KNN crystal with these two components found under the same electric field has great advantages.

We have done a lot of repeated experiments on KNN crystals of these two components and found that these two components have very good recoverability, that is, by adding a small reverse electric field and not exceeding the coercive field, the experiment can be repeated, or the experiment can be repeated after being placed for

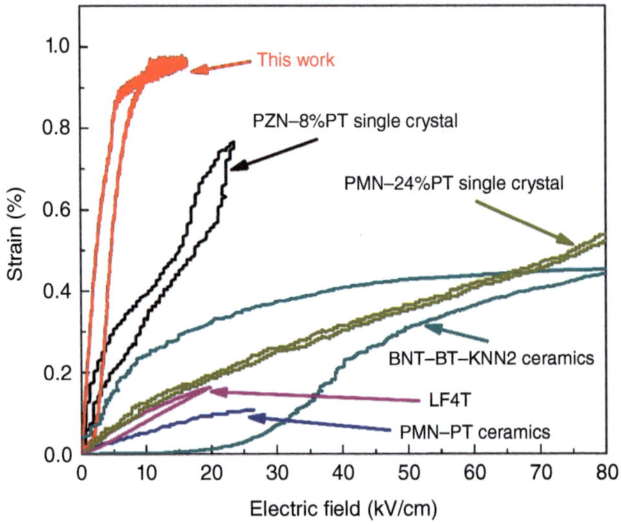

Figure 11.13 Strain–electric field of different materials. Source: Hu et al. [17]. Reproduced with permission of American Association for the Advancement of Science.

about six hours. This phenomenon is very important for the piezoelectric properties of large signals because the general strain is caused by lattice distortion and the movement of electric domains. The lattice distortion is usually manifested as lattice stretching or shrinking, so it is reversible and is an intrinsic contribution. The reorientation of electric domains under the external electric field, that is, domain wall movement, is irreversible and is an extrinsic contribution. So the phenomenon that the strain value reaches 1% found in this paper can be sure that it is the movement of the electric domain, but the recoverability experiment breaks the irreversible rule of the electric domain.

In order to further study the large strain properties, we first use piezo force microscopy (PFM) to make the phase lag loop diagram for single crystals, as shown in Figure 11.14a. We find that there is a appreciable internal electric field in the KNN crystal. It is further confirmed that there is a strong spontaneous polarization in the KNN crystal, so that the strain in the microscopic region is affected by the local macroscopic internal electric field as the electric field changes. When we make a bilateral electrostriction curve for the crystal, it does not show the traditional "butterfly" curve. As shown in Figure 11.14b, the negative strain reaches −0.36%, the positive strain reaches 0.95%, and the total strain (the sum of positive and negative strains) reaches 1.31%. This phenomenon further reflects the fact that the internal electric field formed by spontaneous polarization prevents domain inversion.

Then we measured the electrostriction curve of 10 single side cycles under variable temperature, and the test frequency is 1 Hz, as shown in Figure 11.14c,d. We can see that with the increase in temperature, the maximum strain becomes larger and larger. At 175 °C, the maximum strain is 1.84%. At 200 °C, the strain will suddenly be low, and the crystal has passed the phase transition temperature. The huge strain of 1.84% has exceeded that of 1.7% under the electric phase transition of PZN–8%PT.

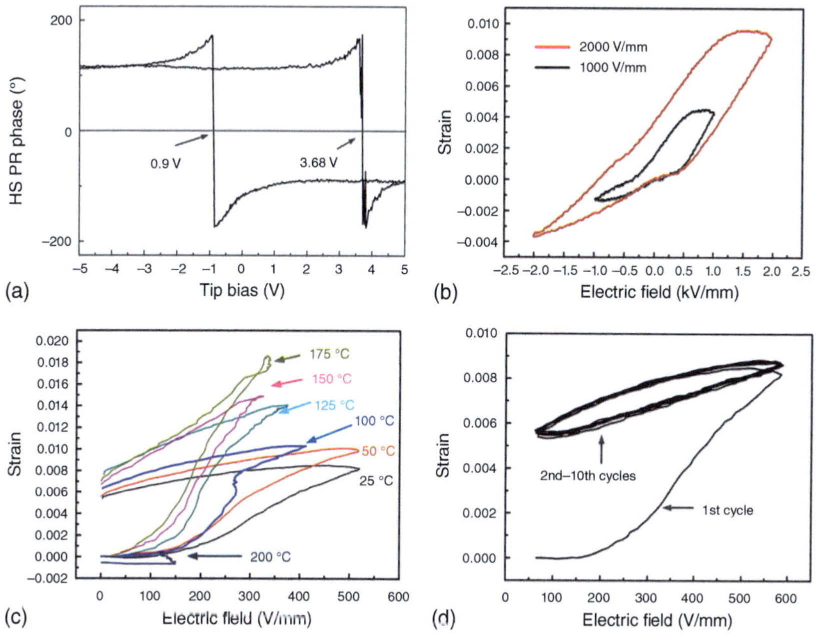

Figure 11.14 The strain versus electric field for the $(K_{0.43}Na_{0.57})NbO_3$ crystal. (a) Phase–voltage hysteresis loop. (b) The strain versus different electric field at 25 °C. (c) The strain versus electric field at different temperatures. (d) The strain versus electric field during 10 cycles.

It should be noted here that the external applied electric field strength is always less than 500 V/mm and can only reach about 350 V/mm at high temperatures, even due to leakage. With the high electric field of PZN–8%PT, it has the advantages of small electric field and large signal.

On the other hand, we can draw from the cyclic electrostriction curve (Figure 11.14d) that the strain of the 2nd–10th cycles is basically unchanged, about 0.40% of the strain, accounting for 43% of the maximum strain. It can be seen from the figure that this part of strain is very effective and has very strong strain recoverability. We summarize the cyclic strain at different temperatures in Figure 11.15. Before 100 °C, KNN crystal had very good recoverability. The converse piezoelectric coefficient S/E reaches 10 000 pC/N. After increasing the temperature, the domain activity is intense due to the influence of the phase transition temperature, and the increase in strain is obviously felt. It can be seen from the 10th cycle curve that although the inverse piezoelectric coefficient will increase to 13 000 pC/N as the strain increases with temperature, it will generally maintain a relatively stable state after several cycles. Therefore, these domains with good recoverability have excellent characteristics in terms of temperature stability.

To sum up, we can find that KNN crystal components of these two components have very good converse piezoelectric properties, not only with large strain and small applied electric field, but also with very good recoverability. From the source of strain contribution, its main contribution comes from the movement of the electric domain. As shown in Figure 11.16, the electric domain disappears obviously after

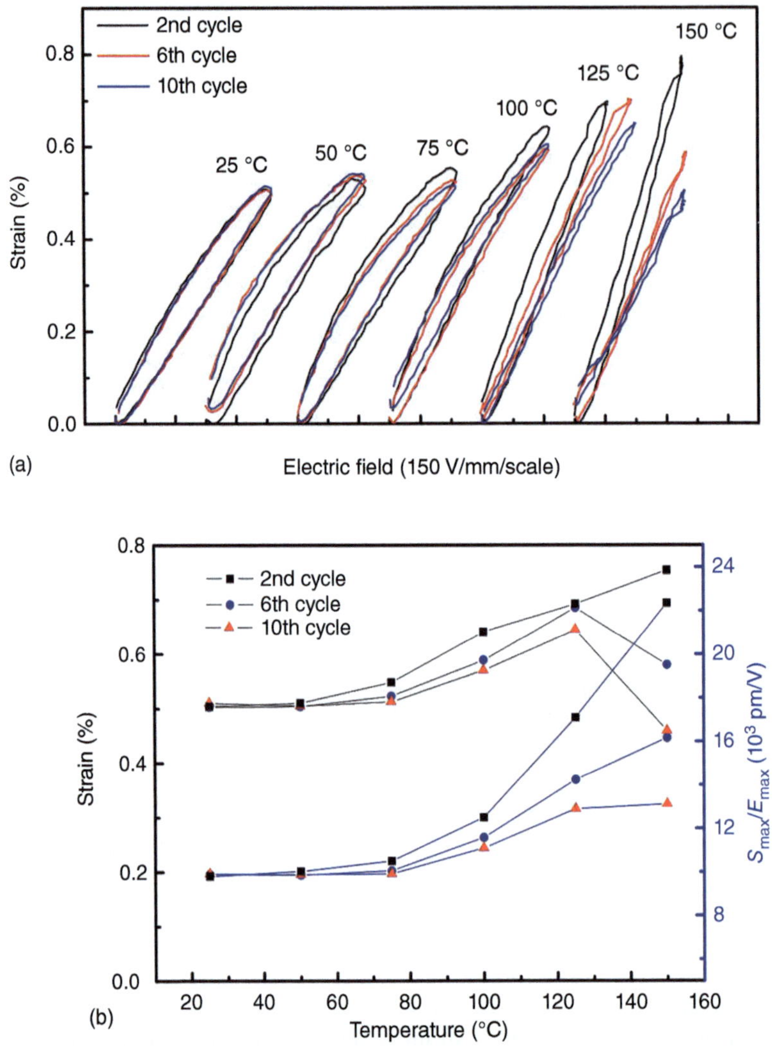

Figure 11.15 (a) The second, sixth, and tenth of strain in $(K_{0.43}Na_{0.57})NbO_3$ crystal at different temperatures, and (b) S_{max}/E_{max}.

1500 V/mm is applied; however, it reappears after removing the electric field. Similar lattice movement can also be observed by single crystal diffraction experiment, as shown in Figure 11.17. When the external electric field is applied to the crystal, the lattice will obviously deflect. The greater the electric field is, the greater the peak value of lattice X-ray along the electric field direction is. However, the lattice recovery process was not observed in the single crystal diffraction experiment. From the cyclic electrostriction curve, we can find that there are two different kinds of domain motions. One is considered as irreversible domain motions, which are generally considered as those with larger sizes, and the other is considered as reversible domain motions, which are understood as those with polar nanodomains.

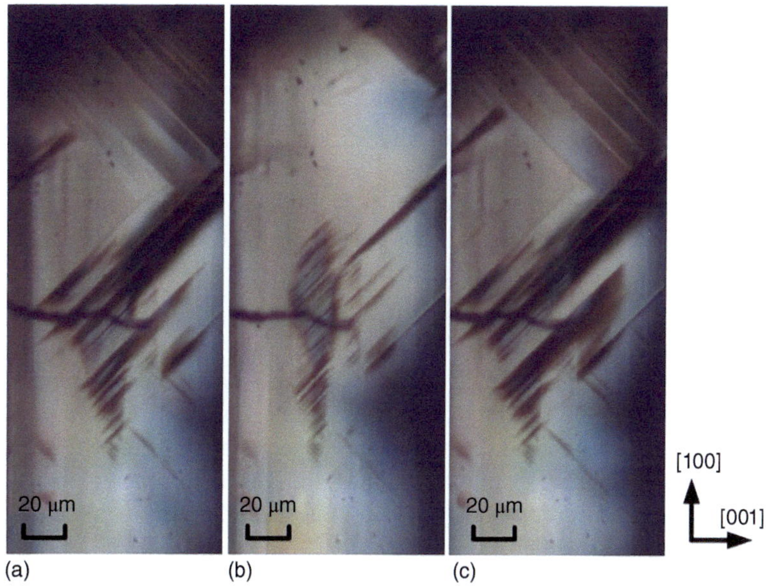

Figure 11.16 The changes of domains in $(K_{0.43}Na_{0.57})NbO_3$ crystal before and after the electric field, (a) before applying the electric field, (b) electric field of 1500 V/mm, (c) remove the electric field.

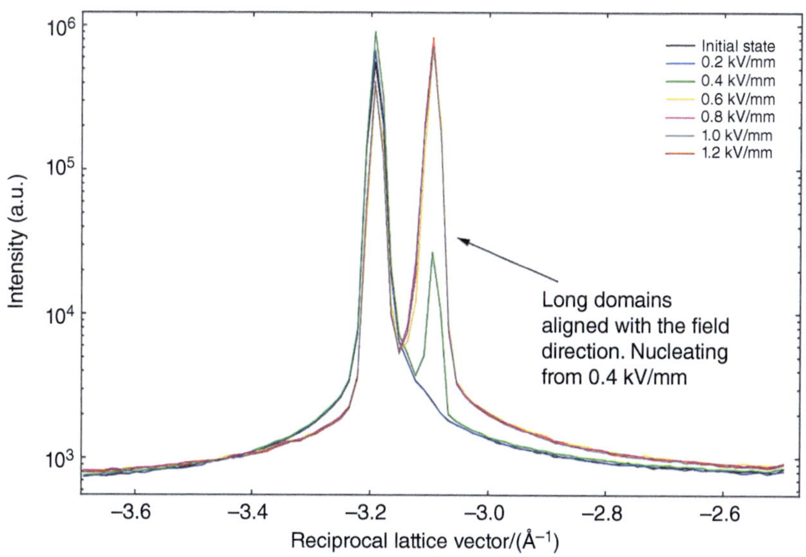

Figure 11.17 (02l) reciprocal space line scan showing pseudo-cubic 022 diffraction peak. Source: Hu et al. [17]. Reproduced with permission of American Association for the Advancement of Science.

11.4 The Electro-Optic Effect and Applications of KTN Crystal

11.4.1 The Electro-Optic Effect of KTN Crystal

The electro-optic effect is a phenomenon in which the refractive index varies with an applied electric field. The frequency of applied electric field is usually much lower than light frequency. According to the different symmetry of the material structure, the refractive index shows different change rules with the electric field as the equation:

$$\Delta n = n - n^0 = aE_0(\omega) + bE_0^2(\omega) \tag{11.8}$$

where the first item shows the linear relationship between n and E called Pockels effect (linear electro-optic effect), which only exists in materials without centrosymmetry, and the second item is the Kerr effect (quadratic electro-optic effect). For KTN single crystal, the Pockels effect mainly contributes to electro-optic property in ferroelectric phase, while the Kerr effect plays a major role near the Curie temperature (ferro-paraelectric phase transition).

The optical indicatrix is a stereoscopic representation of the dielectric tensor, usually used to reveal the propagation of light in a crystal. The relationship between element of inverse dielectric tensor β_{ij} $(i, j = 1, 2, 3)$ and ε_{jk} is shown as follows:

$$\beta_{ij}\varepsilon_{jk} = \delta_{ik} \tag{11.9}$$

At same time, the electric displacement vector and the external electric field satisfy the following relation in crystal:

$$\mathbf{D} = \varepsilon_0 \varepsilon \cdot \mathbf{E} \tag{11.10}$$

In the main axis coordinate system of the crystal, the electric field density can be expressed as follows:

$$w_e = \frac{1}{2}\mathbf{E} \cdot \mathbf{D} = \frac{1}{2\varepsilon_0}\left(\frac{D_1^2}{\varepsilon_1} + \frac{D_2^2}{\varepsilon_2} + \frac{D_3^2}{\varepsilon_3}\right) \tag{11.11}$$

The energy density is determined for the experimental system. In **D**-space, it is expressed in the same way as an ellipsoid. Thus, the optical indicatrix is also called a refraction index ellipsoid. For KTN, the refractive index is $n_i = \sqrt{\varepsilon_i}$ due to the nonmagnetic, which is $\mu_i = 1$. We set the $x_i = D_i/\sqrt{2\varepsilon_0 w_e}$ $(i = 1, 2, 3)$ shows three directions of main axis in crystal, and the equation transformed into:

$$\beta_{11}^0 x_1^2 + \beta_{22}^0 x_2^2 + \beta_{33}^0 x_3^2 = 1 \tag{11.12}$$

where β_{ij}^0 $(i, j = 1, 2, 3)$ is the inverse dielectric tensor element without an external electric field, and the above equation can be changed under the action of external electric field:

$$\left(\beta_{11}^0 + \Delta\beta_{11}\right) x_1'^2 + \left(\beta_{22}^0 + \Delta\beta_{22}\right) x_2'^2 + \left(\beta_{33}^0 + \Delta\beta_{33}\right) x_3'^2 = 1 \tag{11.13}$$

where $\Delta\beta_{ij}$ is the change of refractive index under an external electric field, which can be described as electro-optic effect:

$$\Delta\beta_{ij} = \beta_{ij} - \beta_{ij}^0 = \gamma_{ijk}E_k + s_{ijkl}E_kE_l + \cdots \tag{11.14}$$

where γ_{ijk}, s_{ijkl} are Pockels and Kerr coefficients. KTN crystal has strong linear and quadratic electro-optic modulation. For tetragonal KTN crystals, the linear electro-optic coefficient matrix is as follows:

$$\begin{pmatrix} 0 & 0 & \gamma_{13} \\ 0 & 0 & \gamma_{13} \\ 0 & 0 & \gamma_{33} \\ 0 & \gamma_{51} & 0 \\ \gamma_{51} & 0 & 0 \\ 0 & 0 & 0 \end{pmatrix} \tag{11.15}$$

Then $\beta_{ij} = \beta_m$, $\gamma_{ijk} = \gamma_{mk}$ $(m = 1, 2, 3, 4, 5, 6)$. If only the linear response is considered under the external electric field condition, the changes of β_m is:

$$\begin{pmatrix} \Delta\beta_1 \\ \Delta\beta_2 \\ \Delta\beta_3 \\ \Delta\beta_4 \\ \Delta\beta_5 \\ \Delta\beta_6 \end{pmatrix} = \begin{pmatrix} \beta_1 - \beta_1^0 \\ \beta_2 - \beta_2^0 \\ \beta_3 - \beta_3^0 \\ \beta_4 - \beta_4^0 \\ \beta_5 - \beta_5^0 \\ \beta_6 - \beta_6^0 \end{pmatrix} = \begin{pmatrix} 0 & 0 & \gamma_{13} \\ 0 & 0 & \gamma_{13} \\ 0 & 0 & \gamma_{33} \\ 0 & \gamma_{51} & 0 \\ \gamma_{51} & 0 & 0 \\ 0 & 0 & 0 \end{pmatrix} \begin{pmatrix} E_1 \\ E_2 \\ E_3 \end{pmatrix} \tag{11.16}$$

The tetragonal KTN crystal belongs to uniaxial crystal. It has the relationship of $\beta_1^0 = \beta_2^0 = 1/n_o^2$, $\beta_3^0 = 1/n_e^2$ when the external field is not added. To ensure that the main axis direction of the optical indicatrix under the external field remains unchanged, the external electric field E_3 can be applied only in the direction x_3 ($[001]_C$), then the optical indicatrix with the applied electric field can be expressed as:

$$\left(\beta_1^0 + \gamma_{13}E_3\right)\left(x_1^2 + x_2^2\right) + \left(\beta_3^0 + \gamma_{33}E_3\right)x_3^2 = 1 \tag{11.17}$$

The above equation is the changed optical indicatrix under the spindle condition. To calculate the refractive index change caused by the external field, the variation of refractive index can be shown as:

$$\Delta n_i = -\frac{1}{2}n_i^3\Delta\beta_i \tag{11.18}$$

Therefore, the refractive index change in the direction of the principal axis is obtained as:

$$\begin{cases} \Delta n_1 = -\dfrac{1}{2}n_o^3\gamma_{13}E_3 \\ \Delta n_3 = -\dfrac{1}{2}n_e^3\gamma_{33}E_3 \end{cases} \tag{11.19}$$

At the same time, the effective electro-optic coefficient γ_{eff} can be defined as:

$$\gamma_{eff} = \gamma_{33} - \frac{n_o^3}{n_e^3}\gamma_{13} \tag{11.20}$$

It characterizes the anisotropy difference under the external field of the material, which can be used to realize the polarization conversion, half-wave voltage, and other performance evaluations. Then the difference in value of refractive index anisotropy in this state can be expressed as:

$$\Delta n = \Delta n_3 - \Delta n_1 = -\frac{1}{2}n_e^3\gamma_{eff}E_3 \tag{11.21}$$

For cubic phase KTN crystal, the quadratic electro-optic coefficient matrix is as follows:

$$\begin{pmatrix} s_{11} & s_{12} & s_{12} & 0 & 0 & 0 \\ s_{12} & s_{11} & s_{12} & 0 & 0 & 0 \\ s_{12} & s_{12} & s_{11} & 0 & 0 & 0 \\ 0 & 0 & 0 & s_{44} & 0 & 0 \\ 0 & 0 & 0 & 0 & s_{44} & 0 \\ 0 & 0 & 0 & 0 & 0 & s_{44} \end{pmatrix} \tag{11.22}$$

The change of inverse dielectric tensor caused by electro-optic effect after applying an electric field can be obtained using the same calculation method as linear electro-optic effect as:

$$\begin{pmatrix} \Delta\beta_1 \\ \Delta\beta_2 \\ \Delta\beta_3 \\ \Delta\beta_4 \\ \Delta\beta_5 \\ \Delta\beta_6 \end{pmatrix} = \begin{pmatrix} \beta_1 - \beta_1^0 \\ \beta_2 - \beta_2^0 \\ \beta_3 - \beta_3^0 \\ \beta_4 - \beta_4^0 \\ \beta_5 - \beta_5^0 \\ \beta_6 - \beta_6^0 \end{pmatrix} = \begin{pmatrix} s_{11} & s_{12} & s_{12} & 0 & 0 & 0 \\ s_{12} & s_{11} & s_{12} & 0 & 0 & 0 \\ s_{12} & s_{12} & s_{11} & 0 & 0 & 0 \\ 0 & 0 & 0 & s_{44} & 0 & 0 \\ 0 & 0 & 0 & 0 & s_{44} & 0 \\ 0 & 0 & 0 & 0 & 0 & s_{44} \end{pmatrix} \begin{pmatrix} E_1^2 \\ E_2^2 \\ E_3^2 \\ E_2E_3 \\ E_1E_3 \\ E_1E_2 \end{pmatrix} \tag{11.23}$$

The change of optical volume under the external field response in the direction of x_3 is:

$$\left(\beta_1^0 + s_{12}E_3^2\right)\left(x_1^2 + x_2^2\right) + \left(\beta_3^0 + s_{11}E_3^2\right)x_3^2 = 1 \tag{11.24}$$

The change of refractive index caused by the external field in the direction of x_3 in the spindle direction can be written as:

$$\begin{cases} \Delta n_1 = -\frac{1}{2}n_o^3 s_{12}E_3^2 \\ \Delta n_3 = -\frac{1}{2}n_o^3 s_{11}E_3^2 \end{cases} \tag{11.25}$$

Similarly, an effective quadratic electro-optic coefficient can be defined as:

$$s_{eff} = s_{11} - s_{12} \tag{11.26}$$

According to the above calculation results, linear and quadratic electro-optic coefficients can be obtained by applying the electric field E_3 in the direction of x_3 and testing the refractive index changes in different spindle directions.

11.4.2 The Evolution of Polarization in KTN Crystals Near Curie Temperature

The electro-optic properties of KTN crystals tend to be excellent near the Curie temperature, which is closely related to the structural changes near the phase transition. However, the evolution process of electric domains in this region cannot be clearly explained through the traditional understanding of phase transition. That is, neither the ferroelectric domain structure nor the cubic phase can provide sufficient structural support for the excellent performance of secondary electro-optic properties of KTN crystals. Compared with other perovskite crystal materials, PZT and PMN–PT crystals have certain relaxation properties during phase transition due to the existence of polar nanoscale regions in their interior. The relaxed features mainly exist in the intermediate state, which contains the polar nano-regions, showing the dispersion of frequency in dielectric–temperature curve. It originates from the different response abilities of complex states of domain under an applied electric field. By using the Curie–Weiss law fitting above the Curie temperature, it is found that the KTN crystal also has weak relaxation property, that is, the existence of polar nano-region during the phase transition may be the origin of its excellent performance.

The temperature dependence of the relative dielectric permittivity is a typical method to show the Curie temperature and relaxor characteristics, and the remarkable frequency dispersion appears near the Curie temperature. In KTN crystal, it also shows similar characteristics, as shown in Figure 11.18.

From the spectrum, we can see the obvious anomaly dielectric peak near the tetragonal-cubic phase transition, which corresponds to the change in charge storage capacity of the material driven by the external field caused by the disappearance

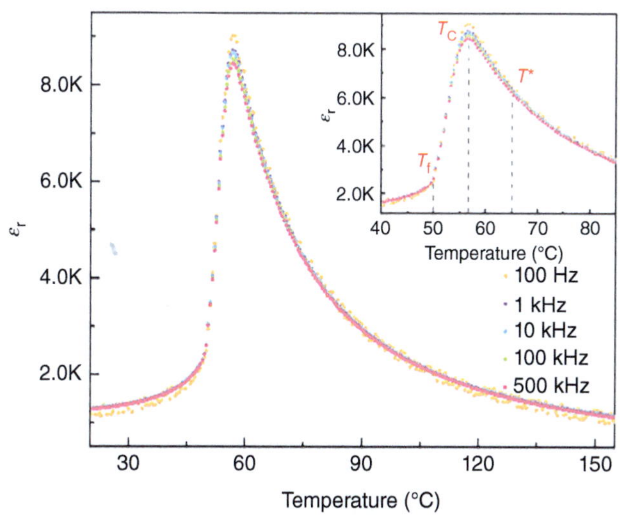

Figure 11.18 The temperature-dependence of the relative dielectric permittivity for KTN crystal. Source: Wang et al. [18]. Reproduced with permission of American Chemical Society.

of ferroelectric domain. It shows the weak relaxation characteristic with frequency dispersion. It is worth noting that the dielectric response of KTN crystals is dispersive at different frequencies of applied electric field at a temperature (freezing temperature $T_f = 50\,°C$) slightly lower than the phase transition temperature. It is due to the fact that the ferroelectric domains undergo the decline of long-range interactions between polar units before disappearing during the heating process. In this temperature range, the overall size of the domain decreases due to the weakening of the long-range ordering force. Then, the response rates of the domains with different sizes show different characteristics under the external electric field, including distortion and switching, resulting in different dielectric peaks at different frequencies.

Figure 11.19 shows that the domain shape does not change significantly when the temperature is lower than 48 °C, and the domain walls begin to decrease significantly when the temperature is higher than 48 °C. At the same time, because all the sizes of the low-frequency ferroelectric domains can fully respond to the field driving, while the large size domains cannot fully follow the high-frequency electric field, the crystals show larger dielectric peak under the low-frequency. However, the method cannot be used to observe the walls of the nanoscale domains. The domain walls disappear completely after 58 °C, indicating that no macroscopic ferroelectric domains exist at this time.

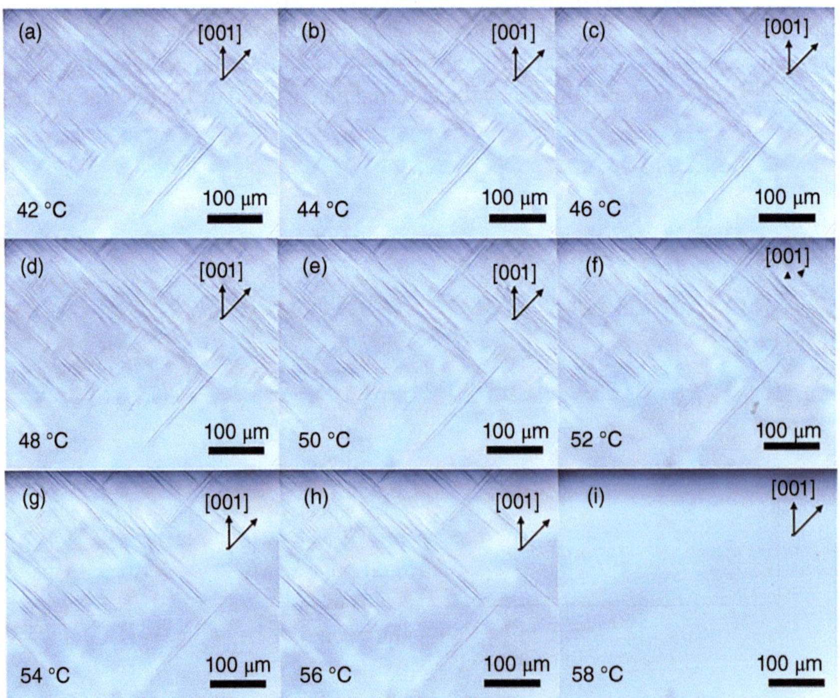

Figure 11.19 Polarizing microscope measurement of KTN crystal phase transition.

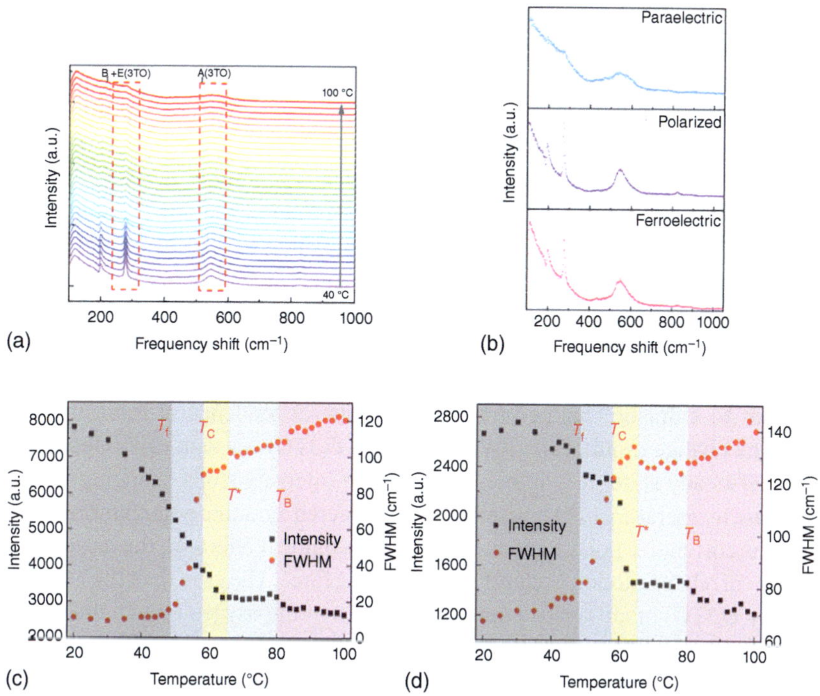

Figure 11.20 The temperature-dependence of the Raman spectra for (a) the region from 150 to 1000 cm^{-1} (the data was measured every 2 °C in the range of 40–100 °C), (b) is the Raman spectrum for paraelectric cubic phase, ferroelectric phase and polarized sample, respectively, (c) B_1 + E(3TO), and (d) A_1(3TO), respectively. Source: Wang et al. [18]. Reproduced with permission of American Chemical Society.

The Raman scattering spectrum in Figure 11.20a shows the temperature-dependence of intensity and full width at half maxima (FWHM). Both the intensity and the FWHM slowly change when the temperature is under T_f. The decrease in intensity shows the weakening of the polarization. The FWHM shows a slow positive correlation with the temperature, discribed by the equation written as:

$$\Gamma\,(\mathrm{T}) = A\left[1 + \frac{2}{e^{\frac{\hbar\omega_0}{2k_B T}} - 1}\right] + B\left[1 + \frac{3}{e^{\frac{\hbar\omega_0}{3k_B T}} - 1} + \frac{3}{\left(e^{\frac{\hbar\omega_0}{3k_B T}} - 1\right)^2}\right] \tag{11.27}$$

Here, A and B are constants and ω_0 represents the harmonic frequency of the optical mode. The change in the FWHM was affected by the expansion of the lattices. The ferroelectric domains change to the relaxational polar regions above the T_f, which induces the intensity to decline rapidly. This leads to the decrease in intensity of the Raman spectrum, which is caused by the further weakening of polarization for both the B_1 + E(3TO) mode and A_1(3TO) mode. The steady intensity of the A_1(3TO) indicates that the intensity of the polarization has a weak correlation with the temperature below the T_f. The temperature mainly acts on the interaction of domains, which breaks the stability of the ferroelectric domains. The FWHM of both the A_1(3TO) mode and the $B_1 + E$(3TO) mode starts to grow rapidly in the vicinity of T_f. This means that the evolution of the domains near the T_f is quasi-static. The interaction of the domains cannot provide a steady consistent orientation, which can be driven by the external electric field. Therefore, the frequency dispersion appears in the temperature-dependence of the relative dielectric permittivity, as shown in the inset of Figure 11.18. All the modes should not be found above the T_C (the space group of *Pm*3*m*), according to the Raman selection rules, as a result of the phase transition. However, it is still active in the vibration mode, as shown in Figure 11.20b.

The polar regions start to shrink and correlate with each other in the short-range due to the specific polarization not being the static solution for Gibbs free energy. This influences the vibration for both the $B_1 + E$(3TO) mode and A_1(3TO) mode, which leads to a drastic change in the temperature-dependence of intensity and FWHM. The change trend of the FWHM for $B_1 + E$(3TO) is a sign of the breaking of the quasi-steady state. It is interesting that the intensity stops changing at the intermediate temperature (T^*), caused by the disordered dynamic polar nanoregions (PNRs). The correlative polar regions translate to dynamic PNRs with the disappearance of the correlated motion under the local driving force. The growth of the FWHM for the $B_1 + E$(3TO) mode verifies that the vibration of the oxygen ions in the plane perpendicular to the $[001]_C$ direction is unstable. The deformation of the octahedral units would not be influenced by the correlated motion of PNRs, which indicates that the correlated motion affects the orientation of PNRs more than the intensity of polarization. This can be confirmed by the steady region for FWHM between the T^* and T_B. And the disappearance of the correlated motion of PNRs means that the disappearance of the frequency dispersion in the temperature-dependence of the relative dielectric permittivity because of the response speed of PNRs is enough to reach the different frequency of ac field. Finally, the intensity of both the $B_1 + E$(3TO) and A_1(3TO) modes declines tardily as a result of the disappearance of PNRs above the T_B due to the high symmetry of the structure.

11.4.3 The Structure Origin of Electro-Optic Effect in KTN Single Crystal

The KTN has linear and quadratic electro-optic effects in tetragonal phase and cubic phase, respectively, which depends on the microstructure. Generally, KTN mainly shows linear electro-optic effect below T_C due to the response of ferroelectric domains with weak quadratic electro-optic effect. However, KTN shows the remarkable quadratic electro-optic effect in the vicinity region above T_C. According to previous research, the PNRs are considered to contribute to the quadratic electro-optic effect.

11.4.3.1 The Relationship Between Linear Electro-Optic Effect and Ferroelectric Domains Structure

In the ferroelectric phase (tetragonal phase), the strong coupling of local dipoles will freeze the PNRs and induce ferroelectric domains. The spontaneous polarization intensity of crystals (P_s) is critical to linear electro-optic properties, and Ps is directly related to ferroelectric domain states. Under T_C, the relationship between the spontaneous polarization intensity of the crystal and the linear electro-optic coefficient is as follows [19]:

$$\gamma_{33} - \gamma_{13} = 2\varepsilon_0(\varepsilon_r - 1)(g_{11} - g_{12})P_s \tag{11.28}$$

where γ_{33} and γ_{13} are the linear electro-optic coefficients, ε_0 is permittivity of vacuum, and ε_r is relative dielectric constant. g_{11} and g_{12} are electro-optic g-coefficients (temperature-independent). Therefore, high polarization intensity P_s is beneficial to improve the electro-optic response of the crystal. However, the complex ferroelectric domain structure in KTN crystal has obvious scattering effect on light, which will greatly affect the application of electro-optic effect. A good polarization treatment can effectively achieve single domain of crystal, inducing stable polarization state with weak light scattering. In order to reflect the electro-optic effect of ferroelectric domain with different states, the state of ferroelectric domain can be changed by controlling the poling time. The two same crystals were kept for 5 and 10 hours, respectively, under the action of 2 kV/mm polarization electric field E_p (complete poling), and the polarization direction was perpendicular to its optical axis. Due to the difference of poling time, the polarization state in the crystal will be different. Therefore, the difference of linear electro-optic effect after two poling progresses can be compared, and the influence of ferroelectric domain on electro-optic effect can be studied by linear electro-optic property evolution. Their effective electro-optic coefficient γ_c is shown in Figure 11.21.

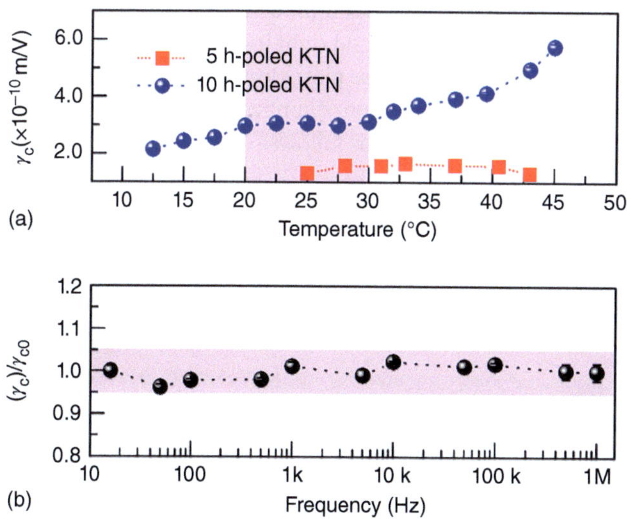

Figure 11.21 Linear electro-optic effect in KTN. (a) Temperature dependences of the linear electro-optic coefficient γ_c after 5 hours (red) and 10 hours (blue) poling processes; (b) Frequency dependence of γ_c in the 10 h-poled KTN (the γ_{c0} is the γ_c measured under the 16 Hz electric field). Source: Tan et al. [20]. Reproduced with permission of Optical Society.

The γ_c is 3.00×10^{-10} m/V at 25 °C after 10 hours poling, which is more than twice the sample with 5 hours poling. It indicates that the full poling is helpful to improve the linear electro-optic properties of the crystal. After 10 hours poling, the formed single domain or single vector domain can effectively suppress light scattering. It also enhances the macroscopic P_s, leading to a large increase of γ_c. In addition, the fully polarized state makes the stable γ_c at 20–30 °C at room temperature for 10 hours poling. The γ_c as high as 5.76×10^{-10} m/V at (T_C −5 °C), which further proves the optimization effect of stable single-domain polarization state on linear electro-optic performance of KTN. According to the frequency response of KTN sample γ_c in Figure 11.21b, the frequency stability of γ_c at room temperature exceeds 95% in the range of 10 Hz–1 MHz in AC electric field. It shows that stable polarization state plays a decisive role in the excellent frequency stability. Based on the above analysis, it can be seen that the ferroelectric domain ordered orientation (single domain) can effectively improve the size, temperature, and frequency stability of the intrinsic γ_c, which is the key to improve the linear electro-optic properties of KTN crystals.

In poled KTN, the inverse piezoelectric effect needs to be considered to further reveal the domain structure contribution to linear electro-optic effect. The inverse piezoelectric effect mainly changes the optical path length by means of sample distortion along the direction of light beam under applied electric field, which influences the results of γ_c. Considering the inverse piezoelectric effect, the optical path length δ satisfies the equation as follows:

$$\delta = \Delta n \cdot L + \Delta L \cdot (n - 1) \tag{11.29}$$

where n and Δn are the refractive index and the change of refractive index under intrinsic electro-optic effect, respectively. L and ΔL are the length of sample along the direction of light beam and the change of length of sample along the direction of light beam under inverse piezoelectric effect. The left and right items show the contribution of the intrinsic electro-optic effect and inverse piezoelectric effect, respectively. The part of inverse piezoelectric effect (the direction of applied electric field and light beam is along $[001]_C$ and $[100]_C$) can be written as follows:

$$\Delta L = d_{31} \cdot EL \tag{11.30}$$

where d_{31} and E are piezoelectric coefficient and applied electric field. According to the measurement, the contribution of inverse piezoelectric effect to effective electro-optic coefficient can be approximated as:

$$\Delta \gamma = \frac{2(n-1)}{n^3} d_{31} \tag{11.31}$$

When considering the anisotropic refractive index n_e and n_o, the effective electro-optic coefficient γ_c combining with the intrinsic electro-optic effect and inverse piezoelectric effect is:

$$\gamma_c = \gamma_c^0 + \frac{2(n_e - n_o)}{n_e^3} d_{31} \tag{11.32}$$

where γ_c^0 is intrinsic electro-optic coefficient. The d_{31} is −42 pm/V at room temperature in same component KTN crystal. Thus, the contribution of inverse piezoelectric effect d_{31} to electro-optic effect γ_c is about 0.09×10^{-10} m/V (without considering the anisotropic refractive index) and 0.005×10^{-10} m/V (considering the anisotropic refractive index). It is known that the inverse piezoelectric effect is much smaller than the intrinsic linear electro-optic effect.

11.4.3.2 The Relationship Between Quadratic Electro-Optic Effect and Polar Nano-regions

The most remarkable advantages of KTN crystal are the large quadratic electro-optic effect of about 10^{-14} m^2/V^2 near Curie temperature, high speed response in nanoseconds magnitude, wide light transmittance range of 400–4000 nm, and high diffraction efficiency (>80%). These properties, especially the large electro-optic coefficient, make KTN crystal one of the most excellent electro-optic materials with great potential for optical switching, deflection devices, and modulator applications.

Generally speaking, the polar microstructure of crystal and the contribution of ionic electrons together affect the macroscopic electro-optic properties of the materials. In recent years, researchers have found that this disordered polarity microstructure is very important for a series of properties of materials, including electro-optic effects, especially near the phase boundary.

In KTN crystals, the displacement of Nb ion provides local polarization. Near Curie temperature, the associated coupling of local dipoles leads to the formation of PNRs, which are in a disordered state due to the substitution of B-sites. The excellent properties of KTN crystals, especially the non-diffracted light transmission and the giant electro-optic effect, are believed to be directly related to the dynamic characteristics of disordered PNRs. Pierangeli and Tian et al. showed that the electro-optic effect in disordered ferroelectrics could reflect the local symmetry breaking induced by PNRs, leading to the specific characteristic of electro-optic effect. The Fröhlich entropy was used to show the ordered/disordered state. Based on the relation between the temperature-dependence of the real part of the dielectric constant and the entropy of the system $S(T)$, the unit volume $S(T)$ can be expressed as:

$$S(T) = S_0(T) + \frac{E^2\varepsilon_0}{2}\frac{\partial \varepsilon_s}{\partial T} = S_0(T) + S_E(T) \tag{11.33}$$

where S_0 is the system entropy without an external field, ε_0 is the vacuum permittivity, ε_S is the real part of the low frequency permittivity, and T is the Kelvin temperature. Fröhlich entropy (s) represents the entropy change under unit volume and unit electric field:

$$s(T) = \frac{S_E}{E^2} = \frac{\varepsilon_0}{2}\frac{\partial \varepsilon_s}{\partial T} \tag{11.34}$$

When $\partial \varepsilon_S/\partial T > 0$, the entropy increased caused by electric field. It means that the applied electric field will disturb the established ordered state and promote the generation of disordered dipoles, which will lead to an increase of entropy in the original ordered dipole system. On the contrary, when $\partial \varepsilon_S/\partial T < 0$, the electric field will create an ordered state of the system because the dipoles get rid of their original random arrangement along the electric field orientation, resulting in a decrease in entropy. Thus, $\partial\varepsilon_S/\partial T > 0$ means that the electric field induces disordered changes in the system with the relatively ordered original system, and $\partial\varepsilon_S/\partial T < 0$ indicates that the electric field-induced system has an orderly transformation with the relatively disordered original system.

During the phase transition, the Fröhlich entropy of the system reflects the evolution of ordered/disordered local polarization states. During the ferro-paraelectric phase transition, especially near T_C, the interaction between the polarized units in the crystal is gradually weakened. The large ferroelectric domains are decomposed into smaller polarized units, which are more significantly affected by the electric field. Therefore, the evolution of microstructure near T_C leads to a drastic change in Fröhlich entropy. Below T_C, stronger interactions of ferroelectric domain promote the ordering of local polarization units, and the applied AC electric field disrupts this ordering with $\partial\varepsilon_S/\partial T > 0$, as shown in Figure 11.22. Above T_C, PNRs, which are susceptible to thermal disturbance, are formed due to the sharply weakened interaction of polarization elements. The sample shows a disordered state, and the applied electric field can drive its ordered vibration, leading $\partial\varepsilon_S/\partial T < 0$. PNRs are generally considered to be the essence of disordered polarization units near T_C, and this field-driven disordered characteristic of PNRs is the key to the dynamic electro-optic effects of KTN crystals.

Figure 11.23 shows the s_{11}–s_{12} under AC+DC electric field. The s_{11}–s_{12} was measured by changing the intensity of biased electric field E_{dc} with same AC electric field E_{ac}. It shows the strong electric-field dependence of quadratic electro-optic effect of KTN crystal. s_{11}–s_{12} first increases with the increase of E_{dc} when the E_{ac} is 4.55 and 11.4 V/mm, respectively. Then, the electro-optic coefficient reaches the maximum when the E_{dc} increases to a certain value. When the E_{dc} is further increased, s_{11}–s_{12} tends to be stable at 11.4 V/mm AC electric field component,

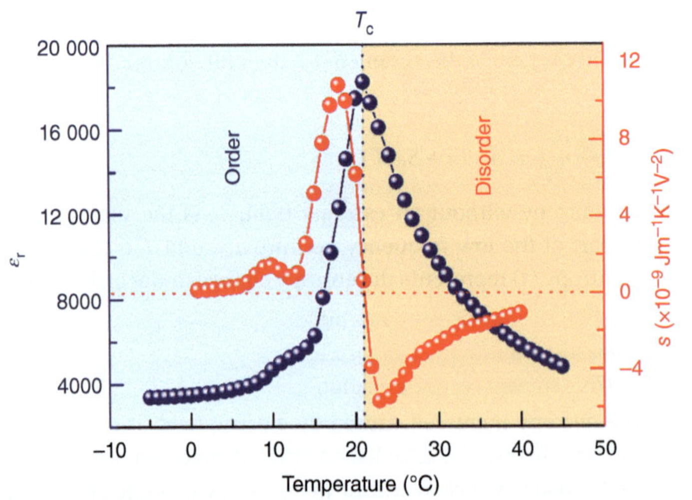

Figure 11.22 Temperature-dependence of ε_r (blue) and Fröhlich entropy (red) at an electric field of 1 kHz. Source: Tan et al. [21]. Reproduced with permission of American Institute of Physics.

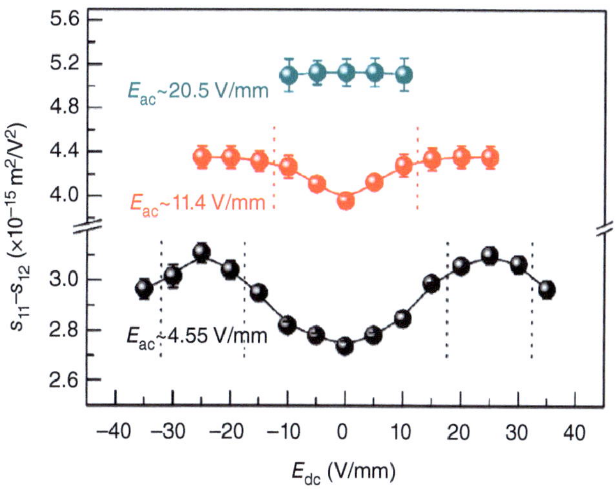

Figure 11.23 $s_{11}-s_{12}$ under biased electric field. Source: Tan et al. [21]. Reproduced with permission of AIP.

while $s_{11}-s_{12}$ decreases when E_{ac} = 4.55 V/mm. When E_{ac} = 20.5 V/mm, $s_{11}-s_{12}$ remains at 5.12 × 10–15 m^2/V^2 in the ±10 V/mm E_{dc} range. To sum up, as shown in Figure 11.23, the quadratic electro-optic effect of KTN crystal shows strong electric field dependence. The field-driving characteristic of PNRs is the main reason for the dependence of electric field on electro-optic effect in KTN crystal.

In KTN, the inherent heterogeneity causes random breaking of (B-site cation)–(oxygen anion) bonds. The broken bonds redistribute the charges and give rise to the local random field, which directly affects the behaviors of the PNRs. Ideal PNRs can be distributed randomly among the equivalent lattice sites in the presence of thermal activation and have both vibration and reorientation. The vibration of a PNR results from the PNR boundary motion, which acts as a polar spring, and the reorientation originates from the redistribution of the PNR spontaneous dipole moment as a polar rigid ellipsoid, as shown in Figure 11.24a.

According to Landau theory, the initial direction of PNR polarization is determined by the local random field. Landau free energy indicates that the local random field deepens the potential well in a specific direction, which constrains the PNRs in the local state and inhibits the performance of the PNRs orientation response mode. Therefore, the initial state PNRs in the KTN crystal presents a non-ergodic state due to the local random field action, and it is difficult to realize the transformation between equivalent states, as shown in Figure 11.24b–d. The relaxation time τ is longer, and the real part of the known polarization can be expressed as:

$$\chi' = \frac{\chi_s}{[1 + (\omega\tau)^2]} \tag{11.35}$$

where χ_s is susceptibility and ω is the frequency of applied electric field. The χ' will decrease with increasing τ, which reduces the quadratic electro-optic effect. However, the metastable PNRs state can be activated to cross the potential barrier, and the orientation dynamics of the activated PNRs can be restored to enhance the χ' and

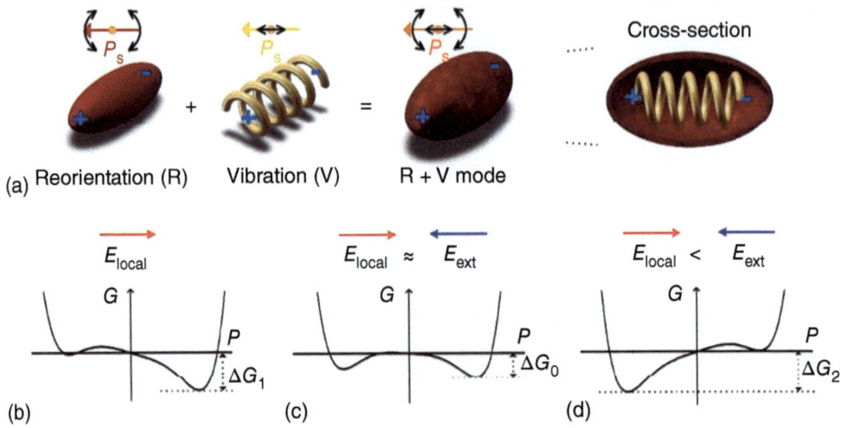

Figure 11.24 (a) Visualized descriptions for the vibrational and reorientational components of the feasible trajectory of the dynamics; (b–d) The orientation states of PNR polarization cloud caused by local E_{local} and external electric field E_{ext}. G and P represent Landau free energy and PNR polarization cloud, respectively. ΔG_1 and $\Delta G_2 > \Delta G_0$ ($\Delta G_0 \approx kT_K$), where k is the Boltzmann constant and kT_K is approximately equal to the PNR thermal kinetic energy. Source: Tan et al. [21]. Reproduced with permission of AIP.

electro-optic effects when the low-frequency electric field is applied strongly enough to the KTN crystal. As a result, the $s_{11}-s_{12}$ is increasing with the increase of E_{ac}.

11.4.4 The Electro-Optic Modulation Devices Based on KTN Crystal

11.4.4.1 High-Speed Large-Angle Beam Scanner Based on Gradient Refractive Index KTN Crystal

Gradient refractive index (GRIN) optical devices have been widely applied in integrated optics, imaging optics, and fiber optics. Because of their intrinsic spatial refractive index distribution, GRIN devices can be used to optimize optical systems and achieve new functions that differentiate them from uniform materials. The electro-optic effect of KTN crystal is closely related to its composition. Thus, the process of crystal composition regulation can be optimized by means of the separation and coagulation effect during KTN crystal growth. Then, the design and preparation of variable GRIN KTN crystals can be achieved by means of the distribution of the electric-optic coefficient induced by the component gradient. According to the method of composition regulation and equal diameter growth, the growth was carried out with different cooling rates in the controllable temperature range, as shown in Figure 11.25a. The inset of Figure 11.25b shows the ε_r gradient along the pulling direction, which indicates possibility of the electro-optic coefficient gradient.

The contribution of quadratic electro-optic coefficient (s_{ij}) was shown in Figure 11.25b in order to reveal the inducing of component gradient on electro-optic

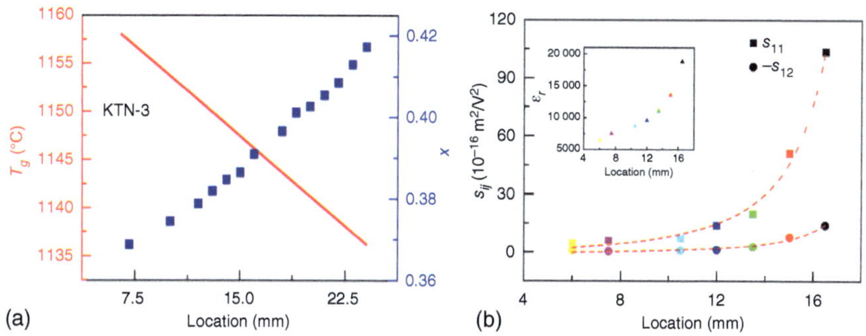

Figure 11.25 (a) Corresponding growth temperatures and composition distributions along the pulling direction of KTN crystal; (b) Distribution of quadratic electro-optic coefficients along the pulling direction at 25 °C in KTN crystal. Source: Tian et al. [4]. Reproduced with permission of Royal Society of Chemistry.

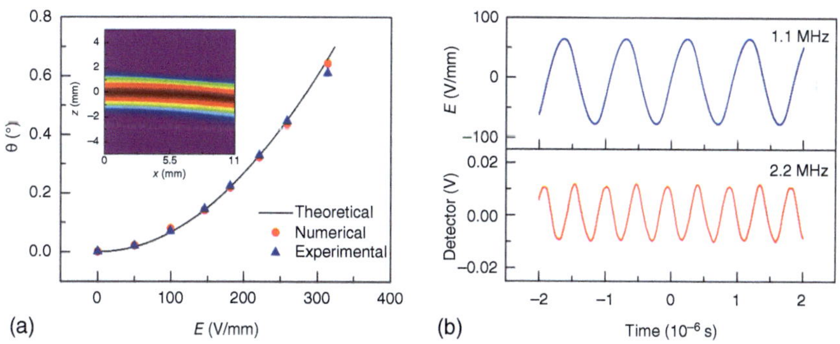

Figure 11.26 (a) Theoretical, numerical, and experimental $E-\theta$ results (the inset figure shows the numerical deflection track when $E = 800$ V/mm). (b) Response rate of the electro-optic deflector. Source: Tian et al. [4]. Reproduced with permission of Royal Society of Chemistry.

effect. It indicates that the s_{ij} changes with the different location of single crystal, which verifies the variation trend of ε_r and design of crystal growth. Based on the GRIN KTN crystals, the transmission optical deflector was achieved. Figure 11.26a shows the good agreement between theoretical and experimental data. The device has a large deflective angle of 0.61° at an applied electric field of 313 V/mm. In addition, the response rate of the electro-optic deflector was also studied. In the test, a slit was placed in front of the optical probe, and an AC electric field (up to 1.1 MHz) was applied to the deflector to make the beam scan continuously through the slit, and the corresponding signal was recorded by an oscilloscope. Figure 11.26b shows that the electro-optic deflector based on GRIN KTN crystal has an optical scanning rate of up to 2.2 MHz, which is much higher than the 500 kHz scanning rate of similar deflectors.

11.4.4.2 Controllable Deflection Devices Based on Photorefractive Effect of Manganese-Doped KTN Crystals

Photorefractive effect is the change of refractive index caused by the spatial distribution of light intensity in materials with electro-optic effect under the condition of light irradiation, which is designable and erasable for controlling the refractive index. The change of refractive index is mainly because of the influence of space charge field excited by light on electro-optic effect. Thus, the designing of irradiation distribution is the key to a multifunction device. In transmission optical deflector based on design of photorefractive effect, the deflection angle θ of the output light is expressed as:

$$\theta = l\left|\frac{d\Delta n(x)}{dx}\right| = n_0^3 l s_{11}[E_0 + E_{sc}(x)]\left|\frac{\partial E_{sc}(x)}{\partial x}\right| \tag{11.36}$$

where E_0 is the external electric field, $E_{sc}(x)$ is the space charge field, s_{11} is the quadratic electro-optic coefficient, n_0 is the refractive index, $\Delta n(x)$ is the distribution of the refractive index along the x axis, and l is the thickness of the sample in the direction of light transmission. To achieve the linear deflective relationship between the input and output signals, the linear gradient distribution of irradiation intensity needs to be built. Figure 11.27a shows the distribution of refractive index after the establishment of space charge field. The E_{sc} maintains good linearity in the middle part of sample, as shown in Figure 11.27.

Based on the induced sample, the deflection function was realized with the transmitted light passing through the middle part of crystal. The center of the beam is extracted to represent the average theoretical deflective angle, which is shown in Figure 11.28a. The red line is the theoretical deflection angle of the center of the beam spot, while the black line denotes the experimental results by measuring the change of spot location. Obviously, the theoretical results agree well with the experimental results, exhibiting a linear relationship between θ versus E. Figure 11.28b shows the frequency response of designed deflector, manifesting the good frequency response of the device. Deflection performance is maintained at 50 kHz.

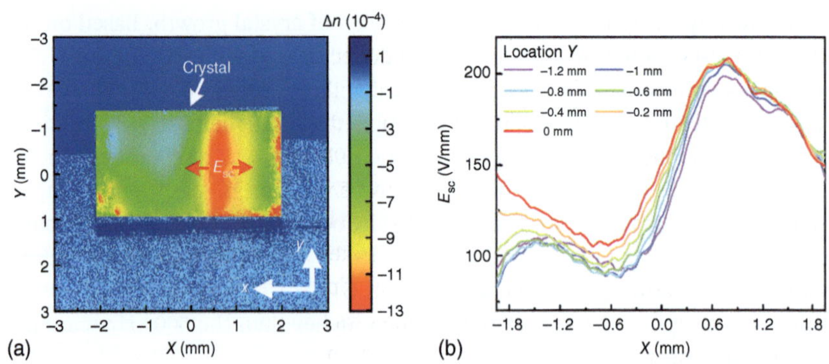

Figure 11.27 Induced results of space charge field. Distribution of space charge field (a) in plane and (b) on different locations of Y direction. Source: Wang et al. [22]. IOP Publishing Ltd / CC BY 4.0.

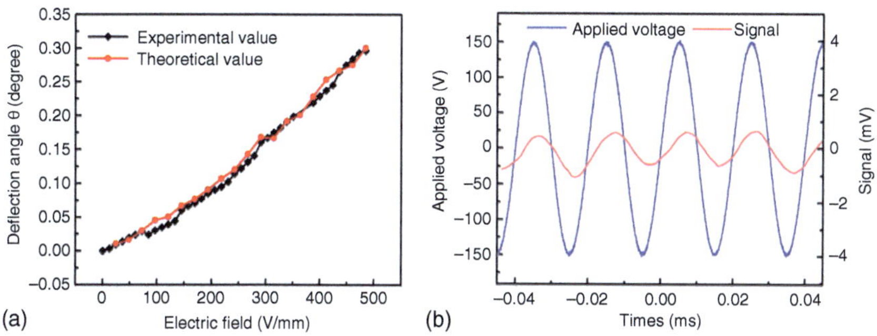

Figure 11.28 (a) Measured and theoretical prediction results of deflection angle. (b) Frequency response characteristics of the deflector. Source: Wang et al. [22]. Optical Society / CC BY 4.0.

11.4.4.3 Static Volumetric Three-Dimensional Display Based on an Electric-Field-Controlled Two-Dimensional Optical Beam Scanner

By exploiting the quadratic electro-optic effect and gradient of composition KTN crystal, a fast response electric-field-controlled two-dimensional optical beam scanner with a wide wavelength range was designed, as shown in Figure 11.29a. The scanner was used to realize a volumetric display based on a two-frequency (850, 1550 nm) two-step-up conversion technique that is used to address the imaging volume.

Er^{3+}-doped (1.0 mol%) tellurite glasses are used as the imaging medium for the display. The optical beam scanning system was used to drive the 850 nm laser to address the large number of voxels required to form a high-resolution image. A long-focal-length cylindrical lens was selected to direct a narrow 1550 nm laser line toward the glass. This line slices through a plane of the glass, and scanning is realized using a traditional galvanometer. The intensities of the incident light beams were 200 mW (850 nm) and 250 mW (1550 nm). Photons from two infrared lasers with different operating wavelengths are incident on the glasses and are absorbed by the Er^{3+} ions. The photons from the first source (which refers to as the addressing source or laser, operating at 1550 nm) are absorbed by Er^{3+} ions in the ground state ($4I_{15/2}$), thereby exciting these Er^{3+} ions to an intermediate energy level ($4I_{13/2}$), where they can remain with a long lifetime. If a photon from the second source (which refers to as the imaging source or laser, operating at 850 nm) arrives while an Er^{3+} ion is in this state, the energy is then absorbed from this photon, and the Er^{3+} ion is excited to a higher state ($2H_{11/2}$). The next action is radiation of the total absorbed energy, minus any internal losses, as the ion returns to the ground state, which generates a visible photon and thus generates a spot of light within the volume of the material. An image is then created by addressing multiple voxels to draw the larger image within the volume of the glass, as shown in Figure 11.29b. It can be seen from the results that the 3D display system based on our scanner has the potential of three-color imagery and high image resolution (nearly 61.5×10^9 voxels per cubic meter) [23].

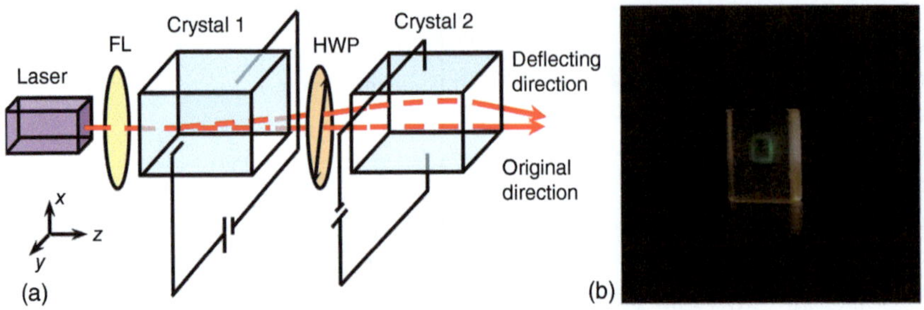

Figure 11.29 (a) Schematic diagram of electric-field-controlled deflection using KTN crystal. HWP, half-wave plate; FL, Fourier lens. (b) Luminescent image of a 3D structure generated in the tellurite glass. Source: Gong et al. [23]. Reproduced with permission of Optica Publishing Group.

11.5 Conclusion

Potassium sodium niobate-based lead-free piezoelectric materials because of their excellent piezoelectric, ferroelectric, and dielectric properties and high Curie temperature, are considered one of the most competitive substitutes for lead-based materials. In this chapter, the growth of KNN and KTN single crystals was explored by the TSSG. High quality and large size KNN and KTN single crystals were grown by optimizing and improving the growth method. The piezoelectric properties and origin of KNN single crystal are studied systematically. KNN single crystals have extremely large piezoelectric properties with an ultrahigh large-signal piezoelectric coefficient d_{33}* of 9000 pm/V, which is superior to the highest value reported in state-of-the-art lead-based single crystals (2500 pm/V). The enhanced electromechanical properties in our crystals are realized by an engineered compositional gradient in the as-grown crystal, allowing significant reversible non-180° domain wall motion. Moreover, our crystals exhibit temperature-insensitive strain performance within the temperature range of 25–125 °C. The enhanced temperature stability of the response also allows the materials to be used in a wider range of applications that exceed the temperature limits of current lead-based piezoelectric crystals.

KTN-based crystal, as a multifunctional material with excellent electro-optic properties, has a broad application prospect in the field of optical communication. The large electro-optic coefficient, high response rate, and broad transmittance make it possible to comprehensively improve the performance of existing modulators. Meanwhile, the controllability of its nonlinearity provides a solution, leading to a completely new paradigm for ultraresolved imaging and microscopy. The tunable composition distribution provides an efficient means for the design and optimization of refractive index gradient devices. The remarkable piezoelectric, electrocaloric, and photovoltaic property of KTN-based crystals are also beneficial for dealing with complex application scenarios and composite control methods. In addition, the gradual maturity of the preparation process of KTN nanoparticles, KTN nanowires, and KTN thin film has made a greater contribution to the integration and functional expansion of KTN devices.

References

1 Chang, Y., Wang, C., Yin, S. et al. (2013). Giant electro-optic effect in nanodisordered KTN. *Opt. Lett.* 38 (22): 4574–4577.

2 Wang, L., Tian, H., Meng, X. et al. (2014). Field-induced enhancement of voltage-controlled diffractive properties in paraelectric iron and manganese co-doped potassium–tantalate–niobate crystal. *Appl. Phys. Express* 7 (11): 112601.

3 Tian, H., Hu, C., Meng, X. et al. (2015). Top-seeded solution growth and properties of $K_{1-x}Na_xNbO_3$ crystals. *Cryst. Growth Des.* 15 (3): 1180–1185.

4 Tian, H., Tan, P., Meng, X. et al. (2015). Variable gradient refractive index engineering: design, growth and electro-deflective application of $KTa_{1-x}Nb_xO_3$. *J. Mater. Chem. C* 3 (42): 10968–10973.

5 Meng, X., Hu, C., Park, D.S. et al. (2020). Ultra-high piezoresponse in tantalum doped potassium sodium niobate single crystal. *Appl. Phys. Lett.* 116 (11): 112902.

6 Tian, H., Tan, P., Meng, X. et al. (2016). Effects of growth temperature on crystal morphology and size uniformity in $KTa_{1-x}Nb_xO_3$ and $K_{1-y}Na_yNbO_3$ single crystals. *Cryst. Growth Des.* 16 (1): 325–330.

7 Chen, K., Xu, G., Yang, D. et al. (2007). Dielectric and piezoelectric properties of lead-free $0.95(K_{0.5}Na_{0.5})NbO_3$–$0.05LiNbO_3$ crystals grown by the Bridgman method. *J. Appl. Phys.* 101 (4): 4103.

8 Ahtee, M. and Glazer, A.M. (1976). Lattice parameters and tilted octahedra in sodium–potassium niobate solid solutions. *Acta Crystallogr. Sect. A: Cryst. Phys. Diffr. Theor. Gen. Crystallogr.* 32 (3): 434–446.

9 Baker, D.W., Thomas, P.A., Zhang, N. et al. (2009). Structural study of $K_xNa_{1-x}NbO_3$ (KNN) for compositions in the range $x = 0.24$–0.36. *Acta Crystallogr., Sect. B: Struct. Sci* 65 (1): 22–28.

10 Saravanan, R., Rajesh, D., Rajasekaran, S. et al. (2012). Crystal growth and dielectric property of $Na_{0.5}K_{0.5}NbO_3$ and Mn-doped $Na_{0.5}K_{0.5}NbO_3$ single crystal grown by flux method. *Int. J. Appl. Phys. Math.* 2 (3): 208.

11 Davis, M., Klein, N., Damjanovic, D. et al. (2007). Large and stable thickness coupling coefficients of $[001]_C$-oriented $KNbO_3$ and Li-modified (K, Na)NbO_3 single crystals. *Appl. Phys. Lett.* 90 (6): 062904.

12 Uršič, H., Benčan, A., Škarabot, M. et al. (2010). Dielectric, ferroelectric, piezoelectric, and electrostrictive properties of $K_{0.5}Na_{0.5}NbO_3$ single crystals. *J. Appl. Phys.* 107 (3): 033705.

13 Lin, D.B., Li, Z.R., Zhang, S.J. et al. (2009). Dielectric/piezoelectric properties and temperature dependence of domain structure evolution in lead free $(K_{0.5}Na_{0.5})NbO_3$ single crystal. *Solid State Commun.* 149 (39): 1646–1649.

14 Hao, T., Chengpeng, H., Xiangda, M. et al. (2015). Dielectric, piezoelectric, and elastic properties of $K_{0.8}Na_{0.2}NbO_3$ single crystals. *J. Mater. Chem. C* 3 (37): 9609–9614.

15 Kizaki, Y., Noguchi, Y., and Miyayama, M. (2006). Defect control for low leakage current in $K_{0.5}Na_{0.5}NbO_3$ single crystals. *Appl. Phys. Lett.* 89 (14): 142910.

16 Lin, D.B., Li, Z.R., Xu, Z., and Yao, X. (2009). Characterization of KNN single crystals by slow-cooling technique. *Ferroelectrics* 381: 1–8.

17 Hu, C., Meng, X., Zhang, M.H. et al. (2020). Ultra-large electric field-induced strain in potassium sodium niobate crystals. *Sci. Adv.* 6 (13): eaay5979.

18 Wang, Y., Meng, X., Tian, H. et al. (2019). Dynamic evolution of polar regions in $KTa_{0.56}Nb_{0.44}O_3$ near the para-ferroelectric phase transition. *Crystal Growth Design* 19 (2): 1041–1047.

19 Wemple, S.H. and DiDomenico, M. (1969). Oxygen-octahedra ferroelectrics. II. Electro-optical and nonlinear-optical device applications. *J. Appl. Phys.* 40 (2): 735–752.

20 Tan, P., Tian, H., Wang, Y. et al. (2018). Impact of dipolar clusters on electro-optic effects in $KTa_{1-x}Nb_xO_3$ crystal. *Opt. Lett.* 43 (20): 5009–5012.

21 Tan, P., Tian, H., Mao, C.Y. et al. (2017). Field-driven electro-optic dynamics of polar nanoregions in nanodisordered $KTa_{1-x}Nb_xO_3$ crystal. *Appl. Phys. Lett.* 111 (1): 012903.

22 Wang, Y., Tan, P., Zhou, Z.X. et al. (2021). Control of photorefractive space-charge field and its deflection application in Mn-doped $KTa_{1-x}Nb_xO_3$ crystal. *New J. Phys.* 23 (1): 013014.

23 Gong, D.W., Wang, C.Z., Wang, X.P. et al. (2019). Static volumetric three-dimensional display based on an electric-field-controlled two-dimensional optical beam scanner. *Appl. Opt.* 5 (26): 7067–7012.